Arthur Aron

**Instructor's Manual
with Handouts
and Transparency Masters**

STATISTICS
FOR
PSYCHOLOGY

Arthur Aron
State University of New York at Stony Brook

Elaine N. Aron
Pacifica Graduate Institute

© 1994 by PRENTICE-HALL, INC.
Simon & Schuster Company
Englewood Cliffs, NJ 07632

0-13-845702-6

Printed in the United States of America

Table of Contents

Introduction

This *Instructor's Manual* accompanies Aron and Aron's *Statistics for Psychology*. It is designed to help you teach the introductory statistics course in psychology using this text.

Section I of this *Manual* includes four chapters (A through D) which discuss general issues in teaching this course: teaching and lecturing about statistics, structuring the course, creating examples based on a class survey of your students, and using the computer with the course.

I have organized these chapters in a detailed outline form, in order to make it easy for you to locate and use just that material pertinent to your needs.

Section II of this *Manual* provides special chapter summaries, lecture outlines, and worked-out examples and other teaching aids to photocopy onto transparencies for use in these lectures. The chapters in this section of the *Manual* correspond to each of the chapters in the text; two or three lectures are outlined for each chapter.

We have also prepared a *Test Bank and Answers to Set II Practice Problems* which is bound separately from this manual.

If at any time you have questions or suggestions regarding the course, please feel free to contact me by electronic mail or post:

Electronic mail: Art.Aron@sunysb.edu

Post: Arthur Aron, Ph.D.
Psychology Department
State University of New York at Stony Brook
Stony Brook, NY 11794-2500

Chapter A
Teaching Statistics

Teaching statistics is a chance to be heroic. Students expect this to be a horrible experience, so if you can make it even bearable, they love you for it.

In this chapter I share with you what I have learned about teaching statistics over the last twenty-two years (as well as what I have learned from reviewing the literature on teaching undergraduate courses and attending various seminars on college teaching). I have organized the chapter into six sections:

I. Overall Teaching Issues
II. Structuring the Lecture
III. Presenting the Lecture
IV. Fitting the Course to the Students
V. Supervising Teaching Assistants
VI. Some Comments to Teaching Assistants

I. Overall Teaching Issues

There are four main issues that make a great deal of difference in the effectiveness of my teaching:

A. Being Organized and Consistent

Students seem reassured by an instructor who has thought out what he or she will do, lets the students know what the plan is, and follows through on the plan. This applies to every aspect of the course: the overall course structure and syllabus, the structure of each lecture, and the content and grading of assignments and exams. If the course includes laboratory or discussion sections, as the overall instructor you must also see that what goes on in the sections is systematically planned and, again, the plan is carried out.

I can not emphasize too strongly that it is not only important to follow an organized plan, but to be sure that the students know, at each point, what this plan is and that you are following it. This is as much an emotional as a rational issue, and applies regardless of whether or not you are feeling organized yourself. My experience is that students need to see order and consistency in you that borders on the compulsive. That way the occasional inevitable error is not taken as a sign of total chaos (something which they may be fearing in themselves).

B. Being Flexible and Responsive to Student Input

On the other hand. Although it is crucial to make and follow a plan, plans do not always work as expected. Teaching requires being alert to whether students are maintaining interest, learning the material, and holding a positive attitude towards the course. There also has to be a sense of compassion coming from you—even mercy—so they can apply that, too, to themselves.

When it is necessary to make changes in the plan, you should plan the changes—perhaps with a discussion with the students (and certainly with any teaching assistants you have)—and then make very clear to everyone just what the changes are and why you have made them. If a change could possibly be interpreted by any students as affecting them adversely, it will be important to address this before they do—preferably by providing some offsetting advantage to those students.

Being responsive to student input also means soliciting it. I encourage students to come to office hours and to send me notes about any suggestions they have about the course (even anonymously, if that is easier for them). In addition, I do two specific things:

1. *Arrive 5-10 minutes early, stay 5-10 minutes late.* I try to be available to students at this time (that is, have all my notes and such prepared and ready before I enter the classroom), and if possible, I actually approach students individually or in small groups and ask them how the class is going. When I know names, that is all the better. But I make a point to speak to some I don't know—not merely the ones who always speak up in class. Also outside of class, if I see a student of mine at the bookstore or at the drinking fountain, I ask him or her about the course.

2. *In about the fourth week of class, I ask for written evaluations.* My procedure is to ask students to take out a sheet of paper, *not* put their name on the paper, and then answer three questions: (a) What do you especially like about the course? (b) What suggestions do you have? and (c) Do you have any other comments? I make sure to collect these in a way that respects their anonymity. I review the evaluations before the next class and report to them on any issues that have come up. I find it is important not to get defensive, to make changes that are really useful, and to let the class also know how many students are really quite satisfied by reading some of the positives (with appropriate modesty, of course). Students are often surprised that not all the other students responded just as they did (social psychologists call this the "false consensus effect"). I think they also learn to recognize the more generic grumbling-against-authority that sometimes crops up in students. Even when there is some truth in the gripe, they can hear when there is more venom or energy in the attack than an instructor deserves, and so I become more human to them—someone who, like them, is sometimes an underdog.

C. Being Scrupulously-Fair

In my experience the one thing that most undermines class morale is when students feel that they are being treated unfairly. In part I try to avert such problems by, again, making the requirements and structure of the course very explicit at the start and sticking to it (point A, above). But problems are also very likely to arise around the grading of exams and assignments, or perhaps around the way you call on students in class, or generally how you are perceived as treating students, both in and out of class. Grades, alas, are often of enormous importance to students, and the one or two points that you know are not important for their final grade can be a source of sleepless nights and tearing of hair to them. And, once again, it matters not only that things be fair, but that students *perceive* them to be fair.

D. Experimenting and Attending to Your Own Strengths and Weaknesses

Your teaching style is probably based on your best learning experiences. And perhaps you have taken to heart advice of people like me. This is fine, but these suggestions may not make the best use of your personal abilities. Everyone I know who has ever won a teaching award comments that they did not really become a good teacher until they found their own voice, their own style. This meant taking some risks, trying out some new and different ways of doing things—and making quite a few mistakes in the process. In fact, mistakes are an inevitable part of teaching even if you don't experiment at all! But excellent teachers watch more closely to see what does and doesn't work for them and let that shape their teaching style. And when teaching statistics, it really helps to pursue excellence.

II. Structuring the Lecture

Most people who teach this course are pretty confident about the material itself. But this does not guarantee a good lecture. In this section I consider the overall structure of the lecture, then focus on what is the heart of any statistics lecture—the explanation of the concepts and the selection and preparation of examples.

A. Overall Structure of the Lecture

The most important part of an effective lecture structure is that there is a structure! Here are specific suggestions for what makes a good structure:

1. *Begin with a statement of this lecture topic and where this topic fits into the overall course.* In this context also review what has been covered in the immediately previous lecture or two.

2. *Next, every class period remind students of the relevance of the course to their larger academic and life goals.* They can't escape learning this if they want to be psychologists. They can't fall asleep, much as they'd like to. In other words, you remotivate them about why they need today's lecture—to understand research they will read in later courses; to prepare them for the more advanced statistics courses which will empower them to do sophisticated research on the issues that matter to them; and to sharpen their conceptual abilities, which are not innate, but only developed by patient persistence. Also, when possible, make the relevance of each procedure clear too, by illustrating its application in studies the students are likely to see as relevant to them or to populations they care about.

3. *Next give an overview of what this lecture will cover.*

4. *Make the focus of the lecture a single theme (or a small number of themes) illustrated with several examples.*

5. *Focus on the material in the text.* In a statistics course you do not bore the students by covering the same material that is in the text. They need and expect the repetition. Indeed, it is crucial that you use the same symbols, methods, formulas, and language as are used in the book. The lecture is made lively by using some different examples, explaining the concepts in your own words and style (but following the same logic as in the book), your give-and-take with them through their questions, and by your overall presentation skills.

6. *Keep each segment discrete and summarize it before going on to the next.*

7. *Plan time for questions at the end of each segment.* I always try to have some extra material to cover if there are very few questions and something near the end that I can leave out if the questions take too long.

8. *End with an overview of what was covered and what is coming in the next class.* It is very important to leave time for this.

B. Explaining Concepts

The most important thing about explaining concepts is to do it *slowly*, step by step, with simple language, and illustrating each step with an example, analogy, or graphic—anything to make the abstract stick. (Abstract concepts of the kind that are so central to statistics are difficult for most students, who find the language itself very strange.) For example, the idea of the variance has three parts: The deviation of each score from the mean, the square of these deviations, and the average of these squared deviations. My experience is that students have no trouble with the first part, but if I move too fast to the next step, the combination of the first with the second, and then of the first two with the third, it all becomes too much of abstraction laid on abstraction.

Keep in mind short-term versus long-term storage—the extra time and the semantic processing that is required to move an idea from one to the other. Also consider primacy and recency effects, which make the middle idea most easily forgotten—unless you emphasize it.

Finally, since the concepts are the most difficult part of a lecture, they should be given when you and the students are freshest and least likely to be rushed—near the beginning.

C. Preparing Good Examples

Working through examples is the heart of most statistics lectures. *For each major point I make in a lecture, I try to have three examples.* I find it most effective to use one example from the book (this keeps continuity) and two new examples. I usually make one new example that is based on responses to a questionnaire the students themselves completed (see Chapter C of this *Instructor's Manual*) and one other example.

Correspondingly, in the lecture outlines provided in this *Instructor's Manual*, for each major theme I provide transparencies for (a) one example from the text, (b) one based on questionnaires of students from one of my classes (which would be replaced by transparency examples from your own class if you send in the data to me), and (c) one new example.

If you want to make your own examples, here are four suggestions:

1. *An example should not involve unnecessary complexities or idiosyncrasies.* The students have enough to do to learn the material without having to deal with examples that are exceptions to rules, include advanced material, involve complicated research designs difficult to understand in their own right, or require learning an unfamiliar theory to understand the rationale of what was done.

2. *The content of the example should be interesting.* If possible, select topics relevant to student interests and having interesting results. I try to take mine from real studies (or if I use made-up data, I try to make up examples leading to conclusions supported by actual studies).

3. *Prepare the example in exactly the same format, with exactly the same formulas, steps, and symbols as the examples in the text.* Students are very confused when steps are left out or when faced with formulas or symbols they have not seen before.

4. *Generally avoid showing alternative ways of working a problem.* This can confuse those students having trouble learning it even one way. (Of course occasionally an alternative approach makes the material come to life.)

III. Presenting the Lecture

Being well prepared is more than half the job. What remains is the way you present. The main issues are the general attitude towards the material you convey, maintaining classroom morale, and style of delivery.

A. Your General Attitude Towards the Material

Students will take their cue from you about the value of what they are learning. Thus, I suggest you follow these three principles:

1. *Show them your own excitement (to the degree you genuinely feel it) about the research process and the use of statistics as part of it.*

2. *Don't belittle statistical methods or the research process.* Sometimes, in order to sound more human, an instructor runs down the subject matter being taught. I have learned that this almost always backfires. Students figure that if I feel this way about it, why should they bother to learn it.

3. *Be accurate with logic, formulas, and numbers.* Mistakes are inevitable, but it is important to take great care, especially when working out problems in class. I have all too often heard students complain about statistics professors who can never get the formulas right. It also drives students crazy when, halfway through a problem they have been laboriously copying down, they have to go back and start over or change every number!

B. Maintaining Classroom Morale

I find that my own day-to-day satisfaction with teaching is primarily a function of the feeling atmosphere in the classroom. And I am also convinced that students' learning is seriously impaired by a classroom ambience that is disinterested, chaotic, or hostile. Here are some things I do to keep classroom morale high:

1. *Pay attention to morale.* Ask students how they are doing and respond to grumbling and any challenges in an open, constructive, even eager way. (I resist the temptation to dismiss the discontented as deviant griping swine who do not appreciate the pearls I have spread before them.) Often just acknowledging students' concerns, even if nothing can be done about them, can turn a classroom situation around. But it is also sometimes possible, and necessary, to make real changes, and sometimes to make some special effort (such as an extra review session on your own time, extra handouts, easing requirements, etc.) to show students you will go out of your way for them. But the main thing is to notice and respond. If things do start to go badly, the longer you let it go on, the harder it gets to do something about it.

2. *Start class on time.*

3. *Keep the difficulty of the material you present (and also the difficulty of assignments and exams) at a level that allows everyone to experience success, even if some say later that you went too slowly.* Remember, your goal is for students to learn—evaluation should always be secondary to this goal. Also, I have found it better not to remind students constantly of how much more there is to learn in statistics and how little of it they are getting in this course, as if this material is really simple and should be a breeze for them. Actually, because of anxiety, for most people it's often the first math courses that seem the hardest.

4. *Avoid saying "clearly," "as is obvious," and the like.*

5. *Be supportive and patient with questions.* A student's attention will never be as consistently on the lecture as yours must be. Students will miss things you said and ask what appear to be stupid questions. It is almost always the case that if one student does not understand something, many others have the same problem but are too shy to speak out. If you are even slightly harsh with a questioner, it will stifle all but the most aggressive students—and also create a fearful (and eventually hostile) climate. Also, don't put the student who asks a question on the hot seat. Repeat the question and give your answer to the whole class.

6. *Don't spoil the experience at the end.* You can do this by getting angry at students' natural tendency to start packing their things a few minutes early. You can do this by expressing your frustration about not having covered everything you

4

intended. And you can do this by trying to cover your last eight points in the last five seconds. (If you are running out of time, recognize it five minutes before the end, leave something out, and allow time to summarize and close graciously.)

7. *Make jokes.*

8. *Let the students know you are a human being.* A personal story now and then goes a long way towards making a connection with your students and adds some random reinforcement during what for some must be very long, dry, difficult lectures.

C. Style of Delivery

Teaching is a performing art. The human performance is what makes it worthwhile for students to attend lectures, as opposed to merely reading the material in a book. Here are some tips for giving a strong performance.

1. *Attend to students' faces to see if they are learning and interested.*

2. *Put variety into your pacing and tone of voice.*

3. *Use your body.* Hand and arm motions, leaning into your points, moving around the room—these all make it easier to follow what you have to say. (You needn't be afraid of being overly dramatic—my experience is that the more outrageous the effort, the more likely students are to pay attention, and that is your goal.) Also, it is generally better to stand if you can, so your motions are less constricted and your level of energy is naturally higher.

4. *Speak to the audience.* Look at individuals, make eye contact, focus on different sections of the audience as you proceed through the lecture. If you are working from detailed notes, look up often—and certainly at the end of each point. Take special care not to spend too much time facing the blackboard or looking down at your notes.

5. *Use simple, clear language.* Remember that even if a student has learned a lot of psychology or math terminology in previous courses, it is not likely to be as familiar to them as to you. If your material is expressed in such language, the student must simultaneously struggle to recall the meaning of the words as well as what you are trying to teach.

6. *Write on the board legibly and in an organized fashion.* (And be sure that the previous material on the board is very thoroughly erased in any area in which you are working.)

7. *Use diagrams.* Many students do not really grasp concepts such as regression or the normal distribution until they see them visually. However, the diagrams we use in statistics require a bit of explanation and time to digest before they can have the needed impact. So take your time in explaining them and return to them often.

8. *Use transparencies (with an overhead projector).* The chapters in Section II of this *Manual* provide material you can copy directly onto a transparency (or onto a handout if you prefer). These are easier for students to read and more attractive than blackboard examples. (Be sure to put the projector back far enough so that those in the back of the room can read the transparency without struggling—if necessary, you may want to enlarge the material on a photocopy machine before making the transparency.) I use one of the special felt-tip pens for writing on transparencies, highlighting points on the transparency as I discuss it. A pointer is also useful. One caution: Many students like to take thorough notes. When you write on the board, they can keep up. But when you project a transparency, you need to allow them ample time to copy it into their notes. Another possibility is to make copies of the transparencies you will use and distribute them to the students at the start of class.

IV. Fitting the Course to the Students

The single biggest difference between new and experienced teachers of statistics is that the latter respond to the reality of the students in front of them. New teachers are more often responding to an unrealistic image of what students ought to be like. Here are some suggestions:

A. Don't Assume Your Students Are Like You

Only a tiny proportion of your class will ever be the sort to teach statistics. Very few will even attend graduate school in psychology. (Of those who do, most will become clinicians, not academics.) In your own first statistics class, you were probably among the most interested, motivated, and capable. And if you think back honestly, even you may have been more interested in your love life than in analysis of variance. So don't expect too much of them—the students themselves become discouraged, if not angry.

In this same light it is important not to assume that most of those taking the course are interested in psychology research. The few who will eventually become researchers usually are inspired to do so later in the major. It is your job to start getting them interested. Presuming they already are could open a large gap between you and them.

B. Assume that Many Will Come to Class Unprepared

On the one hand, it is not fair to those who do prepare to lower the level of your lecture very much, and doing so also reduces the motivation for students to prepare in the future. On the other hand, you don't want to lose half your audience. One solution is to say to your class, "For those of you who were not able to do the reading, you may well have some trouble following the lecture—just do the best you can and hopefully today's class will make it easier to understand the material when you do the reading." Then I give a lecture geared at a level intended to make sense to a poor student who has done the reading and at the same time to a good student who hasn't.

I try not to berate students too much for not having studied, avoiding comments like, "as you know, *if* you've done the reading ..." (Many students have legitimate problems preventing them from keeping up. And it is already the students, not you, who suffers most from being behind.)

C. Monitor How Students Are Doing and Make Adjustments Accordingly

(See point **I**. *B*. above.)

D. Remember that Many Students Are Afraid of (or Loathe) Statistics

Math anxiety, test anxiety, and a general anxiety about school seem to be common among first- and second-year college students. Box 1.2 in the text discusses some things students can do to cope with these anxieties. One of the most important things *you* can do to help students in this regard is to avoid creating any more additional arousal than necessary. Keep the atmosphere of class calm and noncompetitive. Do not emphasize the evaluation aspects and the consequences of poor performance. The arousal-performance curve—particularly for complex, abstract tasks like learning statistics—quickly peaks. Most students are already highly aroused by their fears of this course (and don't be fooled by an apathetic exterior). Any extra arousal only decreases learning. In addition, extra arousal is likely to manifest itself in classroom behavior that undermines class morale.

It is wise not to demand that students pretend to like the material. I show my own enthusiasm, but respect those who dislike or are even hostile to the need for quantitative methods in psychology—I even point to famous psychologists who have said the same things (see our Box 2.1 in the text). Indeed, I usually bring this up on the very first day of class. But in doing so, I also point out that it is unfair of them to create an unpleasant atmosphere and undermine the learning of those students who really want to grasp the material. Further, I note that for everyone who will be a psychology major, whether they like statistics or not, they will have to be able to understand it for their future classes. And this is the place to learn it.

E. Be Aware of Gender Issues

It is helpful to remind the class at the outset that there is no reliable difference in performance in statistics classes as a function of gender (see our Box 1.3 in the text). Students of both genders all too often hold negative stereotypes about women's statistics abilities—stereotypes that interfere with their performance. Some of the subtle ways instructors unintentionally perpetuate these stereotypes include using masculine pronouns—particularly when describing a person carrying out a data analysis—and using examples mainly of research conducted by men. The stereotype may be reinforced if initially your class's women students ask fewer questions. If that happens, it is worth saying something about this explicitly. Also be sure that when women do raise their hands, they are called upon.

F. Be Aware of Ethnic Differences

Students (and instructors) are also affected by ethnic stereotypes. Those belonging to ethnic groups stereotyped as poor at academics may need extra encouragement. They may also need extra help, since they are often the same ones whose secondary schooling provided an impoverished preparation, especially in math and writing. And students in ethnic groups assumed to be good in math may suffer from unrealistically high expectations. Finally, those who have difficulty with English also deserve your extra help and patience if they need it.

V. Supervising Teaching Assistants

When a course uses teaching assistants (TAs), much depends on the quality of their work with the students. In turn, your relation to your TAs is an important part of the course atmosphere and of your own pleasure in teaching. Here are some suggestions:

A. Plan the Section Activities Yourself

Think in advance about what the TAs should be doing in their lab or discussion sections. Often the section activity is simply taking questions on reading and lecture, going over assignments, and reviewing for exams. If this is what you intend, you should make this clear to the TAs and students. But even with this kind of section, it is wise to have the TAs come prepared with extra examples (a good source of these are the practice problems in the text that were not assigned, or problems from the *Test Bank* that will not be used on an exam). Have them organize each meeting into segments to be sure that time is devoted to each of the major topics covered since the last section meeting.

B. Treat Your TAs as Your Teaching Colleagues

TAs are usually graduate students who will be teaching courses themselves before long. Indeed, they may have TA'd this course more times than you have taught it, or may actually have taught it themselves. You can learn from them. And they are likely to work best with you if you show (and feel) respect for what they can contribute. I find it especially important to show this respect when I interact with my TAs in front of our students. I am also careful not to undermine their relations with students when I am talking to students individually.

I meet with my TAs before the term and give them a chance to review and suggest changes in my draft of the syllabus and my plans for the course. Also during the course I discuss with them any changes before I make decisions and announce these to the class.

C. Treat Your TAs as Your Graduate Students

Your TAs are teaching colleagues, but they are also your students, learning how to teach. I provide them with guidance about all the topics in sections I through IV of this chapter (not just having them read it, but discussing my own and their experience with these issues).

D. Meet with TAs Often

I meet with my TAs weekly, having them report on how their sections are going, taking suggestions, and discussing how the course as a whole is going. We also discuss the material coming up for the next week's sections. If an exam is approaching, we discuss its contents and how grading will be organized. I usually prepare an agenda for each meeting. (These meetings serve as an important source of input about how the students are doing in the course—it is striking the things TAs pick up from students that the students will never reveal to me.)

E. Provide Support and Remember Your TAs Are Human Beings.

As graduate students, your TAs have a great many demands on them from their studies and research, as well as undergoing all of the stresses that go along with this period of their life. When a student is unkind to them or if they feel they are not doing an adequate job in their work as a TA, I find it best to provide as much emotional support as I can (a listening and understanding ear) before tackling the specific practical issues of what can be done. Also, I try at the start of each term to learn what special demands each TA may have coming up—such as qualifying exams, proposal dates, or seminar presentations—and to distribute the workload for different parts of the term among the TAs in a way that takes their individual needs into account.

F. Attend Closely to Your TAs' Grading of Exams and Assignments

The grading of exams and assignments are crucial determinants of course atmosphere. Inexperienced TAs (like inexperienced instructors) tend to expect too much and grade too hard. They also tend to forget that undergraduates do not welcome unnecessary "constructive criticism." Thus, before my TAs ever grade an exam or assignment, I discuss with them the importance of providing positive feedback on what is done well and not being harsh or heavy handed in noting errors. I also plan with them, as a group, what the grading standards will be, both to be sure I find the standards appropriate and to be sure that all the TAs are using the same standards.

VI. Some Comments to Teaching Assistants

Being a TA for statistics is one of the most useful teaching assignments, for your own professional development, that you will have as a graduate student. First, it will dramatically strengthen your own mastery of statistics—a crucial tool if you plan a career involving research. Second, it will prepare you to teach statistics as an instructor yourself—an ability that will substantively enhance your value on the academic job market as well as make you a much appreciated colleague in many departments.

You also have the opportunity to be especially loved by the students. Students read the material and hear it in class, but often it does not really sink in until they discuss it in the section with you. The result is that you get the credit for the whole learning experience! (This can be hard on the instructor's ego, but good for yours. And if all the credit is not deserved, much of it is, since this does not happen unless you have done a good job with your part.)

My most important suggestions to you are the same as those I made to the instructor in Sections I through IV, above. Here are a few additional points:

A. Prepare for Sections

Don't just walk in and wait for questions. Even if all you will be doing is answering questions on the book and lecture, put an outline on the board of the topics that are involved (and then ask for questions topic by topic). Think out ahead of time one or two examples of things you anticipate students may be having trouble with. Also think out in advance what aspects of what has been covered are most important for the students to master and what aspects are relatively peripheral. This is something that is not obvious to many students, and a place you can really help.

If the section involves covering material not in the lecture (such as teaching them how to use a computer program or covering additional techniques), then you need to prepare just as you would for teaching any class (see the material in Section II above).

Of course, it is crucial that you have done the same reading as the students and have reviewed it again just before class. It is also important to have attended the lectures. If you do not know what the students have been exposed to, how can you help them make sense of it?

B. Your Relation to Your Students

You are probably only a little older than most of your students, and younger than some. Nevertheless, the students are taking the course and you are the TA precisely because you know more about the material than they do. This does not necessarily make you a better person or even generally more knowledgeable. But it does mean you are the one in the position to teach and they are in the position to learn (you also learn from them, of course). Thus your role should be one of a consultant, a specialist who is there to help. And if it is also your job to evaluate their performance in the course, you are in the best position to do that. Your duty is simply to be as accurate and even-handed as you can.

C. Your Relation to the Course Instructor (Who Is Also a Human Being)

It is the instructor's obligation to treat you with respect and be supportive of you in your role as TA. It is also your responsibility to treat the instructor with respect and to be supportive. In particular, it undermines the course as a whole, as well as the instructor's relation to the students, if you speak negatively to the students about the instructor or the way the course is being taught, or if you complain to them about the way the instructor is treating you or the other TAs. If you feel something is not going well in the course, or the course is not being run properly, you should speak to the instructor about it directly. Or if this is too difficult, send the instructor an anonymous note. You owe that to both the instructor and the students. Also, remember that your instructor is in a political relationship to the rest of the department. If you grumble to

8

your advisor or some other faculty member about something to do with the course (even a casual remark), that could have very large repercussions for the instructor, especially if he or she is untenured.

A Final Word

Statistics is my favorite course to teach. This is partly, as I said at the outset, because I get to be a hero just for making it bearable. But it is also because of the tremendous rewards of seeing students who thought they could never learn statistics become not just passing students, but knowledgeable ones. Or seeing the more confident types, who nevertheless disdain statistics, suddenly discovering its elegance and utility. If you are teaching statistics for the first time, I welcome you to these teaching pleasures.

Chapter B
Structuring the Course

This chapter discusses issues involved in structuring the course, ending with an example course syllabus. Four topics are covered:

I. Selection and Order of Topics
II. Assignments
III. Exams
IV. The Course Syllabus

I. Selection and Order of Topics

A. Selection of Topics

   1. *The core material.* The minimum standard coverage of the introductory statistics course in psychology includes frequency tables and histograms, mean, standard deviation, the correlation coefficient, the normal curve, the *t*-test, and the basics of one-way analysis of variance. In our text this requires Chapters 1-3, 5-7, and 9-11. These chapters can be taught without any of the others in this book (though if Chapter 8 is omitted, students should be told to skip the discussions of power at the ends of Chapters 9-11).

   2. *Power and effect size.* These topics, which in recent years have become an increasingly standard part of the basic course material (for the good reason that psychologists have come to realize their great importance), are introduced in Chapter 8 and brought up again in every subsequent chapter. However, the coverage of these topics in the subsequent chapters is in a discrete section near each chapter's end, permitting the instructor who chooses to do so to omit Chapter 8 easily. (Chapter 8 is one of the more difficult chapters in the book. It can be made easier by not requiring students to learn to compute power, but simply focusing on the concept, on effect size, and how they are related to planning and evaluation of the outcome of studies.)

   3. *Chi-square and nonparametric statistics.* Chi-square tests and nonparametric methods are also very commonly included in the minimum course. These are covered in Chapters 14 and 15. Each can be taught independent of the other and neither requires any background other than the core material.

   4. *Analysis of variance.* As noted above, the basics of one-way analysis of variance are part of the core material. By adding Chapter 12, one introduces the structural model, permitting analysis of unequal Ns. This material also helps the student to grasp the standard computer output for analysis of variance. Chapter 13 discusses factorial designs and two-way analysis of variance. The material in the first part of Chapter 13, which focuses on the logic of interaction effects, only requires Chapter 11 as preparation. However, the students must have learned the Chapter 12 material if you wish to teach the procedures of two-way analysis of variance covered in the second half of Chapter 13.

   5. *Regression.* Chapter 4 covers bivariate regression and gives a conceptual introduction to multiple regression. (This is also one of the more difficult chapters.) This material depends only on Chapters 1, 2, and 3. Some instructors may choose not to assign the material on multiple regression in the second half of the chapter. However, this material is very important for anyone hoping to make sense of journal articles in today's psychology.

   6. *Integration of statistics.* Chapter 16 introduces the general linear model and the links between the major parametric techniques, in the process reviewing most of the previous material covered in the book. This chapter requires the core material plus regression (Chapter 4), and can not be grasped fully without the analysis of variance material in Chapters 12 and 13. However, it does not depend on Chapters 8, 14, or 15.

   7. *Advanced statistical topics.* Chapter 17 offers a conceptual familiarity with many procedures too advanced to be covered in an introductory text, but often encountered in journal articles. In addition to the core material, this chapter requires at least understanding the material on multiple regression in Chapter 4 as well as the material on factorial designs and interaction effects in the first half of Chapter 13. It does not depend on Chapters 8, 12, 14, 15, 16, or the second half of 13.

8. *Reading statistics in journal articles.* Each chapter includes a section on making sense of statistics as they are reported in psychology journal articles. It can be made optional without doing injury to the rest of the course.

9. *Controversies and limitations.* Each chapter includes a discussion of current statistical controversies about the techniques covered. Students usually enjoy these, and they serve to make the student aware of statistics as a lively, growing field. However, again, these sections can be made optional without doing injury to the rest of the course.

10. *Boxes.* Each chapter contains one to three boxes presenting historical or other material intended to show the human side of statistics, or other material we think will interest students. Some instructors may choose to require these, but I prefer leaving them for optional enjoyment.

11. *Chapter appendixes on computational formulas.* The traditional computational formulas for each technique are provided, with a worked-out example, at the end of each chapter for which such formulas are appropriate. These are not at all necessary to any other aspect of the course.

12. *Recommendations about topic coverage.* In my opinion any course should include at least the core material plus power and effect size (Chapter 8), regression (Chapter 4), and the material on factorial designs and interaction effects (first half of chapter 13). My additions, in order of preference, would be chi-square tests (Chapter 14), the material on additional techniques (Chapter 17), and then the material on what to do when normal distribution assumptions can not be met (Chapter 15). When teaching good students, I would then include, in order of preference, the additional analysis of variance material (Chapter 12 and the second half of Chapter 13), and, finally, the integrative general-linear-model material (Chapter 16).

I always require students to know the material in each chapter on reading statistics in journal articles. Whether I require the Controversies sections depends on time and the quality of the students. I never require the boxes—they are intended to be pleasurable study breaks.

In a semester course at an institution with above-average-ability students, I cover the entire book.

In a semester course with average students, I omit Chapter 12, the second half of Chapter 13, Chapter 16, and possibly the fairly difficult material on the computational aspects of power in Chapter 8. If the students are of very minimal ability, I also omit the controversies and might consider leaving out all of Chapter 8 (and the corresponding material on power and effect size in each subsequent chapter), which is of moderate difficulty, and possibly also omitting Chapters 14 and 15. Leaving all this out allows much more time for review.

In a quarter course with above-average students I cover the entire text, except Chapter 13's second half and Chapter 15.

In a quarter course with average students I omit Chapters 12, 15, 16, and the second half of 13, and do not require the Controversies sections or the material on computational aspects of power in Chapter 8. With students of very minimal ability, I would also consider dropping Chapter 8 entirely (and the corresponding material in later chapters), and possibly 14 as well. If necessary, I would also drop Chapters 4 (which is difficult) and 17 (which depends on 4).

B. Order of Topics

    1. Prerequisites for chapters. The following table shows which chapters are prerequisites for other chapters.

Chapter	1	2	3	4	5	6	7	(8)	9	10	11	12	13
1													
2	1												
3	1	2											
4	1	2	3										
5	1	2											
6	1	2			5								
7	1	2			5	6							
8	1	2			5	6	7						
9	1	2			5	6	7	(8)*					
10	1	2			5	6	7	(8)	9				
11	1	2			5	6	7	(8)	9	10			
12	1	2			5	6	7	(8)	9	10	11		
13 (1st ½)	1	2			5	6	7		9	10	11		
13 (2nd ½)	1	2			5	6	7	(8)	9	10	11	12	
14	1	2			5	6	7	(8)	9				
15	1	2	3		5	6	7	(8)	9	10	11		
16	1	2	3	4	5	6	7		9	10	11	12	13
17	1	2	3	4	5	6	7		9	10	11		13 (1st ½)

*Chapter 8 is in parentheses because it can be omitted as a prerequisite to the chapters if students are told to omit the section in each of these chapters on power and effect size.

    2. When to teach correlation and regression. Perhaps the most important implication of the table of chapter prerequisites is that the material on correlation and regression (Chapters 3 and 4) can easily be taught either in the order they appear in the book or later on.

    The advantages of teaching them where they are in the book are (a) the students have just learned Z scores, which makes Chapters 3 and 4 much easier; and (b) correlation and regression are naturally understood as descriptive statistics, the next logical step from the univariate descriptive statistics learned in Chapters 1 and 2.

    On the other hand, if Chapters 3 and 4 are taught after some or all of the inferential statistics technique chapters (5-15), then it is easy to include a discussion of the significance of a correlation coefficient when teaching this topic. Finally, Chapter 4 is relatively taxing. Teaching it early can have advantages spreading out the "hard stuff" and catching students while they are fresh. It can also have disadvantages, discouraging some because of the abrupt jump in difficulty.

    We have included the basic material on testing the significance of a correlation coefficient in Chapter 3, in a second chapter appendix. If you cover Chapter 3 in the usual order, you may want students to go back to that appendix some time after covering Chapter 9. If you decide to cover Chapter 3 later in the course (a good place would be right after the *t*-test chapters), then this appendix should be included when teaching the chapter.

II. Assignments

    For most students to learn statistics, they need actively to work problems. I also find that the more writing about statistics the better, to engrain it in their minds and provide a solid preparation for the next topic.

A. Amount of Assignments

    My policy is to assign about an hour's worth of problems for each class, to be turned in at the start of the next class session. I select a variety of problem types, using the Set I problems. These permit students to check their answers against those in the back of the book, but I do require them to show their work. (If you prefer to assign problems for which the student is not given the answers, use the Set II problems.)

13

B. Marking Assignments

Unless your institution provides more than the usual funding for readers or TAs, it is not practical to provide much feedback to students on their daily assignments. I ask my TAs to look over the assignments and check that all the different parts were done while spot-checking a few more closely. (If time permits, they may also review more closely the work of students having difficulties, or when a student has specifically expressed concern about their answer to a particular problem.) But in general, the TA just marks the assignment as completed, without giving it further evaluation.

C. Assignment Requirements

I require *all* assignments to be completed to pass the course. I consider the working out of these problems as a central part of learning the course material, which exams can only partially evaluate.

D. Late Assignment Policy

I strongly discourage late assignments. Because the course material is largely cumulative, students who turn in assignments late (and hence probably do not learn that material as well) are going to do poorly on learning subsequent material—and bring down the entire level of class discussion.

I have tried several policies to encourage timeliness without imposing draconian discipline (and the attendant disruption to the course atmosphere from the grumbling of the guilty). The most successful has been a rule that a student is allowed two late assignments during the term without penalty. Each additional late assignment requires a short paper summarizing the statistics in a research article provided to the student by the instructor. (I usually require them to read a short research article that uses a topic covered in the course up to that point, and write an explanation to a lay person of the result in the article. A good source of short articles that use fairly simple statistical procedures is *Psychological Reports*.)

I also emphasize that "on time" means the *beginning of the class*—otherwise, students miss class to complete the assignment. However, because there are other reasons students can be late, I usually also add the rule that a paper between 5 minutes and 24 hours late counts as half late.

III. Exams

In my experience, exams (and the grading of them) do the most to determine the atmosphere of the course.

A. Objectives of Exams

Below I describe, in what I consider their order of importance, the four objectives of exams in this course:

1. *Incentive for the student.* An exam encourages a concentrated review of the material. It seems to me that, for better or worse, half the learning that takes place in a course occurs in the 48-hour period preceding each exam. I am especially aware of this from the excellent teaching that is called forth from me by desperate students at the long review sessions I hold.

2. *Feedback to the student.* The graded exam gives students a realistic appraisal of how they are doing in the class. This can encourage students by reminding them just how much they have learned. It can also be a rude awakening for those who have not mastered the material well, hopefully inspiring them to improve the situation. However, I have found that in statistics most students benefit more from the encouragement of seeing what they have learned than from the shock of seeing what they failed to get. Thus, I make it a priority in structuring exams to be sure that the problems provide average students a chance to demonstrate their accomplishments, even if this means the overall class average is relatively high.

3. *Feedback to the instructor.* The distribution of performance of students on the different parts of the exam provides the instructor valuable information about what the students are and are not learning well. One implication is that it is important to make a separate distribution (I find a histogram most helpful) of class performance for each problem or question and to review these carefully and consider what modifications might be made now and in future courses in light of what has been mastered and what is still shaky. Sometimes I find I need to change the schedule and spend an entire class reviewing some concept, instead of proceeding willy-nilly and leaving much of the class hopelessly behind.

4. *Basis for course grade*. Exams are the primary basis for assigning grades. In my opinion this is the least important value of an exam, and when this value comes in conflict with one of the above three, I give those three priority. Nevertheless, when constructing an exam I make an effort to be sure it will yield a distribution of student accomplishment, tap those aspects of student accomplishment that provide an appropriate basis for assigning a course grade, and be structured so that it can be scored reasonably objectively.

B. How Many Tests to Give?

1. *Advantages of giving a large number of exams*. Giving many exams assures that each segment of the course is mastered before proceeding, gives the student and the instructor ongoing feedback on how much has been learned, and provides a more reliable basis for a final grade. Based on these advantages, some statistics instructors give exams weekly—some even give short quizzes as part of each class.

2. *Disadvantages of giving a large number of exams*. Giving many exams constantly focuses the course on evaluation (and possibly quibbling over grades), takes up a lot of class time, and each exam must be prepared, printed, graded, recorded, and handed back. For these reasons, some statistics instructors give only a single midterm plus a final.

3. *My recommendation*: Compromise by giving four to five mid-term exams, one at the end of each major segment of the course, plus a final.

C. Content of Exams

The *Test Bank* which accompanies this text provides several kinds of test questions. Below I describe the pros and cons of each type, some possible mixes and lengths of exams that would be appropriate for different course situations, and the issue of whether exams should be cumulative.

1. *Problem-and-essay questions*. I have found the most useful test question is one in which the student is given a description of a study and its result (including data), then is asked to carry out the appropriate statistical procedure, draw a conclusion, and explain what was done to a person who has never had a course in statistics. (To shorten the task, I usually modify the essay so that it is written to a person who knows some statistics, but not the material covered since the last exam.)
Advantages: Knowing that there will be this kind of question on the exams focuses the students' studying on the core skills I want them to gain. And their performance on this kind of question provides the most direct feedback to them and me of whether they have indeed gained what matters.
Disadvantages: First, such questions take a long time to answer. This means you cannot ask many of them, which in turn means low reliability (if they miss the point of a problem, a huge chunk of the exam is a zero) and a reduced range of topics that can be tested. Second, they are time-consuming to grade, and thus often impractical for large classes. Third, the grading of essay questions necessarily has a subjective component that can be a source of disruptive student quibbling. Finally, such questions put special emphasis on writing ability, and so special burdens on students with poor writing skills or for whom English is the second language.

2. *Problem questions*. These are the standard statistics exam fare—the student is given a description of a study and data (or just data) and carries out the statistical computations. (I always require students to show their work at each stage of the problem.)
Advantages: These questions directly examine whether the student has learned the computation of the statistical procedures. If the problem does not indicate which procedure is to be used, such questions also require the student to determine this. (For example, does this problem require a *t* test for independent or dependent means?) Since so much of a statistics course necessarily focuses on computation (and usually the assignments very heavily so), including such questions appropriately taps what the students have been studying. Answers are usually relatively brief, so several can be included, and they can be graded fairly rapidly and objectively. Finally, such questions minimally impact those not good at English.
Disadvantages: Problem questions do not necessarily get at underlying concepts, but may show only a rote memorization of a formula and the ability to plug numbers into it. Nor are such problems appropriate for testing vocabulary, for example, or controversies, or how to read statistics in a research article. In addition, such problems give great emphasis to computational skills. Thus students who understood the ideas, but are not agile on a calculator or facile with numbers, may not be able to show what they have learned. Finally, although problems of this kind take less time to answer and are easier to grade than problems with essays, they are nevertheless moderately time consuming. Thus, you still can only include a relatively small number of such problems in any one exam, which creates low reliability and limits topic coverage. And for large classes, grading problems is still a substantial task.

3. *Reading-statistics essays.* A unique feature of this text is its emphasis on reading and interpreting statistics as they are presented in psychology journal articles, and the *Test Bank* includes exam questions that evaluate how well students have mastered these skills. These questions consist of a brief description of a study followed by a short excerpt (or table) from the results section. The student is asked to explain what the result means to a person who has never had a course in statistics (or who has only learned the statistics material covered through the last exam).

Advantages: These essays directly assess whether the student has mastered one of the most important objectives of the course. They also tap the student's general mastery of the principles behind the technique. Further, these essays often put special emphasis on evaluating the role of power and effect size in a relatively concrete context.

Disadvantages: These essays share many of the disadvantages of problem/essays: Each takes up a lot of test time, the grading is time consuming and somewhat subjective, and they penalize those with poor writing skills or whose native language is not English.

4. *Multiple-choice questions.* Multiple-choice questions mainly focus on definitions of terms, understanding of concepts, and retention of factual material (such as issues involved in controversies).

Advantages: The grading is economical and objective—especially important in large classes or where the instructor wants to give daily short quizzes. In addition, studying for multiple-choice questions focuses the student on concepts (as opposed to rote computation), and such questions are especially appropriate for testing the material in controversy sections. Finally, a relatively large number of multiple-choice items can be included on an exam, making for higher reliability and giving the opportunity to cover a broad range of topics.

Disadvantages: Multiple-choice questions only minimally tap computational skills (which is usually one of the main things students are being asked to do in their assignments) and do not provide any opportunity for student creativity or expression. Students also often report a sense of mechanization about multiple-choice exams, a feeling that they are being treated as numbers. Finally, in my experience, multiple choice exams seem to be most prone to student quibbling about the right answers.

5. *Fill-in-the-blank items.* These items primarily emphasize definitions of terms and knowledge of factual material.

Advantages: Fill-in items share much of the grading advantages of multiple-choice questions, as well as also being rapid for the student to complete. But fill-in items can be superior when they give the student the sense of a more personal connection with the instructor; and they test recall, not just recognition.

Disadvantages: Fill-in items share most of the limitations of multiple-choice questions, are even less well suited to assessing conceptual material, are somewhat less economical and objective to grade, and can be even more troublesome for those with language deficits.

6. *Recommended mixes and exam length.* The main question is whether it is practical to grade essays and problems. If it is, I prefer a mix of one essay (of either type), two or three problems, and a few fill-in or multiple-choice items to test the factual material. I have found that as long as the students know that the exam will include at least one essay—but do not know on which subtopic it will focus—they study the concepts thoroughly. And having this mix, as opposed to all essay-type problems, gives me a chance to cover several different topics (allowing the students to show their competence at each), and to increase the reliability of the overall exam.

For a 50-minute class period, I give one essay, one other not-very-complex problem, and four fill-ins (which focus entirely on the factual material). For a 75-minute class, I give one essay, one to two other problems (depending on the complexity of the problems), four fill-in and four multiple-choice items.

If it is not practical to include essays and problems, a mix of multiple choice and fill-in items does a quite good job. Depending on the length of the testing time and the ability level of the students, I would use 20 to 40 multiple choice and 5 to 15 fill-ins.

7. *Should exams be cumulative?* Much of the material in a statistics course is intrinsically cumulative. For example, almost every topic after the start requires a mastery of mean and standard deviation. Nevertheless, I have found that students are better able to focus on mastering what they have just learned if I assure them that the upcoming test will cover only that material, with previous material only acting as a foundation for the current topics.

I do give cumulative final exams. In this case, I want them to review the whole course.

D. What to Tell Students in Advance About an Exam

1. *Why tell them anything?* I talk about an exam in advance (a) to reduce their anxiety so that over arousal does not interfere with learning and (b) to focus their attention on what I most want them to master.

2. *What to say?* To achieve the above goals, I usually tell them the kind of exam questions to expect (how many of each type) and the general topics to be covered. If practical, I provide this information on a printed hand-out. For example, if the upcoming exam were on Chapters 9-11, I might tell them that there will be three problems, one on each of the three main types of problems (dependent and independent means *t* tests and one-way analysis of variance), one of which would either include an essay or be a reading and interpreting essay. I would also tell them to expect five fill-in items on the controversies.

3. *Practice exams.* To relieve anxiety for the first exam, I generally give my students a practice exam, with answers. I make up the practice exam to follow precisely the format of the real exam, using items in the *Test Bank* not used on the actual exam. However, I emphasize, both in class and in print on the practice exam, that the specific content will not be identical. (For example, if it were for Chapters 9-11, the practice exam might include a problem/essay for a one-way analysis of variance, but on the actual exam there might be a reading-statistics essay instead, on either kind of *t*-test instead of one-way analysis of variance.)

If you have assigned the *Student's Study Guide* (or made it available as an optional text), providing practice exams is especially unnecessary after the first exam (which gives them the chance to see your particular exam format), though I still give them a list of study question numbers from the study guide that would amount to the equivalent of the upcoming exam.

D. Exam Feedback and Grading

1. *Posting answers.* As soon as the last person in the class has completed the exam, I post the correct answers outside the classroom. This way students can find out immediately how well they have done. More important, students who have made errors do not go home convinced of the correctness of a mistake. This approach also greatly minimizes upset when the tests are eventually returned, as well as easing the pressure on the instructor or TAs to grade the test rapidly.

2. *General grading process.* Unless grading is mechanical (as for multiple-choice), I do one problem at a time for all students. If I have TAs doing the grading, I have them divide up the problems rather than each doing the tests of their own students—this way the grading across students with different TAs is standardized.

3. *Grading essays.* I treat grading essays much like a content analysis. I develop a coding scheme, allotting points for each aspect of the answer and modifying the scheme based on a subset of students' answers—it is often necessary to make further rules along the way. (I go back and regrade the papers in my initial set, to be sure they get the benefit of adjustments made later.) Also as in content analysis, I prefer to have the grading done blind to the student's name, gender, etc. (Identifying tests by student ID number accomplishes this adequately in most cases.) However, although I like to have an organized scheme for grading these essays, with points for each part laid out, I also make adjustments up for someone who has missed some specific point but demonstrated an unusually good grasp of the problem overall.

When grading essay/problems, it is extremely important to focus on the logic, and not to give many points for simply describing the series of computations in words. Unless this grading norm is established on the first test, the essays will be of little use.

Finally, I try to make as great an effort to note really good answers (and parts of answers) as I do to note errors and confusions.

4. *Grading computational problems.* I first check if the final answer is correct. If it is not, I look at the steps of computation and take off only a small number of points for an error along the way, provided the rest of the computations are correct leading to that wrong number. However, I take off more points when an error is sufficiently large that, had the student been thinking about the meaning of the numbers, he or she would have realized the error. (For example, if a correlation coefficient comes out greater than 1.) In fact, I tell students in advance that if they realize that the numbers are not making sense but cannot figure out where their error is, they should note that on the test and I will take off fewer points.

5. *Returning exams.* Prior to returning exams or giving students any information about exam distributions, I look at the outcome myself. If many students have done poorly on a particular problem or question, I reconsider the grading on that item (and may even drop it). If the overall mean is low I may add a constant or percentage to everyone's score. Although this does

not in any way affect final grades if a curve is being used, it can have a dramatic impact on how the students feel about their performance and about the course.

I never return exams at the beginning of a class session. Doing so substantially undermines attention to the lecture or discussion. If exams are to be reviewed in class or a section, I return the exam and do this discussion during the last part of the period. If exams are simply returned at the end of class, I do try to arrange to stay after to talk to those who want to discuss the exam (or have the TAs stay), or I arrange an office hour for me or my TAs shortly after the test.

6. *Minimizing student disagreements about grading of their exams.* The main way I avert disagreements (besides grading carefully) is to post a copy of the exam with the correct answers at the end of the exam so that students are able to determine right away if they disagree with an answer. Any disagreements about the correct answers must be turned in to me, *in writing*, within 48 hours of the exam. (I announce this policy prior to each exam and post it with the correct answers at the back of the room.)

This policy really encourages students to review the correct answers right away, while the material is fresh in their minds. It puts disagreements into writing, where they are less likely to be confrontive. It lessens classroom hostilities. And it gives me a chance to make adjustments prior to finalizing grades. Particularly when multiple choice or fill-in questions are used, this policy averts a great many problems. It also usually leads to at least one changed answer per test (which lets students know you are responsive to their input).

E. Make-Up Exams

In any course it is rare not to have at least one student request to make up a missed exam. However, giving a make-up requires creating a new exam, finding a time and place to administer it, grading it, and adjusting scores to match those on the regular exam. Organizing a time can be particularly difficult if there are several students who need to take such an exam. Also, it can be difficult to determine who genuinely deserves the opportunity and who is making up an excuse because they did not feel prepared.

Having tried different policies, I think the following works best: Students who can demonstrate a legitimate medical or similar reason are allowed to miss one exam during the term. Their grade for that exam will be the average of the other exams (not counting the final).

IV. The Course Syllabus

Here is an example course syllabus. The various policies are based on the issues discussed in this chapter. The schedule assumes a 15-week semester, Monday-Wednesday-Friday course, taught to above-average students. As noted in Section IA, with less able students or when teaching on a quarter system (and also taking holidays into account), less material would be covered. This example syllabus also does not include computer lab sessions—a topic considered in Chapter D of this *Manual*.

University of North America
Winter, 1994
Psychology 2
Introduction to Psychological Statistics
MWF 10 AM - Gosset Hall 105

INSTRUCTOR
Jane Professor, Ph.D.
Office: 308 Gosset Hall
Office Hours: Mondays 11-1
Phone-4208 Email-JPROF@UNA

SYLLABUS

In this course you should gain the following:

1. The ability to understand and explain to others the statistical analyses in reports of psychological research.
2. A preparation for more advanced courses in statistical methods.
3. The ability to identify the appropriate statistical procedure for many basic research situations and to carry out the necessary computations.
4. Further development of your quantitative and analytic thinking skills.

Methods of learning:

1. Reading the assigned material, which includes following the numeric examples closely and writing down questions about anything not entirely clear to you. Reading statistics requires close study and rereading, not just reading through once as you might an ordinary book.
2. Testing your knowledge and reviewing each lecture using your *Student's Study Guide.*
3. Completing the assigned practice problems (and turning them in on time). Statistics is a skill—it is necessary to DO statistics, not just read and understand.
4. Attending lectures, listening closely, asking questions—be sure to have done the reading *first.* DON'T fall behind!
5. Attending discussion sections led by the teaching assistants—be sure to bring questions from the reading with you. This is your chance to get real help with what is not completely clear and to pursue deeply whatever has excited you (yes, there can be exciting things in statistics!).
6. Studying for, taking, and reviewing answers for exams.

Required texts:

1. Aron, A., & Aron, E. N. (1994). *Statistics for psychology.* Englewood Cliffs, NJ: Prentice-Hall.
2. Aron, A. (1994). *Student's study and computer workbookguide to accompany Aron and Aron's Statistics for Psychology.* Englewood Cliffs, NJ: Prentice-Hall.

Basis of evaluation:

1. Five mid-term exams and a final.
2. Completion of assignments on time.
3. Participation in class and sections.

About exams:

1. Each of the five mid-term exams will cover only the material since the last exam (except to the extent that the previous material is necessary for understanding the new). The final is cumulative.
2. *There will be NO make-up quizzes* and *a missed quiz counts as a zero*.
3. Those who provide a *written* medical excuse can drop that quiz. Most other excuses will not be accepted.

About assignments:

1. *All* assignments must be completed by the start of the final exam to pass the course.
2. Assignments are due at the *start* of each class.
3. Assignments turned in between 5 minutes and 24 hours after they are due are 1/2 late.
4. Two late assignments will be allowed.
5. For each additional late assignment you must write a short paper summarizing in your own words the statistical conclusions of a research article assigned by the professor.

Knowledge of mathematics:

The course does not emphasize mathematics. There will be many calculations, but these require nothing more than elementary high-school algebra. The emphasis, instead, is on understanding the LOGIC of the statistical methods. The most important part of each exam will be either (a) a problem in which you use a statistical procedure to analyze the results of a study and then write an essay explaining what you have done to someone who has no knowledge of statistics or (b) a problem in which you are presented with the results of a study and must explain what they mean to a person who has never had a course in statistics.

Calculators:

I *strongly* encourage you to use a hand calculator for doing your assignments, and I *will* permit calculators during tests. I would much prefer you to spend your time developing an understanding of the statistical concepts rather than adding and dividing numbers. A simple calculator that adds, subtracts, divides, multiplies, and takes square roots should be of great help. Since you must show your work on all assignments and exams, calculators that also do statistical calculations will not be of much help, so don't feel any pressure to spend a lot of money. About $10 or less should do.

Wk	Day	Topic	Reading	Assignment Due
		Part I: Descriptive Statistics		(Set I)
1	Mon	Introduction/Administrative	Intro	
	Wed	Frequency Tables & Graphs	Ch 1	
	Fri	Distribution Shapes		Ch 1: 1-2
2	Mon	The Mean	Ch 2	Ch 2: 4,6
	Wed	Variance and Standard Deviation		Ch 2: 1ab
	Fri	*Z* Scores		Ch 2: 1c-e,2,3
3	Mon	Correlation I	Ch 3	Ch 2: 4,6
	Wed	Correlation II		Ch 3: 1ab,2ab,3ab
	Fri	Bivariate Regression	Ch 4	Ch 3: 2c-e,3c,4,5
4	Mon	Multiple Regression		Ch 4: 1,2,4
	Wed	Review		Ch 4: 6,7
	Fri	First Exam		
		Part II: Basics of Inferential Statistics		
5	Mon	Normal Curve & Probability	Ch 5	
	Wed	Hypothesis Testing Logic I	Ch 6	Ch 5: 1-3,5,6
	Fri	Hypothesis Testing Logic II		Ch 6: 1a-d
6	Mon	Distributions of Means	Ch 7	Ch 6: 2,3,5,6
	Wed	Hypothesis Testing with N > 1		Ch 7: 1,2,7
	Fri	Power and Effect Size I	Ch 8	Ch 7: 3,4,6
7	Mon	Power and Effect Size II		Ch 8: 1-3
	Wed	Review		Ch 8: 4-6
	Fri	Second Exam		
		Part III: The t test		
8	Mon	One-Sample *t* test	Ch 9	
	Wed	Dependent Means *t* test		Ch 9: 1,2
	Fri	Independent Means *t* Test I	Ch 10	Ch 9: 3,4
9	Mon	Independent Means *t* Test II		Ch 10: 1,2
	Wed	Review		Ch 10: 3,4
	Fri	Third Exam		
		Part IV: Analysis of Variance		
10	Mon	One-Way Analysis of Variance I	Ch 11	
	Wed	One-Way Analysis of Variance II		Ch 11: 1,6
	Fri	Structural Model	Ch 12	Ch 11: 3,5
11	Mon	Unequal N 1-Way Analysis of Variance		Ch 12: 1
	Wed	Introduction to Factorial Designs	Ch 13	Ch 12: 3,5,6
	Fri	Interaction Effects		Ch 13: 1
12	Mon	Two-Way Analysis of Variance		Ch 13: 2,3,4
	Wed	Review		Ch 13: 6,7
	Fri	Fourth Exam		
		Part V: Additional Topics		
13	Mon	Chi-Square Test of Goodness of Fit	Ch 14	
	Wed	Chi-Square Test of Independence		Ch 14: 1a-c,2
	Fri	Data Transforms/Rank-Order Tests	Ch 15	Ch 14: 3,5
14	Mon	Computer Intensive Methods		Ch 15: 1-3
	Wed	General Linear Model I	Ch 16	Ch 15: 4,5
	Fri	General Linear Model II		Ch 16: 1,2
15	Mon	Advanced Methods I	Ch 17	Ch 16: 4,6
	Wed	Advanced Methods II		Ch 17: 1-4
	Fri	Fifth Exam		Ch 17: 5,6
		Final Exam: [Insert date, time, location]		

Chapter C

Lecture Examples Based on a Questionnaire Administered to Your Students

On the first day of class, I administer a questionnaire to my students, the data from which I use as the basis of lecture examples throughout the course. Students find such examples particularly engaging since they are based on their own and their classmates' responses.

I have made this procedure very easy for you to use. In this chapter I provide the questionnaire I use in a form you can photocopy. Once you have administered it, simply record the data, send it to me, and I will provide by return mail a set of transparencies of lecture examples based on your students' data. (I also provide as part of this manual ready-to-photocopy transparencies of examples based on administering this questionnaire to one of my classes, which you may use for comparison, or instead—particularly if your class is too small to provide adequate power to get interesting results on the various statistical procedures.)

This chapter includes the following two sections:

I. An Example Questionnaire You Can Use
II. How to Record and Send in the Data to Have Lecture Examples Made Up for You

I. An Example Questionnaire You Can Use

The questionnaire on the next page comes from one being used in a research project Elaine Aron and I are conducting on the "highly sensitive person." I have used this questionnaire in my statistics classes as a basis for constructing lecture examples because the questions are so varied and interesting to most students. The questions were selected primarily for their usefulness in creating good class examples and give a somewhat skewed impression of the research project. This questionnaire is *not* a measure of the trait.

Permission is hereby granted to anyone who is using the Aron and Aron text in their course to reproduce this questionnaire for administration to the students in that course.

I distribute the questionnaire to students at the very start of the first day of class, while students are arriving and getting settled. After most of the students are done, I ask them to fold their questionnaires in half (to help maintain anonymity) and pass them to the aisles, where I collect them. About 10 minutes before the end of class, I collect any remaining questionnaires and explain the purpose of the questionnaires:

> This questionnaire consists of items taken from a longer questionnaire being used in a research program on "highly sensitive people" conducted by Elaine Aron and Arthur Aron, the authors of your textbook. I will use your responses in my lectures as data for examples of the various statistical techniques you will be learning in the course. Since this questionnaire has been used in other statistics classes around North America, we will also be able to compare the responses of our class to those of students at other colleges.
>
> The research program for which these questions were developed focuses on people who are very sensitive to sensory stimulation, so that they become uncomfortably overaroused sooner than others by noise, a lot to look at, and so forth. It is a trait that appears to be inherited, to occur in at least twenty percent of the population, and to be about equally common in men and women. (It is not the same as introversion or shyness, although it can lead to these because people are one common source of stimulation in our lives.) This is *not* a measure of sensitivity, but a means of studying its aspects and effects. The best measure of being a highly sensitive person on this questionnaire is item 28, the one that defines it and asks how true it is of you.
>
> The last question is from research on adult attachment style conducted by Cindy Hazan and Philip Shaver (1987). It is intended to assess your typical style of relating to intimate others.

Questionnaire

Your answers to this questionnaire are completely anonymous Do not write your name on it.

You are not required to complete this questionnaire, and you should not answer any questions that you feel uncomfortable about or prefer not to answer for any reason.

Your answers and those of your classmates in this course will provide data for examples in lectures throughout the term on each of the various statistical methods you will be learning. The data from this class will also be sent to a data base for use in a large personality-psychology research project.

Please answer each question, honestly and accurately, according to the way you personally feel, using the following scale:

1	2	3	4	5	6	7
Not at All			Moderately			Extremely

1   Are you introverted?
2   Are you easily overwhelmed by strong sensory input?
3   Do you make a point to avoid violent movies and TV shows?
4   Do you find yourself thinking about some movies the next day?
5   Do you avoid crowds (at malls, carnivals, fairs, etc.)?
6   Are you made uncomfortable by loud noises?
7   Did you tend to fall in love in your early school years (from 5 to 12 years old)?
8   Do you tend to fall in love very hard?
9   Were you prone to hide as a child (under beds or tables, in closets, bushes, etc.)?
10 When you must compete or be observed while performing a task, do you become so nervous or shaky that you do much worse than you would otherwise?
11 Would you characterize your childhood as troubled?
12 Were you close to your father?
13 Was your father involved in your family during your childhood?
14 Were you close to your mother?
15 Was your mother fond of infants and small children (liking to hold and cuddle them, have them around her)?
16 Was alcoholism a problem in your immediate family while you were growing up?
17 When you have a lot to do in a short amount of time, do you get "rattled"?
18 Would you prefer to live out in the country with not many people around?
19 Were you sexually or physically abused as a child?
20 To what extent are you a "morning person"?
21 Do you find yourself needing to withdraw during busy days, into bed or into a darkened room or any place where you can have some privacy and relief from stimulation?
22 Are you a tense or worried person by nature?
23 Are you prone to fears?
24 Do you cry easily?
25 Do you startle easily?
26 Do you like having just a few close friends (as opposed to a large circle of friends)?
27 Do you make it a high priority to arrange your life to avoid upsetting or overwhelming situations?
28 A "highly sensitive person" has been defined as someone who is highly introverted and/or easily overwhelmed by sensory stimulation. To what extent are you a highly sensitive person?

Background Questions. Gender _____ Age _____ Number of older siblings _____

[ ] Check here if you have taken this questionnaire before.

Please read the following three alternatives and decide which best describes your feelings. Then CHECK ONE.

[ ] I find it relatively easy to get close to others and am comfortable depending on them and having them depend on me. I don't often worry about being abandoned or about someone getting too close to me.
[ ] I am somewhat uncomfortable being close to others: I find it difficult to allow myself to depend on them. I am nervous when anyone gets too close, and often love partners want me to be more intimate than I feel.
[ ] I find that others are reluctant to get as close as I would like. I often worry that my partner doesn't really love me or won't want to stay with me. I want to merge completely with another person and this desire sometimes scares people away.

© 1994 by Elaine N. Aron (Extracted from HSP Questionnaire)

24

II. How to Record and Send in the Data to Have
Lecture Examples Made Up For You

As soon as I receive the data from your administration of the questionnaire, I will quickly analyze your data and send you a complete set of lecture examples made up for copying onto transparencies. In most cases, I can send these to you within two days of receiving your data.

Please send us your data in a computer file of one of the following types:

1. *IBM type (MS-DOS) computer:* Save your data in what is called "Text," "DOS," or "ASCII" format. (Almost all word processing programs have an option for saving your data in this format.) WordPerfect 5.1 (or earlier) is also acceptable. You can use any standard disk size (3-1/2" or 5-1/4", either double or high density). Mail your disk to me at the following address:

> Arthur Aron, Ph.D.
> Psychology Department
> State University of New York at Stony Brook
> Stony Brook, NY 11794-2500

2. *Any system that you can use to send electronic mail.* (Most college mainframe computers have this capacity. Also, it is often possible to transfer a file from your personal computer, including Macintosh computers, to the college mainframe.) Send the file to me directly at the following electronic mail address:

> aron@psych1.psy.sunysb.edu

Along with the data, please send a note including:

1. Your return address.

2. Confirmation that you are using the Aron & Aron text in this course.

3. Permission to use your data as part of an international data base being assembled on the highly sensitive person. (Your agreement to this use of the data is not required, but will be much appreciated. We realize that at some institutions you may need approval from a human subjects committee if the data will be used for any purpose other than class demonstration.) Also, please indicate if there is anything about the administration to your class that would make the data not suitable for inclusion in this data base.

The file itself should be typed so that there is one line per student. Within each line, use the layout described in the following table:

Format for Entering Each Line of Data (One Questionnaire per Line)

Column	Entry (what you type into that column)
1-4	Subject Number (a number of your choice)
5	
6	Question 1 Answer (that is, the number they give from 1 through 8)
7	
8	Question 2 Answer
9	
10	Question 3 Answer
11	
12	Question 4 Answer
13	
14	Question 5 Answer
15	
16	Question 6 Answer
17	
18	Question 7 Answer

Format for Entering Each Line of Data (Continued)

Column	Entry (what you type into that column)
19	
20	Question 8 Answer
21	
22	Question 9 Answer
23	
24	Question 10 Answer
25	
26	Question 11 Answer
27	
28	Question 12 Answer
29	
30	Question 13 Answer
31	
32	Question 14 Answer
33	
34	Question 15 Answer
35	
36	Question 16 Answer
37	
38	Question 17 Answer
39	
40	Question 18 Answer
41	
42	Question 19 Answer
43	
44	Question 20 Answer
45	
46	Question 21 Answer
47	
48	Question 22 Answer
49	
50	Question 23 Answer
51	
52	Question 24 Answer
53	
54	Question 25 Answer
55	
56	Question 26 Answer
57	
58	Question 27 Answer
59	
60	Question 28 Answer
61	

Format for Entering Each Line of Data (Continued)

Column	Entry (what you type into that column)
62-63	Age
64	
65-66	Number of older siblings (if just one digit, put a 0 first—for example, if the number is 1, put "01," and if the number is 0 or it is left blank, enter "00")
67	
68	Put an "x" if they have ever taken the test before, or leave blank otherwise.
69	
70	Final question (enter a 1 if they checked the first paragraph, a 2 if they checked the second paragraph, a 3 if they checked the third paragraph)

NOTE: "Column" refers to the number of spaces across the page, starting from the left margin. The first space on the left edge is column number 1. If nothing is shown under "Entry," leave the column empty. Also, if a student has failed to answer a question, leave the column blank for that item.

Chapter D
Using the Computer in the Statistics Course

This chapter examines issues and provides suggestions regarding using the computer in your course. I have divided the material into three sections:

I.   **Advantages and Disadvantages of Using the Computer with this Course**
II.   **Computer Software Packages Compatible with This Course**
III. **Facilities and Equipment Needed**

I. Advantages and Disadvantages of Using the Computer with this Course

A. Advantages

1. *It familiarizes students with the way data are analyzed in actual psychological research.* This serves as an experiential path, giving the student a better sense of what it is like to be inside the research process. In addition, it prepares the student for advanced courses that use these methods and gives them tools they can use in analyzing student projects of their own.

2. *It provides students the opportunity to conduct analyses quickly.* This permits them to carry out exercises involving variations on data sets that illustrate various statistical principles. It also encourages them to try out alternative data analysis schemes of their own.

3. *It carries a sense of excitement.*

B. Disadvantages

1. *It is time consuming.* Teaching students to use the computer to analyze data usually takes considerable teaching effort. Even today, many students are not even familiar with using the computer for word processing. So they must be made familiar with such issues as handling the keyboard, using disks, and creating and saving files. (If a mainframe is used, accounts must be created and they must be taught about logging on and off and such.) Most time consuming of all is teaching them the editing on whatever system is used. Finally, of course, the software program itself must be taught.

2. *It can be confusing to students.* Learning statistics is hard enough for many students. Learning to use the computer at the same time can make the experience overwhelming. I often find that even half-way through the course a few students are still struggling with editing their files when they should be working on learning the concepts and methods.

3. *Facilities and equipment are required.*

C. Circumstances Under Which I Recommend Using the Computer

Presuming appropriate facilities and equipment were available, I would use the computer in my course under any of four conditions:

1. *The course includes a substantial laboratory component.* Ideally, this would include about 2 hours of lab time each week (with appropriate unit credit for a laboratory course provided to students).

2. *A small class of very good students.*

3. *Nearly all students are already comfortable using a computer system of the type to be used in the course.*

4. *The course is expected to provide students the skills needed to analyze substantial data of their own.*

II. Computer Software Packages Compatible with This Course

The supplements that accompany the Aron & Aron text include instructions and examples for SPSS/PC+ Studentware and MYSTAT. Perhaps the major considerations in deciding which package to use is its availability and use in other courses in your institution. In this section we consider special advantages and disadvantages of each package, along with a discussion of the possibility of using other packages with this course.

A. SPSS

SPSS is one of the most widely used statistical packages in psychology, available at nearly every university, so that familiarity with it is likely to be particularly useful to the student who goes on to graduate school. The version available for students is specially designed to make this high powered program accessible to the computer novice.

B. MYSTAT

MYSTAT is also well designed for student use and is inexpensive for each student to own a copy. Its main disadvantage is that SYSTAT (MYSTAT's parent program) is not nearly as widely used among psychology researchers as SPSS, so that familiarity with SYSTAT is not as likely to be directly useful to students who will be going on to graduate school or taking more advanced statistics classes. Also, MYSTAT could not be used if you are relying on a computer laboratory using terminals to a mainframe system.

C. Other Software Packages

With some effort, you could modify the material in our supplementary materials to create your own instructions and examples to use with other statistical packages, such as SAS or BMDP. It is also possible that some other instructor has done this already. So if you are considering using some other software package, contact me (by post or electronic mail) to find out if such material has already become available. And if you do prepare such material yourself, please let me know so that I can arrange for future instructors to contact you.

III. Facilities and Equipment Needed

Whatever package you use, each student needs access to the package and the appropriate computer system for sufficient time to complete assignments, practice, and attempt some creative uses of their own.

If a personal-computer based system is used and copies of the package are made available to students, some students will be able to work at home on their own machines. However, do not overestimate the number of students who are likely to have the appropriate system for the package you are using. Many students do not have any computer system; many have dedicated word processing systems; some have unusual systems that will not run your software.

If students do not each purchase copies of the program (in which case, manuals are included), it is important to arrange for several copies of the manual to be available at the computer lab.

Introduction to the Student
and
Chapter 1
Displaying the Order in a Group of Numbers

Instructor's Summary of Chapter

Difficulty of course. We have never had a student who could pass other college-level psychology courses who could not also pass this course—though for many students this course requires more work.

Reasons for psychology students to learn statistical methods: Reading the psychology research literature, conducting research, and developing analytic and critical thinking.

How to gain the most from this course: Attend to the concepts (not just the numbers), master each concept before going on to the next, keep up with reading and assignments, study especially intensely during the first half of the course, and study with other students.

Descriptive Statistics summarize and make understandable a group of numbers collected in a research study.

Frequency Tables organize the numbers into a table in which each of the possible values is listed along the left from highest to lowest, accompanying each value by the number of cases that have that value.

Grouped frequency tables are used when there are a large number of different values. The frequencies are given for intervals which include a range of values. An interval size should be selected so that the total number of intervals is between 5 and 15; the interval size is a common, simple number; and each interval starts with a multiple of the interval size.

Histograms and frequency polygons. A histogram is a kind of bar graph in which the height of each bar represents the frequency for a particular value or interval. In a frequency polygon, a line connects dots, where the height of each dot represents the frequency for a particular value or interval.

Distribution shapes. The general shape of the histogram or frequency polygon can be unimodal, bimodal, multimodal, or rectangular; symmetrical or skewed; or kurtotic in relation to the normal curve.

Misuse of graphs. Graphs presented to the general public sometimes mislead the eye by methods such as failing to use equal intervals and exaggerating proportions.

How the procedures of this chapter are reported in research articles. When frequency tables appear in research articles, it is usually to compare distributions and often involves frequencies (and percentages) for various categories. Histograms and frequency polygons rarely appear in articles, though the shapes of distributions are occasionally described in words.

Box 1.1. Important Trivia for Poetic Statistics Students. Summarizes the major historical sources of statistical methods.

Box 1.2. Math Anxiety, Statistics Anxiety, and You: A Message for Those of You Who Are Truly Worried about This Course. Summarizes research and thinking on various kinds of anxiety associated with studying statistics and methods for coping with these anxieties.

Box 1.3. Gender, Ethnicity, and Math Performance. Reviews research and thinking on gender and ethnic differences in math and statistics performance, emphasizing the lack of evidence for differences in underlying abilities.

Lecture 1.1: Introduction to the Course

Materials

Lecture outline
Transparencies of syllabus
Questionnaires
Syllabi
Enrollment forms (as appropriate to your institution)

Outline for Blackboard

[Name and number of course and name of instructor]

I. **Complete Questionnaires**
II. **Why Study Statistics?**
III. **What Will You Learn in this Course?**
IV. **Introductions**
V. **Course Structure and Requirements**
VI. **Administrative Matters**
VII. **Review this Class**

Instructor's Lecture Outline

I. **Complete Questionnaires**
   NOTE: The questionnaire and description of its content are given in Chapter C of this *Manual*.
   A. Distribute questionnaires as students enter classroom.
   B. Collect questionnaires when nearly all are done (the remainder can finish during class).
   C. Explain briefly content of questionnaire and how it will be used for data for examples throughout the course (see material in Chapter C of this *Manual*).

II. **Why Study Statistics?**
   A. It is required for psychology majors! But why is it required?
   B. Statistical methods are essential tools used in most psychological research (including in clinical and other areas of applied psychology). Therefore:
      1. This course prepares you for later psychology courses which usually require reading research articles.
      2. This course prepares you for more advanced statistics courses which equip you to use statistics in research you conduct yourself.
   C. This course often meets a general education requirement in quantitative reasoning. But why is there such a requirement, and how does psychological statistics fulfill this requirement?
      1. Psychological statistics involves abstract logical and numeric methods.
      2. Mastering these methods develops your ability to think clearly and very precisely about these kinds of abstractions—something every educated person ought to be able to.

III. What Will You Learn in This Course?
A. How to *understand* statistical methods. Note: The course is not very math-oriented, but is very logic-oriented.
1. You will write essays describing statistical procedures as well as carrying them out.
2. We will emphasize "definitional formulas," which express the concepts, rather than "computational formulas," which ease computation but obscure the concepts.

B. Hand out syllabi and systematically go through goals and topics. [An example syllabus is included in Chapter B of this *Manual*.]

IV. Introductions
A. Introduce yourself.
B. Introduce any teaching assistants.
C. Ask students about themselves using the following categories (if a small class, each introduces self; if a large class, ask for numbers of students in each category):
1. Area of psychology that most interests you.
2. Year.
3. If not a psychology major, what is your major?
4. Have you taken statistics before?
5. Have you had introductory psychology?
6. What other psychology courses have you taken?

V. Course Structure and Requirements: Read and discuss each section of syllabus—be sure to discuss any aspects involving using a computer and any discussion or laboratory sections.

VI. Administrative Matters
A. Subject pool requirements.
B. Instructor's and teaching assistants' office hours.
C. Organizational matters such as enrollment, etc., as required by your institution.

VII. Review this Class: Use blackboard outline.

Lecture 1.2: Frequency Tables

Materials

Lecture outline
Transparencies 1.1 through 1.5
(If using transparencies based on your class's questionnaires and they have arrived already, replace 1.2 and 1.5 with 1.2R and 1.5R.)
Questionnaires (for those who missed first class)
Syllabi (for those who missed first class)

Outline for Blackboard

I. **Organizational Matters**
II. **Roles of Statistics in Psychological Research**
III. **Frequency Tables**
IV. **Grouped Frequency Tables**
V. **Review this Class**

Instructor's Lecture Outline

I. **Organizational Matters**
   A. Be sure each student has a syllabus; answer questions on course structure, etc.
   B. Arrange for those who missed the first class to complete the questionnaire.
   C. Complete any remaining administrative matters.

II. **Roles of Statistics in Psychological Research**
   A. Describe data—"descriptive statistics." Focus of beginning part of course and foundation of rest of course.
   B. Make inferences based on data—"inferential statistics." Focus of most of course after beginning, but builds on beginning material.

III. **Frequency Tables**
   A. General question: Given a set of numbers, how can we make sense of them? Show TRANSPARENCY 1.1 top (example of unordered stress ratings data from text) and discuss.
   B. Show TRANSPARENCY 1.1 bottom (frequency table for stress-ratings data from text) and discuss:
      1. Key terms:
         a. Frequency: Number of cases in a given category or of a particular score.
         b. Frequency distribution: The pattern of frequencies over different categories or scores.
      2. Points in constructing a frequency table:
         a. Go from highest to lowest.
         b. Meaning of symbols at top.
         c. All cases are included.
         d. Sometimes cumulative frequency is included as an additional column.
   C. Second example: Show TRANSPARENCY 1.2 or 1.2R (frequency table for Highly Sensitive Scale using class questionnaire) and discuss.

D. Third example: Show TRANSPARENCY 1.3 top (frequency table from horn-honking study) and discuss.

IV. Grouped Frequency Tables
A. Needed to make large distributions more comprehensible.

B. Procedure.
  1. Subtract lowest from highest score to find range.
  2. Divide range by reasonable interval size (use 2, 3, 5, 10, or multiples of these). *All intervals must be same size.*
  3. List intervals from highest to lowest, down the page. *Be sure they do not overlap.*
  4. Determine number of cases in each interval. (Work from ordinary frequency table if you have one already.)
  5. Double check total count to be sure none are missed.

C. Go through three examples from before, now making grouped frequency tables.
  1. First example: Show TRANSPARENCY 1.4 (stress-ratings grouped frequency table example from text) and discuss, noting that data are not really appropriate for a grouped frequency table as there are already so few intervals.
  2. Second example: Show TRANSPARENCY 1.5 or 1.5R (grouped frequency table for Highly Sensitive Scale using class questionnaire) and discuss.
  3. Third example: Show TRANSPARENCY 1.3 bottom (grouped frequency table from horn-honking study) and discuss.

V. Review this Class: Use blackboard outline.

Lecture 1.3: Describing a Distribution Graphically

Materials

Lecture outline

Transparencies 1.6 through 1.16

> (If using transparencies based on your class's questionnaires and they have arrived already, replace 1.7, 1.14, and 1.16 with 1.7R, 1.14R, and 1.16R.)

Outline for Blackboard

I. Review

II. Histograms

III. Frequency Polygons

IV. Shapes of Distributions

V. Review this Class

Instructor's Lecture Outline

I.   Review
   A. Descriptive statistics.
   B. Frequency tables.
   C. Grouped frequency tables.

II.   Histograms
   A. Purpose: Provide a picture of the distribution.
   B. Show TRANSPARENCY 1.6 (stress-ratings histogram) and use it to explain steps of constructing a histogram.
      1. Begin with frequency table (raw or grouped).
      2. Make a scale of intervals (lowest to highest) along a line at the bottom of the page.
      3. Make a scale of frequencies along a line rising from left edge of bottom line.
      4. Place a box for each case above its interval.
   C. Second example: Show TRANSPARENCY 1.7 or 1.7R (Highly Sensitive Scale histograms) and discuss.
   D. Third example: Show TRANSPARENCY 1.8 (horn-honking example histogram).

III.   Frequency Polygons: Explain principle and relation to histograms using TRANSPARENCY 1.9 (stress-rating example from text).

IV.   Shapes of Distributions
   A. Unimodal, bimodal, and rectangular.
      1. Show TRANSPARENCIES 1.10 and 1.11 (examples of these three shapes from text) and discuss.
      2. Show TRANSPARENCY 1.6 (histograms of stress-rating examples) and discuss.
   B. Symmetric versus skewed.
      1. Show TRANSPARENCIES 1.12 and 1.13 (examples of these shapes from text) and discuss
      2. Show TRANSPARENCY 1.14 or 1.14R (examples of skewed distributions from class questionnaire ratings) and discuss.

C. Normal versus kurtotic distributions.
1. Show TRANSPARENCY 1.15 (examples of these shapes from text) and discuss.
2. Show TRANSPARENCY 1.7 or 1.7R (Highly Sensitive Scale histogram) and discuss as example of an approximately normal distribution.
3. Show TRANSPARENCY 1.16 or 1.16R (examples of skewed distributions from class questionnaire ratings) and discuss.

V. **Review this Class:** Use blackboard outline.

4, 7, 7, 7, 8, 8, 7, 8, 9, 4, 7, 3, 6, 9, 10, 5,
7, 10, 6, 8, 7, 8, 7, 8, 7, 4, 5, 10, 10, 0, 9, 8,
3, 7, 9, 7, 9, 5, 8, 5, 0, 4, 6, 6, 7, 5, 3, 2, 8,
5, 10, 9, 10, 6, 4, 8, 8, 8, 4, 8, 7, 3, 8, 8, 8,
8, 7, 9, 7, 5, 6, 3, 4, 8, 7, 5, 7, 3, 3, 6, 5, 7,
5, 7, 8, 8, 7, 10, 5, 4, 3, 7, 6, 3, 9, 7, 8, 5, 7,
9, 9, 3, 1, 8, 6, 6, 4, 8, 5, 10, 4, 8, 10, 5, 5,
4, 9, 4, 7, 7, 7, 6, 6, 4, 4, 4, 9, 7, 10, 4, 7, 5,
10, 7, 9, 2, 7, 5, 9, 10, 3, 7, 2, 5, 9, 8, 10,
10, 6, 8, 3

TABLE 1-1
Number of Individuals Using Each Value of the Stress Rating Scale

Stress Rating	Frequency
10	14
9	15
8	26
7	31
6	13
5	18
4	16
3	12
2	3
1	1
0	2

Note. Data from Aron, Paris, & Aron (1993).

Aron/Aron
STATISTICS FOR PSYCHOLOGY

© 1994 by Prentice-Hall, Inc.
A Paramount Communications Company
Englewood Cliffs, New Jersey 07632

TRANSPARENCY 1.2

Describing scores with a frequency table.
(Scores from class questionnaire.)

Highly Sensitive Scale--sum of three questions:

    Are you introverted?
    Are you easily overwhelmed by sensory input?
    To what extent are you a highly sensitive person?

1	2	3	4	5	6	7
Not at All			Moderately			Extremely

```
14,15,12,17,18,19,10,15,13,8,17,8,13,
10,14,12,14,11,12,13,4,14,6,9,10,12,
11,15,12,16,11,15,15,13,9,12,14,6,10,
12,11,12,16,4,14,11,7,18,9,13,8,10,9,
12,10,12,14,15,10,13,13,12,14,12,9,13,
5,10,9,9,13,10,10,6,5,10,13,12,8,8,12,
12,14,13,12,14,10,6,15,11,16,13,11,9,9
```

Scale Score	Frequency
19	1
18	2
17	2
16	3
15	7
14	10
13	12
12	16
11	7
10	12
9	9
8	5
7	1
6	4
5	2
4	2

Aron/Aron
STATISTICS FOR PSYCHOLOGY

© 1994 by Prentice-Hall, Inc.
A Paramount Communications Company
Englewood Cliffs, New Jersey 07632

TRANSPARENCY 1.3

Interpersonal hostility measured as delay in
seconds for 29 cars before honking horn at stalled
car after the light has changed to green.
(Fictional data based on Kenrick & McFarland, 1986)

3.5, 2.0, 0, 5.0, .5, 1.0, 4.0, 3.5
3.0, 1.5, 1.5, 2.0, 2.5, 3.0, 3.0,
3.5, 4.5, 2.0, 2.5, 4.5, 4.0, 3.5,
3.0, 2.5, 2.5, 3.5, 3.5, 4.0, 3.0

Frequency Table

		X	f
5.0	/	5.0	1
4.5	//	4.5	2
4.0	///	4.0	3
3.5	//////	3.5	6
3.0	/////	3.0	5
2.5	////	2.5	4
2.0	///	2.0	3
1.5	//	1.5	2
1.0	/	1.0	1
0.5	/	0.5	1
0.0	/	0.0	1

Grouped frequency table (Interval = 1)

Interval	f
5.0 - 5.9	1
4.0 - 4.9	5
3.0 - 3.9	11
2.0 - 2.9	7
1.0 - 1.9	3
0.0 - 0.9	2

Aron/Aron
STATISTICS FOR PSYCHOLOGY

© 1994 by Prentice-Hall, Inc.
A Paramount Communications Company
Englewood Cliffs, New Jersey 07632

TABLE 1-1
Number of Individuals Using Each Value of the Stress Rating Scale

Stress Rating	Frequency
10	14
9	15
8	26
7	31
6	13
5	18
4	16
3	12
2	3
1	1
0	2

Note. Data from Aron, Paris, & Aron (1993).

TABLE 1-3
Grouped Frequency Table for Stress Ratings

Stress Rating Interval	Frequency
10–11	14
8–9	41
6–7	44
4–5	34
2–3	15
0–1	3

Note. Data from Aron, Paris, & Aron (1993).

TRANSPARENCY 1.5

Example of frequency table and grouped frequency table.
(Scores from class questionnaire.)

Highly Sensitive Scale--sum of three questions:

    Are you introverted?
    Are you easily overwhelmed by sensory input?
    To what extent are you a highly sensitive person?

Frequency Table		Grouped Frequency Tables			
		Interval = 2		Interval = 3	
X	f	Interval	f	Interval	f
19	1	18-19	3	18-21	3
18	2	16-17	5	15-17	12
17	2	14-15	17	12-14	38
16	3	12-13	28	9-11	28
15	7	10-11	19	6-8	10
14	10	8-9	14	3-5	4
13	12	6-7	5		
12	16	4-5	4		
11	7				
10	12				
9	9				
8	5				
7	1				
6	4				
5	2				
4	2				

Aron/Aron
STATISTICS FOR PSYCHOLOGY

TRANSPARENCY 1.6

Figure 1.3

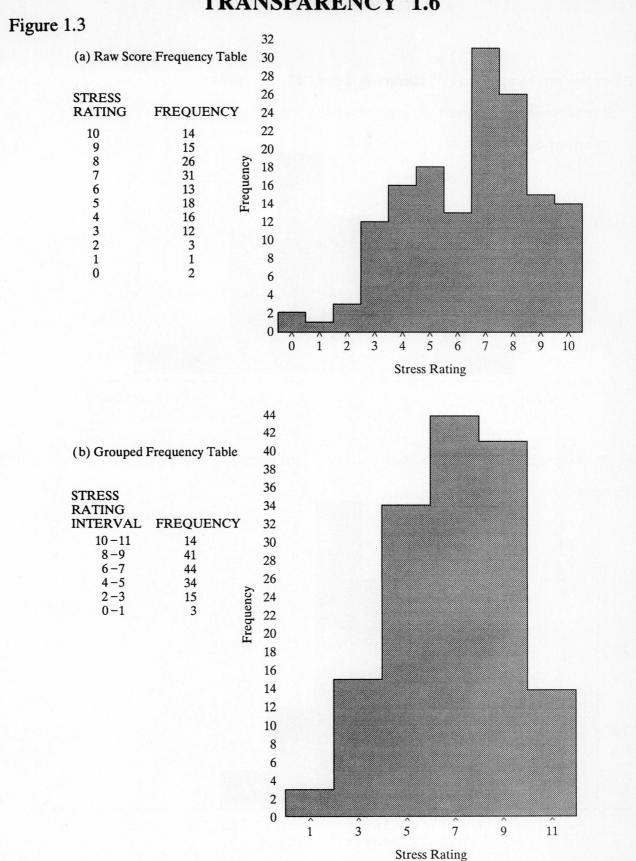

(a) Raw Score Frequency Table

STRESS RATING	FREQUENCY
10	14
9	15
8	26
7	31
6	13
5	18
4	16
3	12
2	3
1	1
0	2

(b) Grouped Frequency Table

STRESS RATING INTERVAL	FREQUENCY
10–11	14
8–9	41
6–7	44
4–5	34
2–3	15
0–1	3

Aron/Aron
STATISTICS FOR PSYCHOLOGY

© 1994 by Prentice-Hall, Inc.
A Paramount Communications Company
Englewood Cliffs, New Jersey 07632

TRANSPARENCY 1.7

Histogram examples. (Scores from class questionnaire.)

Histogram (Grouped Frequencies, Interval Size = 2)

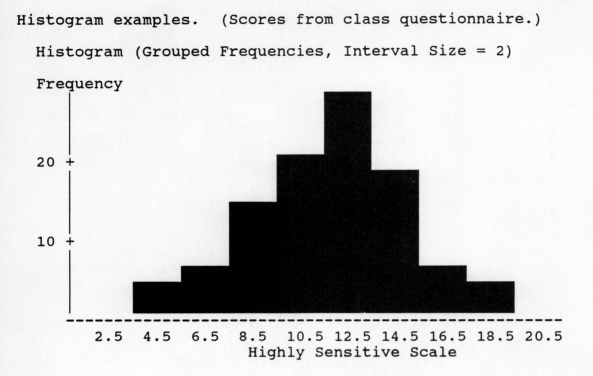

Histogram (Grouped Frequencies, Interval Size = 3)

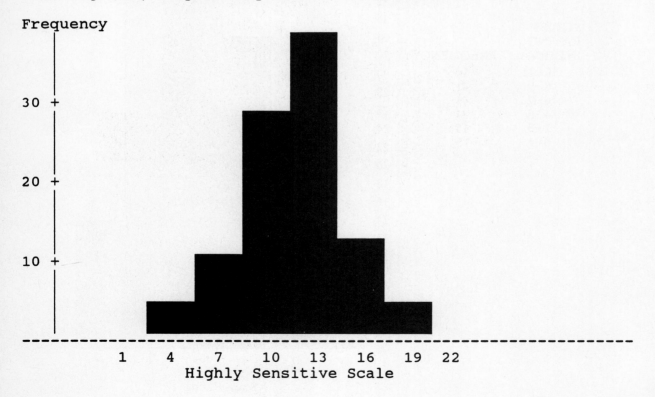

TRANSPARENCY 1.8

Interpersonal hostility measured as delay in seconds for 29 cars before honking horn at stalled car after the light has changed to green. (Fictional data based on Kenrick & McFarland, 1986.)

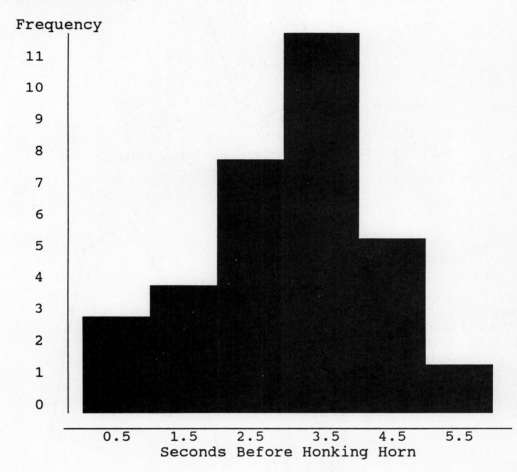

TRANSPARENCY 1.9

Figure 1.6

(a) Frequency Table

STRESS RATING	FREQUENCY
10	14
9	15
8	26
7	31
6	13
5	18
4	16
3	12
2	3
1	1
0	2

(b) Grouped Frequency Table

STRESS RATING INTERVAL	FREQUENCY
10 –11	14
8 –9	41
6 –7	44
4 –5	34
2 –3	15
0 –1	3

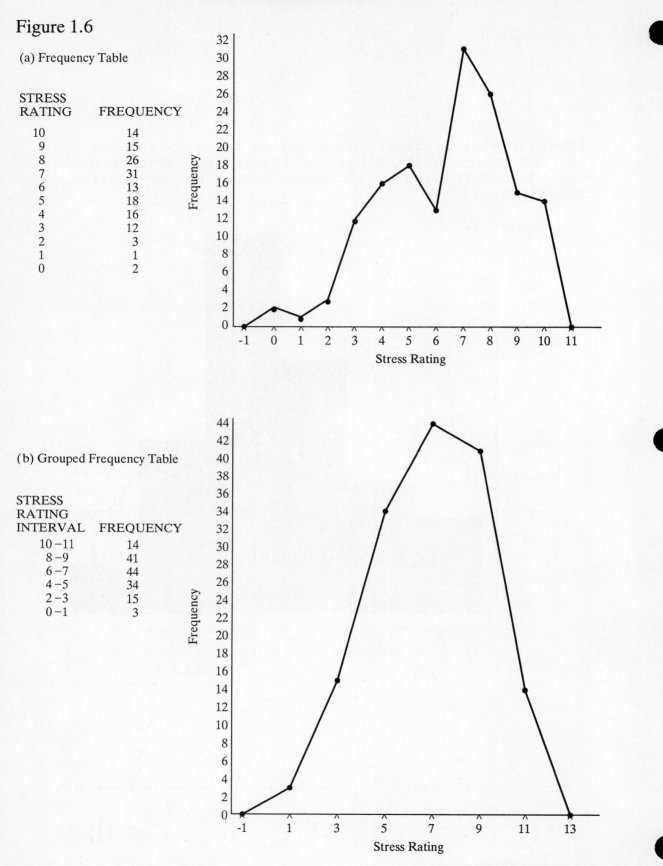

Aron/Aron
STATISTICS FOR PSYCHOLOGY

TRANSPARENCY 1.10

Figure 1.9

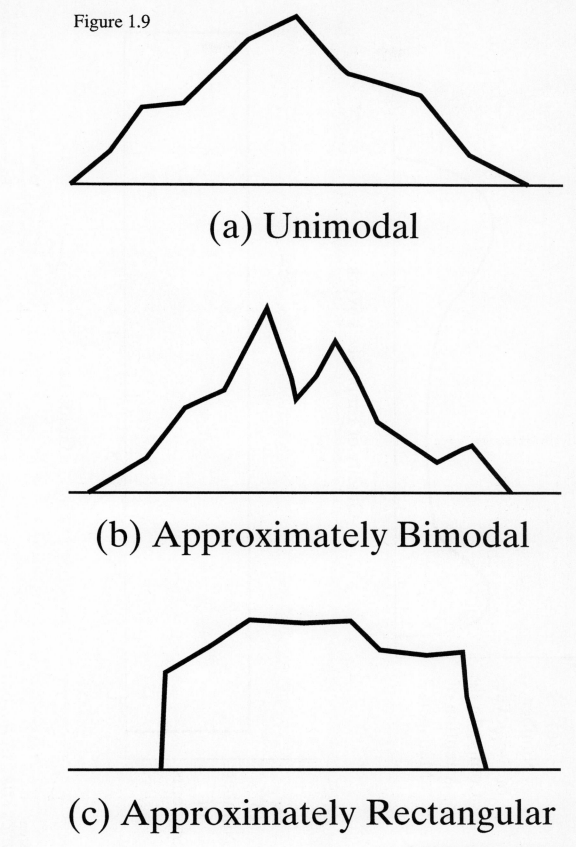

(a) Unimodal

(b) Approximately Bimodal

(c) Approximately Rectangular

Aron/Aron
STATISTICS FOR PSYCHOLOGY

TRANSPARENCY 1.11

Figure 1.10

(a)

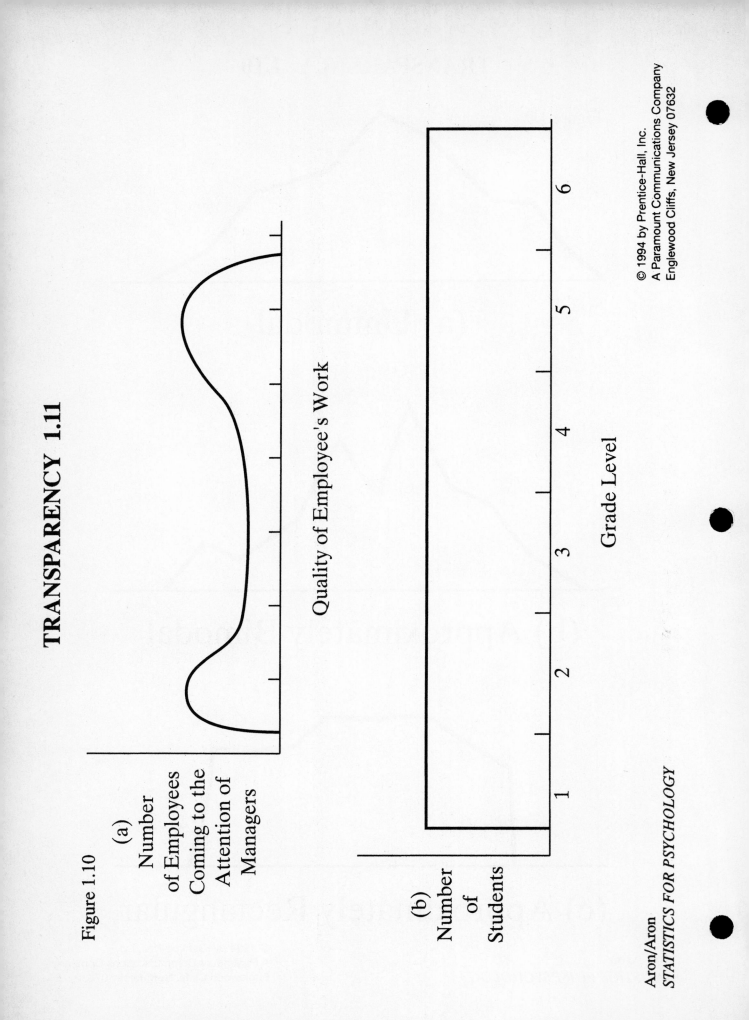

Number
of Employees
Coming to the
Attention of
Managers

Quality of Employee's Work

(b)

Number
of
Students

Grade Level

Aron/Aron
STATISTICS FOR PSYCHOLOGY

© 1994 by Prentice-Hall, Inc.
A Paramount Communications Company
Englewood Cliffs, New Jersey 07632

TRANSPARENCY 1.12

Figure 1.11

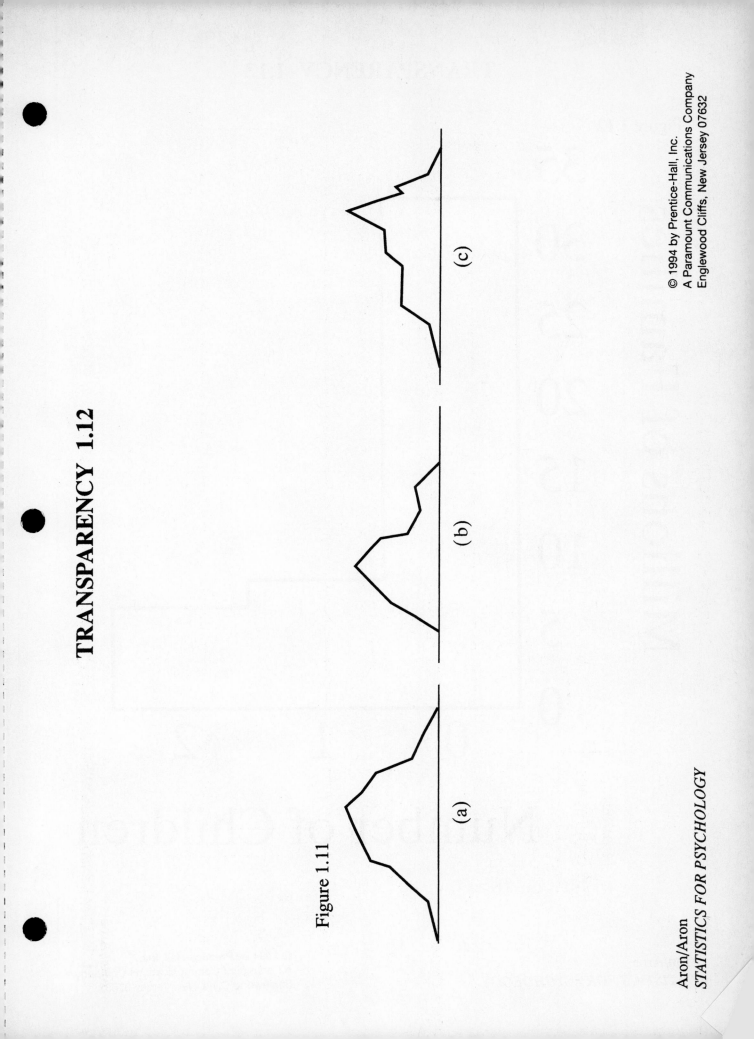

(a)

(b)

(c)

© 1994 by Prentice-Hall, Inc.
A Paramount Communications Company
Englewood Cliffs, New Jersey 07632

Aron/Aron
STATISTICS FOR PSYCHOLOGY

Figure 1.12

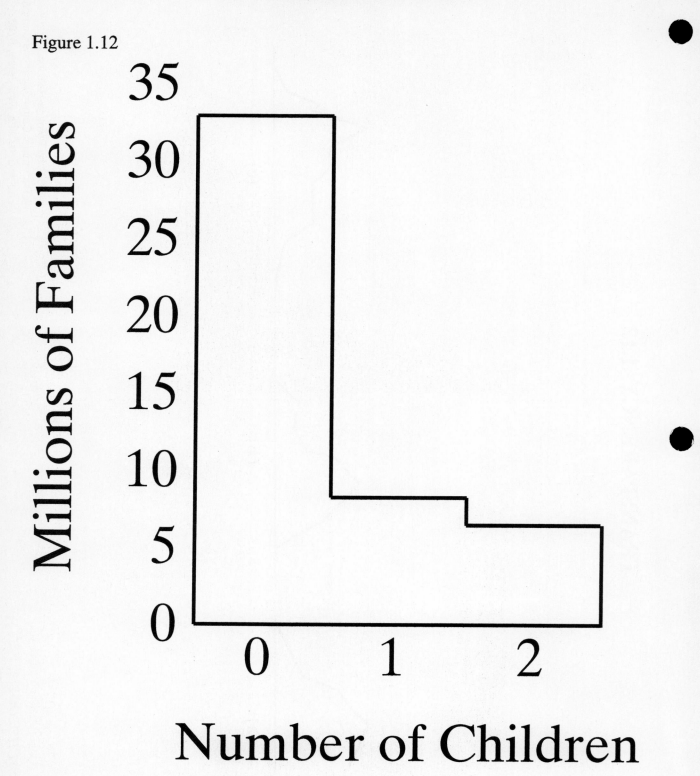

Aron/Aron
STATISTICS FOR PSYCHOLOGY

TRANSPARENCY 1.14

Examples of skewed distributions.
(Scores from class questionnaire.)

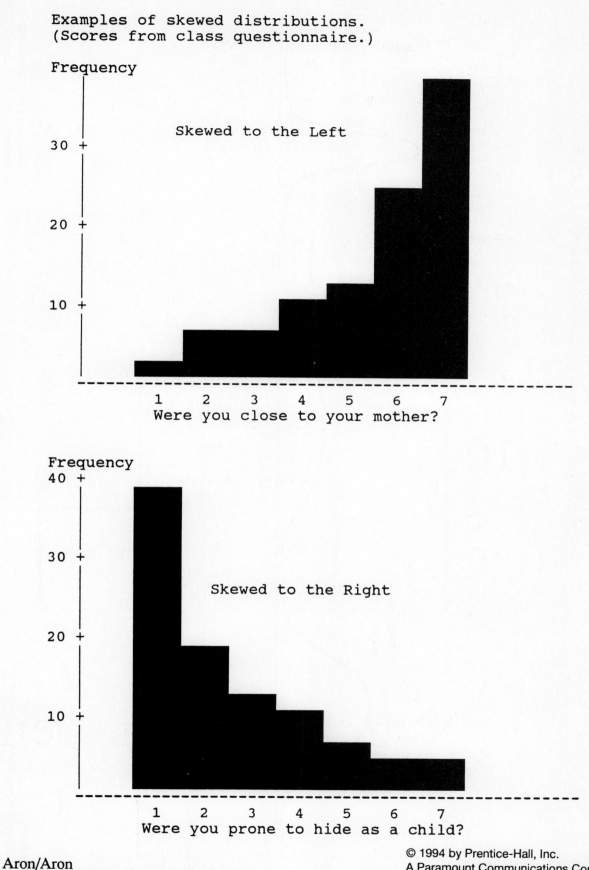

TRANSPARENCY 1.15

Figure 1.14

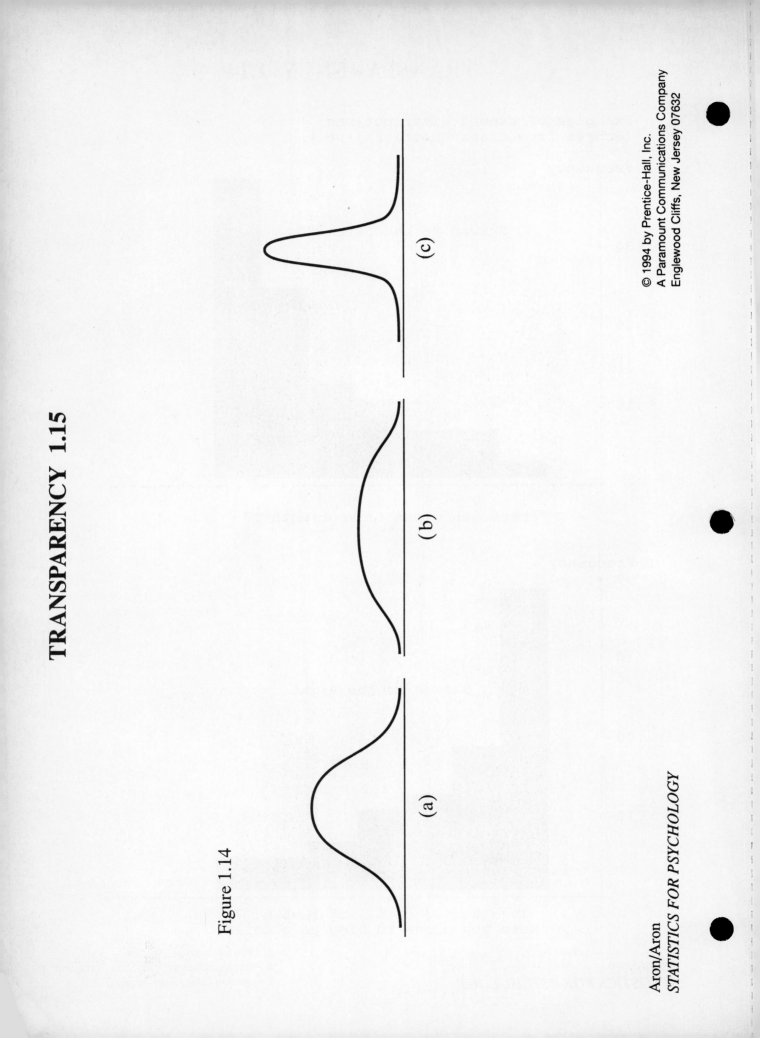

(a)

(b)

(c)

Aron/Aron
STATISTICS FOR PSYCHOLOGY

TRANSPARENCY 1.16

Examples of kurtotic distributions
(Scores from class questionnaire.)

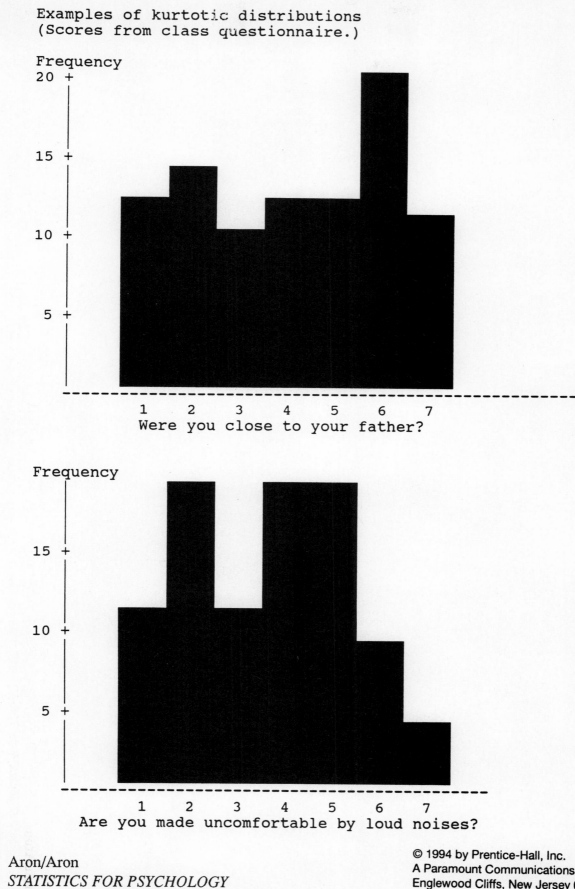

Chapter 2
The Mean, Variance, Standard Deviation, and *Z* Scores

Instructor's Summary of Chapter

Central tendency. The mean is the ordinary average—the sum of the scores divided by the number of scores. Expressed in symbols, $M = \Sigma X/N$. Less commonly used indicators of central tendency of a distribution include the mode, the most common single value, and the median, the value of the middle case if you were to line up all the cases from highest to lowest.

Variation. The spread of the scores in a distribution can be described by the variance—the average of the squared deviation of each score from the mean: $SD^2 = \Sigma(X-M)^2/N$. Alternatively: $SD^2 = SS/N$, where SS is the sum of squared deviations. The standard deviation is the square root of the variance: $SD = \sqrt{SD^2}$. It can be best understood as an approximate measure of the average amount that scores differ from the mean.

Z Scores. A *Z* score is the number of standard deviations a raw score is from the mean of a distribution: $Z = (X-M)/SD$. Among other applications, *Z* scores permit comparisons of scores on different scales.

Controversy: The value of statistics. Some psychologists—especially those associated with behaviorism, humanistic psychology, phenomenology, and qualitative methods—mistrust statistical processes because in the process of creating averages, knowledge about the individual case is lost. Even holding these viewpoints, however, many acknowledge that statistical analysis does play an important role, but argue that when studying any particular topic, careful study of individuals should always come first.

How the procedures of this chapter are described in research articles. Means and standard deviations (but not *Z* scores) are commonly reported in research articles in the text, in tables, or in graphs.

Box 2.1. The Sheer Joy (Yes, Joy) of Statistical Analysis. When several social psychologists were interviewed, many reported that their most enjoyable day-to-day research activity was analyzing data.

Lecture 2.1: The Mean

Materials

Lecture outline
Transparencies 1.1, 1.6, and 2.1 through 2.6
(If using transparencies based on your class's questionnaires, replace 2.2 through 2.5 with 2.2R
through 2.5R.)

Outline for Blackboard

I.   **Review**
II. **Describing the Average**
III. **The Mean**
IV. **Formulas and Symbols:** $M = \Sigma X / N$
V.   **Examples**
VI. **Median and Mode**
VII. **Review this Class**

Instructor's Lecture Outline

I.   **Review**
A. Idea of descriptive statistics and importance for their own right and as a foundation for the rest of
the course.
B. Describing a distribution using frequency tables.
C. Describing a distribution graphically.

II. **Describing the Average**
A. Principle: Summarize a distribution of scores as a single number.
B. Show TRANSPARENCIES 1.1 and 1.6 (stress-ratings example from text).
C. Previously we summarized this group of numbers into a table and graph. Now we want to
summarize it into a single number.

III. **The Mean**
A. Mean is the arithmetic average—sum of scores divided by number of scores.
B. Example calculation: Show TRANSPARENCY 2.1 top (stress-ratings mean computation).
C. Mean as balance point: Show TRANSPARENCY 2.1 bottom.

IV. **Formula and Symbols**
A. Show TRANSPARENCY 2.1 top (stress-rating mean computation) and discuss each symbol.
B. Emphasize value of symbols in statistics and importance of mastering them.

V.   **Examples**
A. First example computation: Show TRANSPARENCY 2.1 and discuss.
B. Second example computation: Show TRANSPARENCIES 2.2 and 2.3 or 2.2R and 2.3R (prone-
to-fears from class questionnaire) and discuss.

C. Third example computation: Show TRANSPARENCIES 2.4 and 2.5 or 2.4R and 2.5R (Highly Sensitive Scale, from class questionnaire) and discuss.

VI. Median and Mode

A. Problem with the mean is that it is highly influenced by extreme scores. Show TRANSPARENCY 2.6 (horn-honking and other examples) and discuss computation of mean with and without additional extreme score.

B. Median.
  1. An alternative to the mean for describing the typical value of a distribution.
  2. The median is the middle score.
  3. Computation:
     a. Organize scores from lowest to highest.
     b. Count to middle score.
     c. If an even number of scores, take the average of the middle two.
  4. Show TRANSPARENCY 2.6 (horn-honking study and other examples) again and discuss:
     a. Computation of median.
     b. How median is not affected by extreme score in horn-honking study.
     c. How median is not affected by extreme scores as shown in example of 1-1-1-1-1-8-9-9-9-9-9 versus 7-7-7-7-7-8-50-50-50-50-50 (that is, both have same median but not same mean).
     d. Computation of median versus mean in feudal village example.

C. Mode.
  1. Another alternative to the mean for describing the typical value of a distribution.
  2. The mode is the value with the most scores.
  3. Computation:
     a. Make a frequency table.
     b. Find the value with the highest frequency.
  4. In horn-honking study it is 3.5.
  5. When mode is most useful:
     a. In combination with mean or median to give a fuller description of a distribution.
     b. To describe a modal interval from a grouped frequency distribution.
     c. There are relatively few values.
     d. The values are categories (that is, this is a distribution over a nominal rather than numeric variable).

VII. Review this Class: Use blackboard outline.

Lecture 2.2: The Variance and Standard Deviation

Materials

Lecture outline

Transparencies 2.1 through 2.9

(If using transparencies based on your class's questionnaires, replace 2.2 through 2.5 with 2.2R through 2.5R.)

Outline for Blackboard

   I.   **Review**
  II.   **Variation**
 III. **Deviation Scores and Squared Deviation Scores**
 IV. **The Variance**
  V.   **Formulas and Symbols:** $SD^2 = \Sigma(X\text{-}M)^2/N = SS/N$
 VI. **The Standard Deviation:** $SD = \sqrt{SD^2}$
VII. **Review this Class**

Instructor's Lecture Outline

I.   **Review**

    A. Idea of descriptive statistics and importance for their own right and as a foundation for the rest of the course.

    B. Describing a distribution using frequency tables and graphs.

    C. Describing a distribution's typical value as the mean: $M = \Sigma X/N$. Show TRANSPARENCY 2.1 (computation of mean for stress ratings) and discuss.

II.   **Variation**

    A. A distribution can be characterized by how much the scores in it vary from each other—they could all be bunched closely together or very spread out, or anywhere in between.

    B. Knowing the mean and variation in a distribution gives a much more complete sense of how scores are distributed than the mean alone.

    C. Examples of possible situations in which the mean SAT is 600 for entering students at a particular college.

        1. All have almost exactly 600—there is little variation.

        2. About half have SAT's of 400 and half have SAT's of 800—there is a very great deal of variation.

        3. About equal numbers having SAT's of 500, 550, 600, 650, 700—a moderate amount of variation.

    D. In general, the amount of variation and mean are independent of each other.

    E. Show TRANSPARENCY 2.7 (distributions with various means and variances from the text) and discuss.

III. **Deviation Scores and Squared Deviation Scores**

    A. One way of determining the variance numerically focuses on the extent to which scores differ from the mean.

    B. A deviation score is the score minus the mean—if a score is 28 and the mean is 25, the deviation score is 3.

    C. Over an entire distribution positive and negative deviation scores balance each other out.

    D. For this and other more complicated reasons, we emphasize squared deviations.

E. Show TRANSPARENCY 2.2 or 2.2R (prone-to-fears example from class questionnaire) and discuss computation of deviation and squared deviation scores.

F. Show TRANSPARENCY 2.4 or 2.4R (Highly Sensitive Scale from class questionnaire) and discuss computation of deviation and squared deviation scores.

IV. The Variance

A. A widely used measure of variation in a distribution.

B. The average of the squared deviation scores.

C. It can be thought of as the average "deviation area"—show TRANSPARENCY 2.8 (figure from text showing squared areas).

D. It is not the average amount that scores differ from the mean; it is the average amount of *squared* differences of scores from the mean. So it will typically be much larger than the average amount that scores differ from the mean.

V.   Formulas and Symbols: $SD^2 = \Sigma(X-M)^2/N = SS/N$

A. Show TRANSPARENCY 2.2 or 2.2R (prone-to-fears example from class questionnaire) and discuss the meaning of each symbol in the computation of variance.

B. Show TRANSPARENCY 2.4 or 2.4R (Highly Sensitive Scale from class questionnaire) and discuss the computation of the variance.

VI.   The Standard Deviation: $SD = \sqrt{SD^2}$

A. The square root of the variance.

B. *Approximately* the average amount that scores differ from the mean in a particular distribution.

C. The most widely used descriptive statistic for describing the variation in a distribution.

D. Show TRANSPARENCY 2.1 (stress-ratings example from text) and discuss computation of standard deviation.

E. Show TRANSPARENCY 2.2 or 2.2R (prone-to-fears example from class questionnaire) and discuss computation of standard deviation at bottom.

F. Show TRANSPARENCY 2.3 or 2.3R (histogram for data in Transparency 2.2) and discuss standard deviation as a distance along the base of the histogram.

G. Show TRANSPARENCY 2.4 or 2.4R (Highly Sensitive Scale from class questionnaire) and discuss computation of standard deviation at bottom.

H. Show TRANSPARENCY 2.5 or 2.5R (histogram for data in Transparency 2.4) and discuss standard deviation as a distance along the base of the histogram.

I. Show TRANSPARENCY 2.6 (horn-honking study) focusing on computation of standard deviation at bottom.

VII. Review this Class: Use blackboard outline and TRANSPARENCY 2.9 (review of mean, variance and standard deviation).

Lecture 2.3: Z Scores

Materials

Lecture outline
Transparencies 2.9 through 2.13
   (If using transparencies based on your class's questionnaires, replace 2.12 with 2.12R.)

Outline for Blackboard

I.   **Review**
II. **Describing a Score in Relation to the Distribution**
III. **Computing the Z Score:** $Z = (X-M)/SD$
IV. **Transforming Z scores to Raw Scores:** $X = (Z)(SD) + M$
V. **Review this Class**

Instructor's Lecture Outline

I.   **Review**
  A. Idea of descriptive statistics and importance for their own right and as a foundation for the rest of the course.
  B. Describing a distribution using frequency tables and graphs.
  C. Describing a distribution's typical value and variation. Show TRANSPARENCY 2.9 (review of mean, variance, and standard deviation) and discuss.

II.   **Describing a Score in Relation to the Distribution**
  A. So far we have described distributions, now we turn to describing a single score's location in a distribution.
  B. Knowing an individual's score gives little information without knowing where that score stands in relation to the entire distribution.
  C. Knowing the mean of the distribution allows you to tell whether a score is above or below the average in that distribution.
  D. Example: An individual has a score of 26 on a leadership test.
     1. What does that indicate? Is the person a particularly good leader? A particularly poor leader? About average?
     2. If the mean on this test is 20, then you know that the person is above average in leadership (compared to other people who have taken this test).
  E. Knowing the standard deviation of the distribution allows you to tell how much above or below the average that score is in relation to the spread of scores in the distribution.
  F. Leadership test example.
     1. Suppose the standard deviation is 3.
     2. The person with a score of 26 is two standard deviations above the mean of 20.
     3. Thus, the person's score is about twice as much above the mean as is the average difference in the scores from the mean.
     4. Show TRANSPARENCY 2.10 (graphic illustration of this example) and discuss how the SD serves as a kind of unit of measure.
     5. (Note that in this example the overall distribution is a normal curve and that when this is the case this approach becomes especially useful, as the student learns in Chapter 5.)

60

G. Second example: Individual scores 84 on a test of planning ability.
    1. If mean is 90 and standard deviation is 12, this person's score is 1/2 standard deviation below the mean.
    2. Thus, the person's score is below average, but not by a lot—about half as much as is the average difference in the scores from the mean.
    3. Show TRANSPARENCY 2.11 (graphic illustration of this example) and discuss.
H. The number of standard deviations a score is above or below the mean is called its Z score.
I. Show TRANSPARENCY 2.11 again and discuss relation of Z scores and raw scores—they are two different ways of measuring the same thing.
J. Z scores provide a helpful way to compare scores on measures that are on completely different scales. For example, if a person scored 26 on leadership and 84 on planning, we can say that the person scores much higher than average on leadership and slightly lower than average on planning.

III. Computing the Z Score: $Z = (X-M)/SD$
A. Show TRANSPARENCY 2.10 (leadership test) and discuss computation of Z score using the formula.
B. Show TRANSPARENCY 2.11 (planning test) and discuss computation of Z score using the formula.
C. Show TRANSPARENCY 2.12 or 2.12R (Highly Sensitive Scale from class questionnaire) and discuss computation of Z scores.
D. Show TRANSPARENCY 2.13 (horn-honking study) and discuss computation of Z scores.

IV. Transforming Z scores to Raw Scores: $X = (Z)(SD) + M$
A. Show TRANSPARENCY 2.10 (leadership test) and discuss conversion of Z scores to raw scores at bottom.
B. Show TRANSPARENCY 2.11 (planning test) and discuss conversion of Z scores to raw scores at bottom.
C. Show TRANSPARENCY 2.12 (Highly Sensitive Scale from class questionnaire) and discuss conversion of Z scores to raw scores at bottom.
D. Show TRANSPARENCY 2.13 (horn-honking study) and discuss conversion of Z scores to raw scores at bottom.

VII. Review this Class
A. Go through blackboard outline.
B. Emphasize importance of knowing Z scores thoroughly as preparation for next topics.

TRANSPARENCY 2.1

Ratings of stress of statistics class students.
(Data from Aron et al., 1993)

Sum of stress ratings = 975
Mean is 151 / 975 = 6.46

$$M = \frac{\Sigma X}{N} = \frac{975}{151} = 6.46$$

$$SD^2 = SS / N = 797.5 / 151 = 5.28$$
$$SD = \sqrt{SD^2} = \sqrt{5.28} = 2.30$$

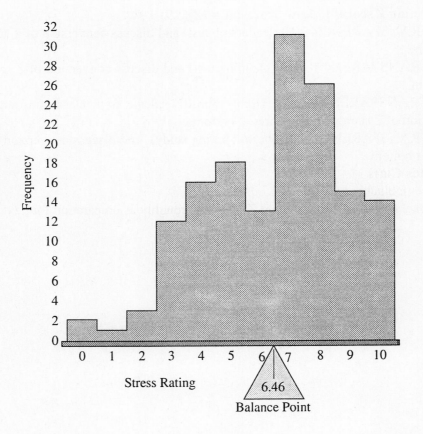

Aron/Aron
STATISTICS FOR PSYCHOLOGY

© 1994 by Prentice-Hall, Inc.
A Paramount Communications Company
Englewood Cliffs, New Jersey 07632

TRANSPARENCY 2.2

Computing the mean, variance, and standard deviation.
(Scores from class questionnaire.)

Are you prone to fears?

1	2	3	4	5	6	7
Not at All			Moderately			Extremely

```
1,4,2,5,6,6,2,5,3,4,5,2,3,2,4,4,5,3,2,1,4,
4,3,2,2,4,1,2,7,6,5,7,5,2,4,4,3,4,3,5,5,4,
4,4,3,3,2,2,4,1,4,3,3,5,6,7,5,6,3,1,5,1,4,
5,5,3,2,3,3,5,3,3,4,5,5,6,5,3,2,2,3,4,1,3,
2,7,2,4,1,7,5,3,4,7,2,3,7,3,2,4,2,2,2,3,4
```

RESPONDENT	RATING		MEAN		DEVIATION	DEVIATION SQUARED
1	1	−	3.64	=	−2.64	6.9696
2	4	−	3.64	=	0.36	0.1296
3	2	−	3.64	=	−1.64	2.6896
4	5	−	3.64	=	1.36	1.8496
5	6	−	3.64	=	2.36	5.5696
6	6	−	3.64	=	2.36	5.5696
7	2	−	3.64	=	−1.64	2.6896
8	5	−	3.64	=	1.36	1.8496
9	3	−	3.64	=	−0.64	0.4096
10	4	−	3.64	=	0.36	0.1296
11	5	−	3.64	=	1.36	1.8496
.	.		.		.	.
.	.		.		.	.
102	2	−	3.64	=	−1.64	2.6896
103	2	−	3.64	=	−1.64	2.6896
104	3	−	3.64	=	−0.64	0.4096
105	4	−	3.64	=	0.36	0.1296

$$\Sigma X = 382 \qquad\qquad SS = 270.2476$$

$$M = \Sigma X/N = 382/105 = 3.64$$

$$SD^2 = \Sigma(X-M)^2/N = SS/N$$
$$= 270.2476/105 = 2.57$$

$$SD = \sqrt{SD^2} = \sqrt{2.57} = 1.60$$

Aron/Aron
STATISTICS FOR PSYCHOLOGY

© 1994 by Prentice-Hall, Inc.
A Paramount Communications Company
Englewood Cliffs, New Jersey 07632

Mean and standard deviation shown on a histogram.
(Scores from class questionnaire.)

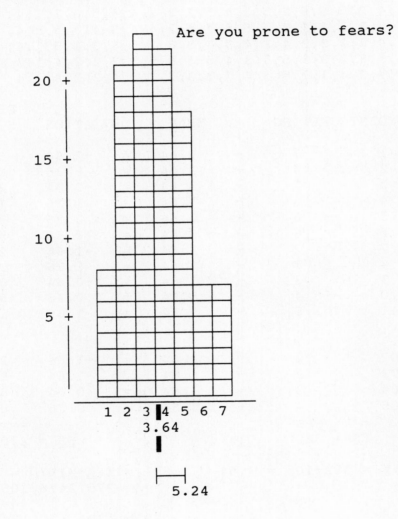

Are you prone to fears?

Aron/Aron
STATISTICS FOR PSYCHOLOGY

TRANSPARENCY 2.4

Computing the mean, variance, and standard deviation.
(Scores from class questionnaire.)

Highly Sensitive Scale (sum of three items):
    Are you introverted?
    Are you easily overwhelmed by sensory input?
    To what extent are you a highly sensitive person?

```
      1     2     3     4     5     6     7
   Not at All      Moderately      Extremely
```

```
14,15,12,17,18,19,10,15,13, 8,17, 8,13,11,
10,14,12,14,11,12,13, 4,14, 6, 9,10,12,16,
11,15,12,16,11,15,15,13, 9,12,14, 6,10,13,
12,11,12,16,12,12,14, 5,13, 9,16,13, 4,14,
11, 7,18, 9,13, 8,10, 9,11,12,10,12,14,15,
10,13,13,12,14,12, 9,13, 9, 5,10, 9, 9,13,
10,10, 6, 5,10,13,12, 8, 8,12, 9,12,14,13,
12,14,10, 6,15,10,15,14,10, 6,17,15,11,12, 6
```

RESPONDENT	SCALE SCORE		MEAN		DEVIATION	DEVIATION SQUARED
1	14	–	11.57	=	2.43	5.9049
2	15	–	11.57	=	3.43	11.7649
3	12	–	11.57	=	0.43	0.1849
4	17	–	11.57	=	5.43	29.4849
5	18	–	11.57	=	6.43	41.3449
6	19	–	11.57	=	7.43	55.2049
.	.		.		.	.
.	.		.		.	.
111	11	–	11.57	=	-0.57	0.3249
112	12	–	11.57	=	0.43	0.1849
113	6	–	11.57	=	-5.57	31.0249

$\Sigma X = 1307$ $SS = 1131.7537$

$M = \Sigma X / N = 1307/113 = 11.57$ $SD^2 = \Sigma(X-M)^2/N = SS/N$
$$= 1131.7537/113 = 10.02$$

$$SD = \sqrt{SD^2} = \sqrt{10.02} = 3.17$$

Aron/Aron
STATISTICS FOR PSYCHOLOGY

© 1994 by Prentice-Hall, Inc.
A Paramount Communications Company
Englewood Cliffs, New Jersey 07632

TRANSPARENCY 2.5

Mean and standard deviation shown on a histogram.
(Scores from class questionnaire.)

Histogram of Highly Sensitivity Scale scores

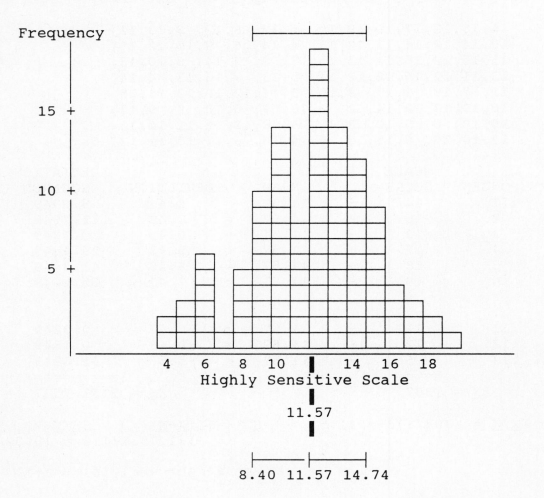

TRANSPARENCY 2.6

Measures of central tendency.
(Fictional data based on Kenrick & McFarland, 1986.)

Seconds honking at stalled car ($\underline{N}$ = 29):

3.5, 2.0,   0, 5.0, .5, 1.0, 4.0, 3.5
3.0, 1.5, 1.5, 2.0, 2.5, 3.0, 3.0,
3.5, 4.5, 2.0, 2.5, 4.5, 4.0, 3.5,
3.0, 2.5, 2.5, 3.5, 3.5, 4.0, 3.0

$\Sigma\underline{X}$ = 82.5 $\underline{N}$ = 29 $\underline{M}$ = $\Sigma\underline{X}$ / $\underline{N}$ = 82.5 / 29 = 3.18

With one additional case of 13 seconds:

$\Sigma\underline{X}$ =82.5+13=95.5 $\underline{N}$=29+1=30 $\underline{M}$=$\Sigma\underline{X}$/$\underline{N}$=95.5/30 = 2.85

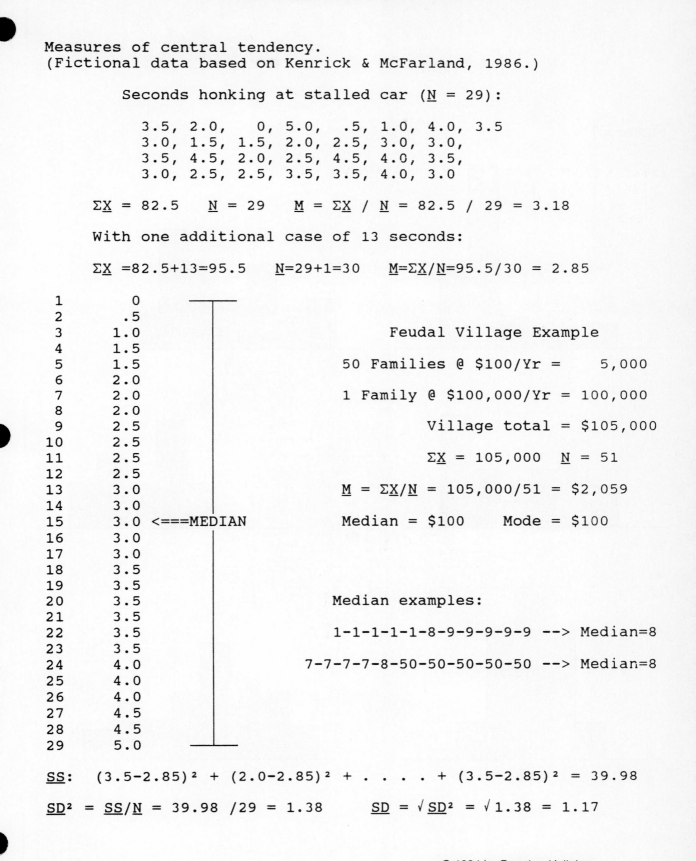

1	0
2	.5
3	1.0
4	1.5
5	1.5
6	2.0
7	2.0
8	2.0
9	2.5
10	2.5
11	2.5
12	2.5
13	3.0
14	3.0
15	3.0 <===MEDIAN
16	3.0
17	3.0
18	3.5
19	3.5
20	3.5
21	3.5
22	3.5
23	3.5
24	4.0
25	4.0
26	4.0
27	4.5
28	4.5
29	5.0

Feudal Village Example

50 Families @ $100/Yr =   5,000

1 Family @ $100,000/Yr = 100,000

Village total = $105,000

$\Sigma\underline{X}$ = 105,000 $\underline{N}$ = 51

$\underline{M}$ = $\Sigma\underline{X}$/$\underline{N}$ = 105,000/51 = $2,059

Median = $100   Mode = $100

Median examples:

1-1-1-1-1-8-9-9-9-9-9 --> Median=8

7-7-7-7-8-50-50-50-50-50 --> Median=8

$\underline{SS}$: (3.5-2.85)² + (2.0-2.85)² + ... + (3.5-2.85)² = 39.98

$\underline{SD}$² = $\underline{SS}$/$\underline{N}$ = 39.98 /29 = 1.38 $\underline{SD}$ = $\sqrt{\underline{SD}^2}$ = $\sqrt{1.38}$ = 1.17

Aron/Aron
STATISTICS FOR PSYCHOLOGY

TRANSPARENCY 2.7

Figure 2.9

(a)

(b)

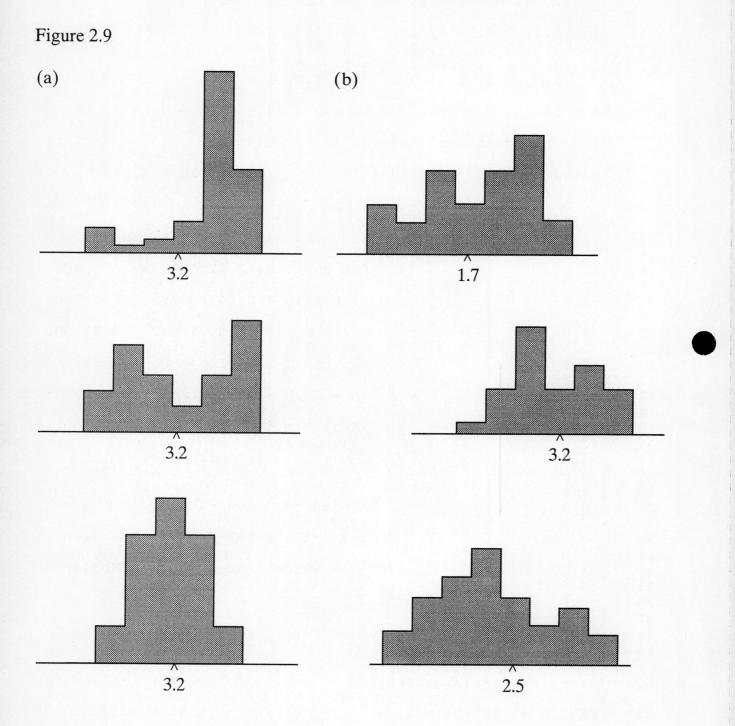

3.2

1.7

3.2

3.2

3.2

2.5

Aron/Aron
STATISTICS FOR PSYCHOLOGY

Figure 2.10

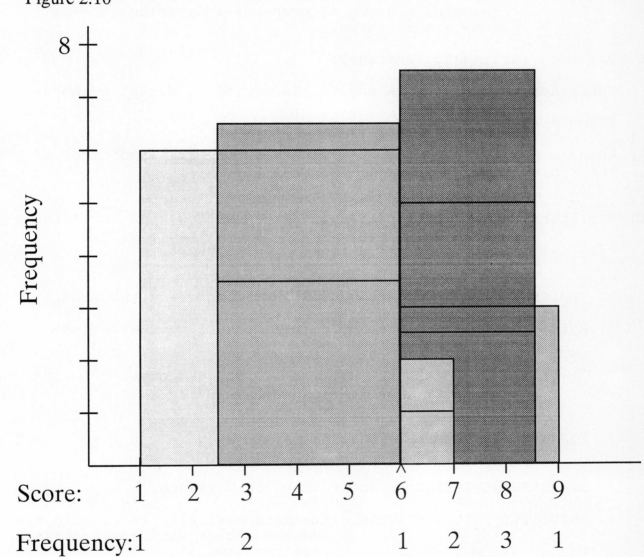

Score: 1 2 3 4 5 6 7 8 9

Frequency: 1   2   1   2   3   1

TRANSPARENCY 2.9

Review of Mean, Variance, and Standard Deviation

A. Principle: Describe (these are descriptive statistics) by
              reducing a group of numbers--a distribution--to
              some simple terms.

B. The mean (arithmetic average):

   RULE: Add up the numbers and divide by the number of numbers

   FORMULA AND SYMBOLS: $\underline{M} = \Sigma\underline{X}/\underline{N}$.

   Mean is most widely used, and generally the best indicator of
   the typical score.

C. Variation: A single number that describes how much variation
              there is in a group of numbers—that is, how
              spread out or narrow a distribution is.

   1. VARIANCE is average of squared deviations from the mean.

      RULE: a. Subtract each score from mean to get deviation
               score.
            b. Square each deviation score.
            c. Add up all the squared deviation scores.
            d. Divide by number of cases to get average of
               squared deviation scores.

      FORMULA AND SYMBOLS: $\underline{SD}^2 = \Sigma(\underline{X}-\underline{M})^2 / \underline{N} = \underline{SS} / \underline{N}$.

   2. STANDARD DEVIATION is square root of variance.

      INTUITIVE INTERPRETATION: Standard deviation is roughly the
                                 average amount each score differs
                                 from the mean.

      RULE: Compute variance and take the square root.

      FORMULA AND SYMBOLS: $\underline{SD} = \sqrt{\underline{SD}^2}$.

Aron/Aron
STATISTICS FOR PSYCHOLOGY

© 1994 by Prentice-Hall, Inc.
A Paramount Communications Company
Englewood Cliffs, New Jersey 07632

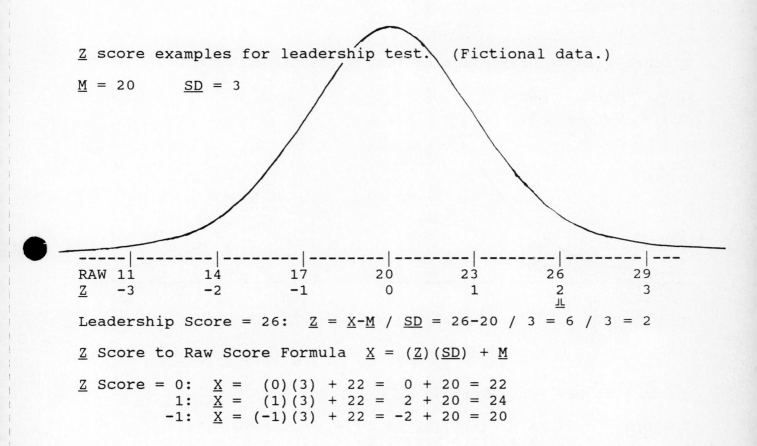

Z score examples for leadership test.   (Fictional data.)

M = 20   SD = 3

```
RAW 11        14        17        20        23        26        29
Z   -3        -2        -1         0         1         2         3
                                                      ⊥L
```

Leadership Score = 26: Z = X-M / SD = 26-20 / 3 = 6 / 3 = 2

Z Score to Raw Score Formula X = (Z)(SD) + M

```
Z Score = 0:   X =   (0)(3) + 22 =   0 + 20 = 22
          1:   X =   (1)(3) + 22 =   2 + 20 = 24
         -1:   X =  (-1)(3) + 22 =  -2 + 20 = 20
```

Aron/Aron
STATISTICS FOR PSYCHOLOGY

© 1994 by Prentice-Hall, Inc.
A Paramount Communications Company
Englewood Cliffs, New Jersey 07632

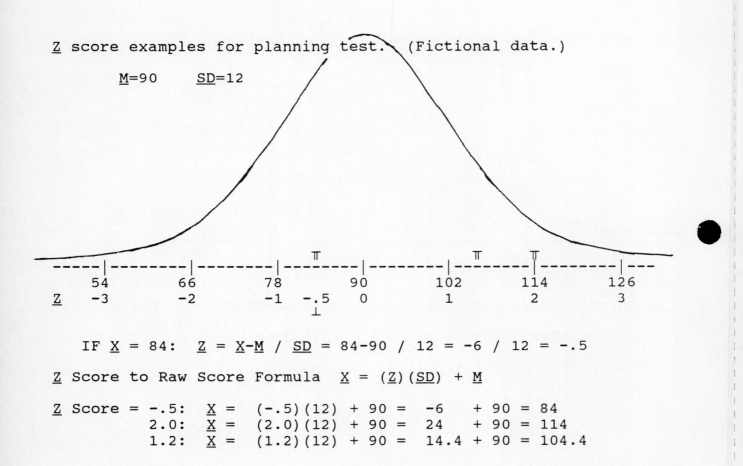

Z score examples for planning test.   (Fictional data.)

M=90   SD=12

$$IF \underline{X} = 84: \quad \underline{Z} = \underline{X}\text{-}\underline{M} / \underline{SD} = 84\text{-}90 / 12 = \text{-}6 / 12 = \text{-}.5$$

Z Score to Raw Score Formula $\underline{X} = (\underline{Z})(\underline{SD}) + \underline{M}$

$$
\begin{array}{llll}
\underline{Z}\text{ Score} = \text{-}.5: & \underline{X} = & (\text{-}.5)(12) + 90 = & \text{-}6 & + 90 = 84 \\
2.0: & \underline{X} = & (2.0)(12) + 90 = & 24 & + 90 = 114 \\
1.2: & \underline{X} = & (1.2)(12) + 90 = & 14.4 & + 90 = 104.4
\end{array}
$$

© 1994 by Prentice-Hall, Inc.
A Paramount Communications Company
Englewood Cliffs, New Jersey 07632

TRANSPARENCY 2.12

Z score examples for Highly Sensitive Scale.
(Scores from class questionnaire.)

Highly Sensitive Scale

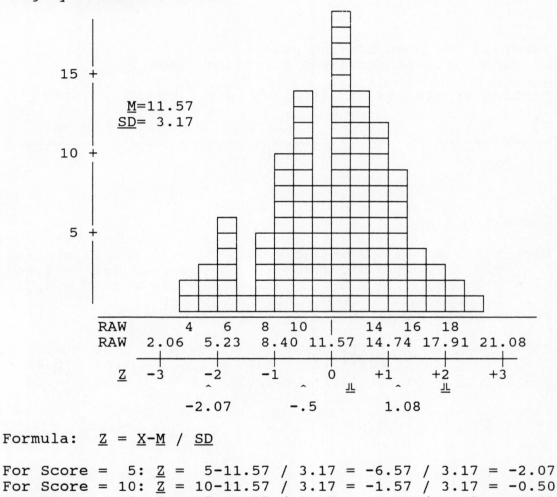

M=11.57
SD= 3.17

| RAW | | 4 | | 6 | | 8 | | 10 | | | | 14 | | 16 | | 18 | | |
| RAW | 2.06 | | 5.23 | | | 8.40 | | 11.57 | | 14.74 | | 17.91 | | 21.08 |

Formula: Z = X-M / SD

For Score = 5: Z = 5-11.57 / 3.17 = -6.57 / 3.17 = -2.07
For Score = 10: Z = 10-11.57 / 3.17 = -1.57 / 3.17 = -0.50
For Score = 15: Z = 15-11.57 / 3.17 = 3.43 / 3.17 = 1.08

Z Score to Raw Score Formula: X = (Z)(SD) + M

Z Score = 2.00: X = (2.00)(3.17) + 11.57 = 6.34 + 11.57 = 17.91
         -2.07: X = (-2.07)(3.17) + 11.57 = -6.56 + 11.57 = -5.01
          0.38: X = (.38)(3.17) + 11.57 = 1.20 + 11.57 = 12.77

Aron/Aron
STATISTICS FOR PSYCHOLOGY

© 1994 by Prentice-Hall, Inc.
A Paramount Communications Company
Englewood Cliffs, New Jersey 07632

TRANSPARENCY 2.13

$\underline{Z}$ score examples for horn honking study.
(Fictional data based on Kenrick & McFarland, 1986.)

Seconds honking at stalled car ($\underline{N}$ = 29) $\underline{M}$ = 2.85, $\underline{SD}$ = 1.17

```
Raw   0.51        1.68         2.85         4.02         5.19         6.36
   ----|---------|---------|---------|---------|---------|-------
   Z    -2          -1           0            1            2            3
                                 ^                         ^                        ^
                                .30                      1.84                      3.55
```

Formula: $\underline{Z} = \underline{X}-\underline{M}$ / $\underline{SD}$

Time to Honk = 2.5 s: $\underline{Z}$ = 2.5-2.85 / 1.17 = -.35 / 1.17 = -.30
             = 5.0 s: $\underline{Z}$ = 5.0-2.85 / 1.17 = 2.15 / 1.17 = 1.84
             = 7.0 s: $\underline{Z}$ = 7.0-2.85 / 1.17 = 4.15 / 1.17 = 3.55

$\underline{Z}$ Score to Raw Score Formula: $\underline{X}$ = ($\underline{Z}$)($\underline{SD}$) + $\underline{M}$

$\underline{Z}$ Score = 1 : $\underline{X}$ =   (1)(1.17) + 2.85 = 1.17 + 2.85 = 4.02
             0 : $\underline{X}$ =   (0)(1.17) + 2.85 =   0   + 2.85 = 2.85
           -2.11: $\underline{X}$ = (-2.11)(1.17) + 2.85 = -2.47 + 2.85 = -.38

Aron/Aron
STATISTICS FOR PSYCHOLOGY

© 1994 by Prentice-Hall, Inc.
A Paramount Communications Company
Englewood Cliffs, New Jersey 07632

Chapter 3
Correlation

Instructor's Summary of Chapter

Scatter diagrams describe the relation between two variables by representing each individual's pair of scores as a dot on a two-dimensional graph of the two variables.

Linear, Curvilinear, Positive, and Negative Correlations. The general pattern of dots in the scatter diagram describe a linear correlation when they roughly fall around a straight line (they describe a positive linear correlation when that line goes upward to the right and a negative linear correlation when it goes downward to the right). It is a curvilinear correlation when they follow a line other than a simple straight line, and no correlation when they do not follow any kind of line.

The correlation coefficient (r), the standard numeric index of linear correlation, is the average of the cross products of Z scores. r approaches 1 when there is a strong positive linear correlation, because positive Z scores are multiplied by positive and negative Z scores by negative; r approaches -1 when there is a strong negative linear correlation, because positive Z scores are multiplied by negative and negative by positive; and r is zero when there is no linear correlation, because positive and negative Z score cross products cancel each other out.

Hypothesis testing. A correlation is said to be significant when hypothesis testing procedures (covered in later chapters and the second appendix to Chapter 3) applied to your data support the hypothesis that a correlation exists in the larger group which your data were intended to represent.

Issues in interpreting the correlation coefficient. (a) The proportionate reduction in error (r^2) is considered the best indicator of degree of relationship. (b) If X and Y are correlated, X could be causing Y, Y causing X, or a third variable causing both. (c) A computed correlation coefficient can underrepresent the true correlation in the general group if the smaller group selected for study is restricted in range or if the scores are obtained with unreliable measures.

Controversies and Recent Developments: What is a Large Correlation? Many psychologists argue that r is an overestimate of the importance of an association and psychologists do tend to overestimate the degree of association implied by a given r. On the other hand, Rosnow and Rosenthal have shown that in some cases even very small rs can have great practical importance.

How the procedures of this chapter are described in research articles. Correlational results are usually presented either in the text with the value of r or in a correlation matrix.

Box 3.1. Galton: Gentleman Genius. Briefly describes the background and character of Francis Galton and what led him to the work on statistics that eventually resulted in the correlation coefficient.

Box 3.2. Illusory Correlation: When You Know Perfectly Well.... Summarizes social psychological research showing that people tend to mistakenly link infrequent (and therefore highly memorable) traits or processes. or mistakenly link variables simply as a result of social prejudices.

Lecture 3.1: The Scatter Diagram and the Logic of Correlation

Materials

Lecture outline
Transparencies 3.1 through 3.9
(If using transparencies based on your class's questionnaires, replace 3.2, 3.5, 3.6 and 3.8 with 3.2R, 3.5R, 3.6R, and 3.8R.)

Outline for Blackboard

I.   **Review**
II. **The Scatter Diagram**
III. **Patterns of Association**
    A. **Linear**
    B. **Positive linear**
    C. **Negative linear**
    D. **Curvilinear and No Relationship**
IV. **Review this Class**

Instructor's Lecture Outline

I.   Review
    A. Idea of descriptive statistics.
    B. Describing distributions by frequency tables and graphs.
    C. Describing distributions by their mean and standard deviation.
    D. Describing a score within a distribution: The Z score as the number of standard deviations above or below the mean.

II.   The Scatter Diagram
    A. Consider the example from the text of a study of stress level of managers who supervise different number of employees.
        1. Each person has a score on both variables. Are they related?
        2. To see, we can make a graph—show TRANSPARENCY 3.1 (number supervised and stress level from text) here and throughout the discussion of lecture outline points IIA and IIB.
        3. Note that the general pattern of dots goes up—the more supervised, the more stress. (Draw the approximate regression line with a transparency pen.)
        4. Line is *positive*, highs with highs and lows with lows.
        5. This is just the general trend—almost none of the dots fall exactly on the line and many are a distance away.
        6. A chart like this is called a scatter diagram.
        7. A scatter diagram is a visual description of the relation between two variables.
    B. Making a scatter diagram.
        1. Shape is square.
        2. One variable goes on each axis:
          a. If it is clear that one is supposed to be causal (independent), that variable goes on the horizontal axis.
          b. If it is not clear which causes which, it does not matter which variable goes on which axis.
        3. Make the scale go from lowest to highest possible value on the variables being measured.
        4. Place dots in their appropriate location.
        5. Discuss how to handle two dots in the same place.

C. Example 2: Show TRANSPARENCY 3.2 or 3.2R (scatter diagram of being highly sensitive and being prone to fears, from class questionnaire) and discuss.

D. Example 3: Show TRANSPARENCY 3.3 (scatter diagram of fictional data based on Orme-Johnson & Haynes, 1981) and discuss.

E. Example 4: Show TRANSPARENCY 3.4 (scatter diagram of fictional data based on Blankstein & Toner, 1987) and discuss.

III. Patterns of Association

A. Linear correlation.
   1. Describes situation where the pattern of dots fall roughly in a straight line.
   2. Examples:
      a. Show TRANSPARENCY 3.1 (supervisor's stress).
      b. TRANSPARENCY 3.2 or 3.2R (Highly Sensitive Scale and prone to fears).

B. Positive linear correlation.
   1. When the pattern of dots goes up from left to right.
   2. Where highs on one variable go with highs on the other, lows on one with lows on the other, and middle scores on one with middle scores on the other.
   3. Examples:
      a. Show TRANSPARENCY 3.1 (supervisor's stress).
      b. Show TRANSPARENCY 3.2 or 3.2R (Highly Sensitive Scale and prone to fears).
      c. Show TRANSPARENCY 3.3 (EEG coherence and creativity).
      d. Show TRANSPARENCY 3.5 or 3.5R (prone to fears with worried, from class questionnaire).
      e. Show TRANSPARENCY 3.6 or 3.6R (father involved and close to father, from class questionnaire).

C. Negative (or inverse) linear correlation.
   1. When the pattern of dots goes down from left to right.
   2. When highs on one variable go with lows on the other.
   3. Examples:
      a. Show TRANSPARENCY 3.7 (fictional study of marital adjustment and overtime hours worked).
      b. Show TRANSPARENCY 3.4 (Blankstein & Toner), noting that this is not a very strong association.
      c. Show TRANSPARENCY 3.8 or 3.8R (fond of mother and fall in love hard, from class questionnaire), again noting that this is not a very strong association.

D. No relationship and curvilinear.
   1. No relationship: Show TRANSPARENCY 3.9 (creativity and shoe size, from text) and discuss.
   2. Curvilinear: Show TRANSPARENCY 3.10 (children's rate of substituting digit symbols and their motivation, from text).

IV. Review this Class: Use blackboard outline.

Lecture 3.2: The Correlation Coefficient

Materials

Lecture outline
Transparencies 2.10, 3.1 through 3.4, 3.7, 3.11, 3.12

Outline for Blackboard

I.   **Review**
II. **Purpose of the Correlation Coefficient**
III. **Logic of Computing the Correlation Coefficient**
IV. **Formulas and Symbols:** $r = \Sigma(Z_X Z_Y) / N$
V.   **Interpreting the Correlation Coefficient**
VI. **Review this Class**

Instructor's Lecture Outline

I. **Review**
    A. Idea of descriptive statistics.
    B. Describing a single distribution:
        1. Frequency tables and graphs.
        2. Mean and standard deviation.
    C. Describing a score within a distribution.
        1. The Z score as the number of standard deviations above or below the mean.
        2. Computation: $Z = (X-M)/SD$.
        3. Example: Show TRANSPARENCY 2.10 (Z score examples for leadership test) and discuss.
    D. Describing the relation between scores in two distributions.
        1. The scatter diagram: Show TRANSPARENCY 3.1 (manager's stress example from text) and discuss.
        2. Directions of relationships:
            a. Positive.
            b. Negative.
            c. Curvilinear.

II. **Purpose of the Correlation Coefficient**
    A. Describes the degree and direction of linear correlation.
    B. Such a measure would be especially useful if it has the following properties:
        1. Numeric on a standard scale for all uses.
        2. A positive number for positive correlations and a negative number for negative correlations.
        3. A zero for no correlation.
        4. A +1 for a perfect positive linear correlation (and a -1 for a perfect negative linear correlation).
        5. The closer a correlation is to perfect, the closer it is to +1 (or -1 if it is negative) and the further from 0.

III. **Logic of Computing the Correlation Coefficient**
    A. Interpreting correlation in terms of patterns of highs and lows.
        1. A correlation is strong and positive if highs on one variable go with highs on the other, and lows with lows.
        2. A correlation is strong and negative if lows with highs, and highs with lows.
        3. There is no correlation if sometimes highs go with highs and sometimes with lows.
    B. But what is "high" and "low"?
        1. How can we compare scores as high and low that are on completely different scales in completely different distributions?

78

2. Solution is to use Z scores.

3. A high score in a distribution always has a positive Z score.

4. A low score in a distribution always has a negative Z score.

C. Multiplying the Z scores of each person's two scores produces the following effects:

  1. With a perfect positive correlation:

  a. Highs always go with highs—multiplying two highs will be multiplying two positive Z scores. Thus the result is always positive.

  b. Lows go with lows—multiplying two lows will be multiplying two negative Z scores. Thus the result is always positive.

  2. With a perfect negative correlation:

  a. Highs on the first variable always go with lows on the second—multiplying a high by a low will be multiplying a positive Z score times a negative Z score. Thus the result is always negative.

  b. Lows on the first variable always go with highs on the second—multiplying a low by a high will be multiplying a negative Z score times a positive Z score. Thus the result is always negative.

  3. With no correlation:

  a. Sometimes highs go with highs and lows with lows, making positive products.

  b. Sometimes highs go with lows and lows with highs, making negative products.

  c. The result over all is that the positives and negatives cancel each other out.

  4. In between degrees of correlation:

  a. To the extent a correlation is positive, there will be more cases of highs with highs and lows with lows than the reverse, making the overall result somewhat positive.

  b. To the extent a correlation is negative, there will be more cases of highs with lows and lows with highs than of both the same, making the overall result somewhat negative.

D. Each multiplication of a person's two Z scores is called a *cross-product of Z scores*.

E. The average of the cross-products of Z scores is called the correlation coefficient.

  1. With a perfect positive correlation, the average of the cross-products of Z scores is +1.

  2. With a perfect negative correlation, the average of the cross-products of Z scores is -1.

  3. With no correlation, the average of the cross-products of Z scores is 0.

  4. With a positive correlation that is not perfect, the average of the cross-products of Z scores is between 0 and +1.

  5. With a negative correlation that is not perfect, the average of the cross-products of Z scores is between 0 and -1.

F. Show TRANSPARENCY 3.1 (manager's stress example from text) and discuss how the average of the cross-products of Z scores comes out near +1.

G. Show TRANSPARENCY 3.7 (fictional study of marital adjustment and overtime hours worked) and discuss how the average of the cross-products of Z scores comes out near -1.

H. Show TRANSPARENCY 3.4 (Blankston & Toner) and discuss how the average of the cross-products of Z scores comes out near to 0.

IV. **Formulas and Symbols:** $r = \Sigma(Z_x Z_y) / N$

A. Show TRANSPARENCY 3.3 (EEG coherence and creativity) and discuss each symbol in the formula.

B. Discuss computation using the formula and show computations in the following:

  1. TRANSPARENCY 3.2. or 3.2R (highly sensitive and prone to fears, from class questionnaire).

  2. TRANSPARENCY 3.4 (Blankston & Toner).

  3. TRANSPARENCY 3.7 (marital adjustment and overtime hours).

V. Interpreting the Correlation Coefficient

A. When comparing correlations, use the square of the correlation coefficient, not r itself.

B. A correlation between two variables does not indicate the direction of causality.

    1. Show TRANSPARENCY 3.11 (directions of causality from text) and discuss the three possible directions of causality.

    2. Show TRANSPARENCY 3.2 or 3.2R (Highly Sensitive Scale and prone to fears, from class questionnaire) and discuss possible directions of causality:

    a. Being highly sensitive could make one prone to fears.

    b. Being prone to fears could make one highly sensitive.

    c. Some third factor, such as genetically endowed temperament, could make a person both highly sensitive and prone to fears.

    3. Show TRANSPARENCY 3.7 (marital satisfaction and overtime hours example) and discuss possible directions of causality:

    a. Being dissatisfied with one's marriage could make one choose to work more overtime hours.

    b. Working more overtime hours could make one less satisfied with one's marriage.

    c. Some third factor, such as being very poor, could make one choose to work more overtime hours (to make needed money) and also put excessive stress on the marriage.

C. Restriction in range.

    1. Principle: If a subgroup of people is studied that represents a specific level of one variable, this could result in the correlation underrepresenting what it would be if everyone had been studied.

    2. Show TRANSPARENCY 3.12 (scatter diagram showing effect of restriction in range from text) and discuss.

D. Unreliability of measurement.

    1. Principle: If either of the variables is measured with low reliability, the correlation computed is lower than it would be if the variables were more precisely measured.

    2. Reliability is the extent to which a measure is consistent.

    a. Loosely speaking, an unreliable measure is "sloppy": for example, a questionnaire with a lot of ambiguous items, or a rating scale that allows for lots of subjective interpretation that could change with the mood of the rater.

    b. For more explanation of reliability, refer students to Appendix A and Chapter 17 in the text.

    3. An unreliable measure reduces the correlation because it adds a random element. This random element makes the dots less likely to fall closely to the line; or to put it another way, highs and highs (or highs with lows if it is a negative correlation) will not go together as consistently as they would if the measure did not include the random element.

VI. Review this Class: Use blackboard outline.

TABLE 3-1
Employees Supervised and Stress Level (Fictional Data)

Employees Supervised	Stress Level on Questionnaire
6	7
8	8
3	1
10	8
8	6

Figure 3.4

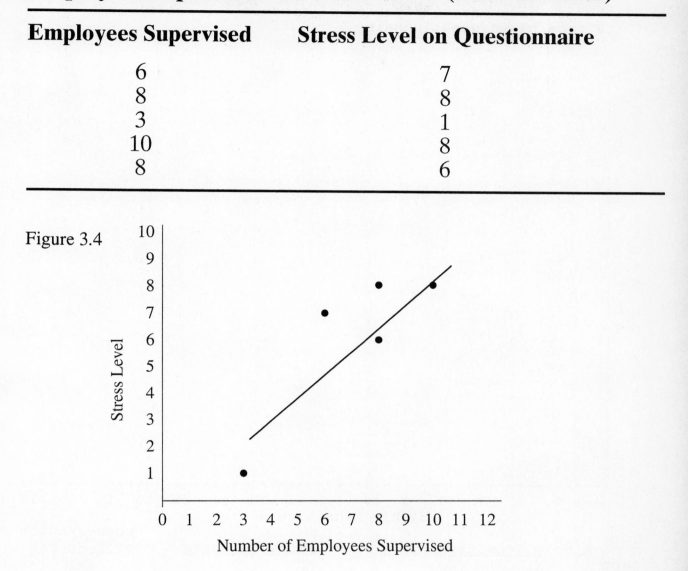

Number of Employees Supervised

TABLE 3-2
Calculations for the Correlation Coefficient for the Managers' Stress Study (Fictional Data)

Number of Employees Supervised (X)				Stress Level (Y)				Cross-Products
X	$X-M$	$(X-M)^2$	Z_X	Y	$Y-M$	$(Y-M)^2$	Z_Y	$Z_X Z_Y$
6	−1	1	−.42	7	1	1	.38	−.16
8	1	1	.42	8	2	4	.77	.32
3	−4	16	−1.69	1	−5	25	−1.92	3.24
10	3	9	1.27	8	2	4	.77	.98
8	1	1	.42	6	0	0	0.00	0.00

(= 35 SS = 28   (= 30 SS = 34 $\Sigma Z_X Z_Y = 4.38$
M = 7 SD^2 = 5.60 M = 6 SD^2 = 6.80 r = .88
        SD = 2.37 SD = 2.61

Aron/Aron
STATISTICS FOR PSYCHOLOGY

© 1994 by Prentice-Hall, Inc.
A Paramount Communications Company
Englewood Cliffs, New Jersey 07632

TRANSPARENCY 3.2

Scatter diagram and computation of correlation coefficient.
(Data from class questionnaire.)

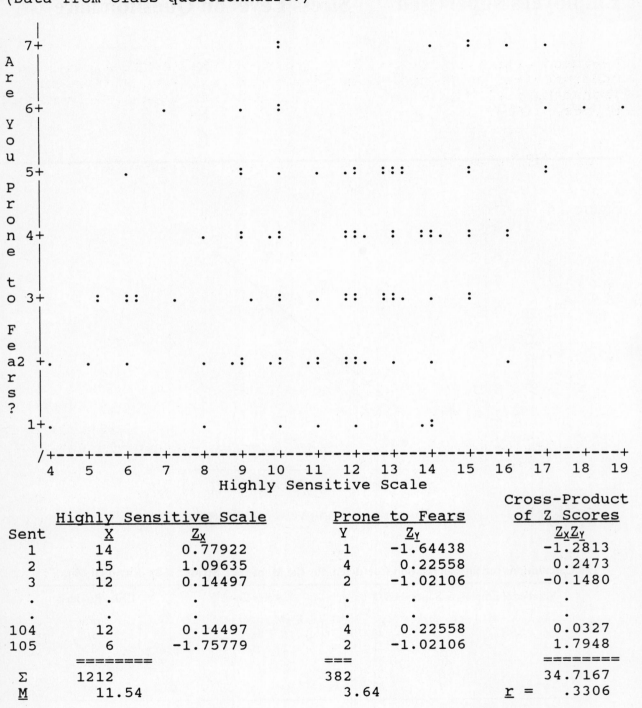

Highly Sensitive Scale

	Highly Sensitive Scale			Prone to Fears		Cross-Product of Z Scores
Sent	X	Z_X		Y	Z_Y	$Z_X Z_Y$
1	14	0.77922		1	-1.64438	-1.2813
2	15	1.09635		4	0.22558	0.2473
3	12	0.14497		2	-1.02106	-0.1480
.	.	.		.	.	.
.	.	.		.	.	.
104	12	0.14497		4	0.22558	0.0327
105	6	-1.75779		2	-1.02106	1.7948
	========			===		========
Σ	1212			382		34.7167
M	11.54			3.64		r = .3306

TRANSPARENCY 3.3

Measured creativity using the Torrence Tests and EEG alpha-band coherence in the frontal area in subjects who were practicing Transcendental Mediation. (Fictional data based on Orme-Johnson & Haynes, 1981.)

EEG Coherence	Creativity Test Score
.2	2
.5	6
1.0	7
.8	5
.5	5

CREATIVITY (0–8) plotted against EEG COHERENCE (.0–1.)

	EEG Coherence					Creativity Test Score				Cross-Product of Z Scores
	X	dev	dev²	Z_X		Y	dev	dev²	Z_Y	$Z_X Z_Y$
	.2	-.4	.16	-1.45		2	-3	9	-1.79	2.60
	.5	-.1	.01	- .36		6	1	1	.60	- .22
	1.0	.4	.16	1.45		7	2	4	1.19	1.73
	.8	.2	.04	.73		5	0	0	0	0
	.5	-.1	.01	- .36		5	0	0	0	0
Σ	3.0		.38			25		14		4.11
	.6	$SD^2 =$	.076			5	$SD^2 =$	2.8	$r = \Sigma(Z_X Z_Y)/N$	
		$\underline{SD} =$	.276				$\underline{SD} =$	1.673	$\underline{r} = 4.11/5 = .82$	

Aron/Aron
STATISTICS FOR PSYCHOLOGY

© 1994 by Prentice-Hall, Inc.
A Paramount Communications Company
Englewood Cliffs, New Jersey 07632

TRANSPARENCY 3.4

Social desirability measured with the Marlowe-Crowne Social
Desirability Scale and test anxiety with the Sarason Test Anxiety
Scale. (Fictional data based on Blankstein & Toner, 1987.)

Marlowe-Crown	Sarason Test Anxiety
33	20
25	15
25	25
18	29
20	20
10	20
20	25

Scatter diagram:

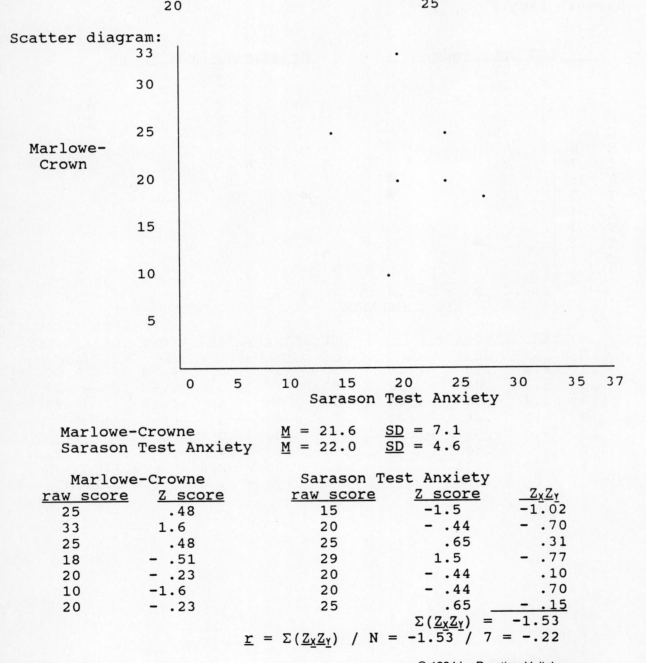

Marlowe-Crowne	M = 21.6	SD = 7.1
Sarason Test Anxiety	M = 22.0	SD = 4.6

Marlowe-Crowne		Sarason Test Anxiety		
raw score	Z score	raw score	Z score	$Z_X Z_Y$
25	.48	15	-1.5	-1.02
33	1.6	20	- .44	- .70
25	.48	25	.65	.31
18	- .51	29	1.5	- .77
20	- .23	20	- .44	.10
10	-1.6	20	- .44	.70
20	- .23	25	.65	- .15

$$\Sigma(Z_X Z_Y) = -1.53$$

$$\underline{r} = \Sigma(\underline{Z}_X \underline{Z}_Y) \ / \ N = -1.53 \ / \ 7 = -.22$$

Aron/Aron
STATISTICS FOR PSYCHOLOGY

© 1994 by Prentice-Hall, Inc.
A Paramount Communications Company
Englewood Cliffs, New Jersey 07632

TRANSPARENCY 3.5

Histogram example.
(Data from class questionnaire.)

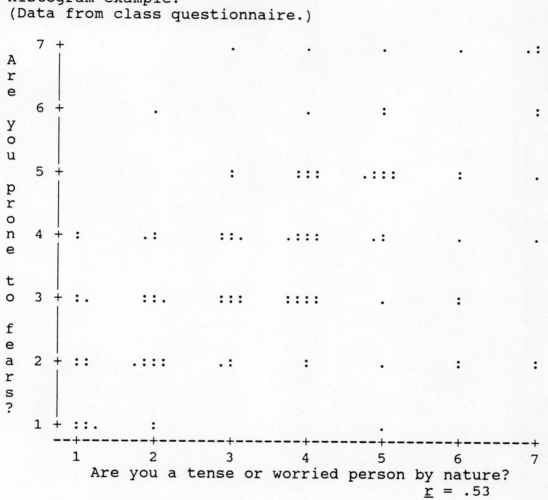

Are you prone to fears?

Are you a tense or worried person by nature?

r = .53

TRANSPARENCY 3.6

Histogram example.
(Data from class questionnaire.)

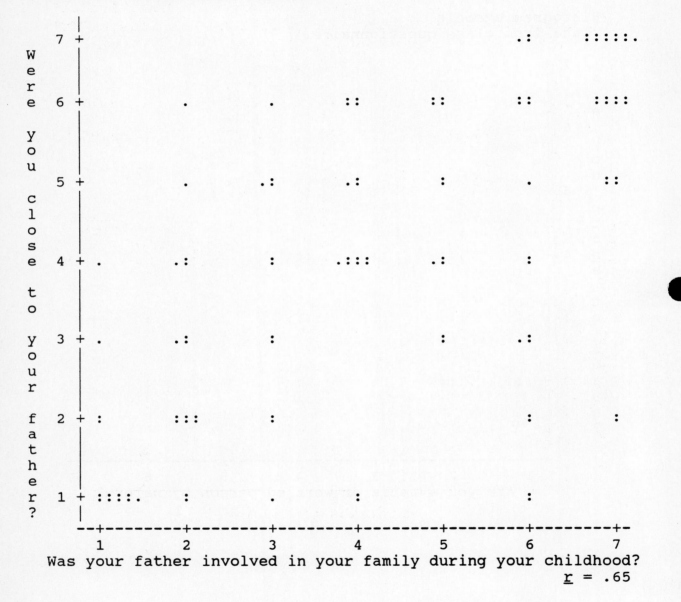

Was your father involved in your family during your childhood?

r = .65

TRANSPARENCY 3.7

Marital satisfaction and overtime hours. (Fictional Study)

Marital Satisfaction		Overtime Hours Typical Month		Cross-Products of Z Scores
X	Z_X	Y	Z_Y	$Z_X Z_Y$
8	.67	4	-.74	-.50
9	1.01	4	-.74	-.75
4	-.67	16	1.04	-.70
8	.67	6	-.44	-.29
2	-1.34	24	2.22	-2.97
3	-1.01	9	0	0
9	1.01	0	-1.33	-1.34
2	-1.34	14	.74	-.99
6	0	8	-.15	0
9	1.01	5	-.59	-.60
Σ 60		90		-8.05
M 6		9		
SD= 2.98		SD= 6.75		

$$\underline{r} = \Sigma(\underline{Z}_X \underline{Z}_Y)/N = -8.05 / 9 = -.81$$

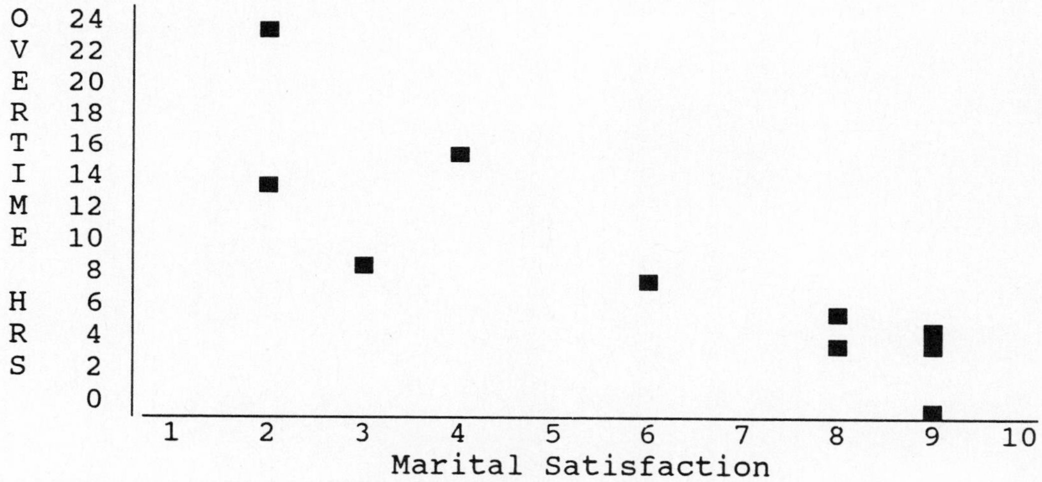

Scatter plot: Y-axis "OVERTIME HRS" with values 0, 2, 4, 6, 8, 10, 12, 14, 16, 18, 20, 22, 24. X-axis "Marital Satisfaction" with values 1 through 10.

Aron/Aron
STATISTICS FOR PSYCHOLOGY

TRANSPARENCY 3.8

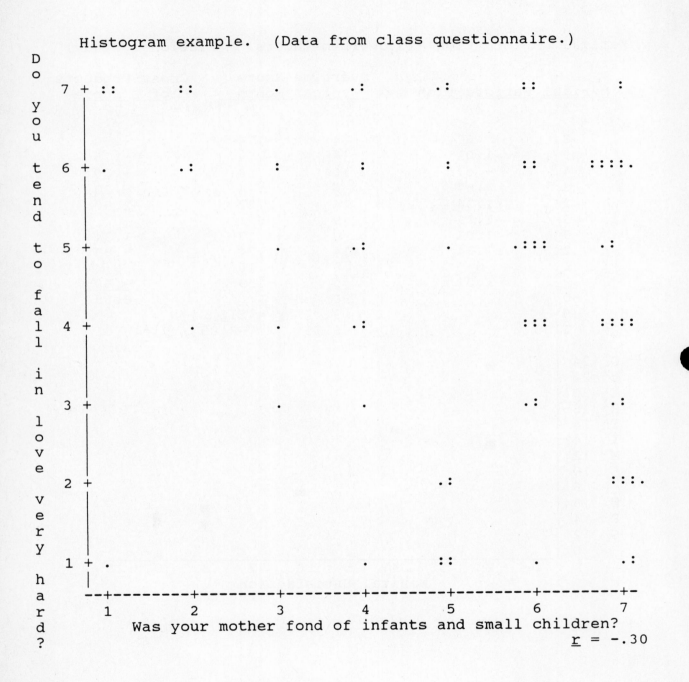

Histogram example. (Data from class questionnaire.)

Aron/Aron
STATISTICS FOR PSYCHOLOGY

Figure 3.8

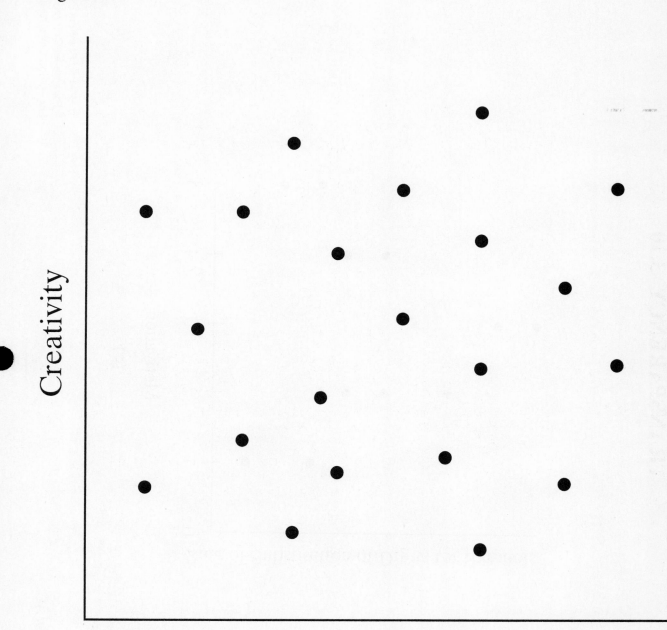

Creativity

Shoe Size

Aron/Aron
STATISTICS FOR PSYCHOLOGY

TRANSPARENCY 3.10

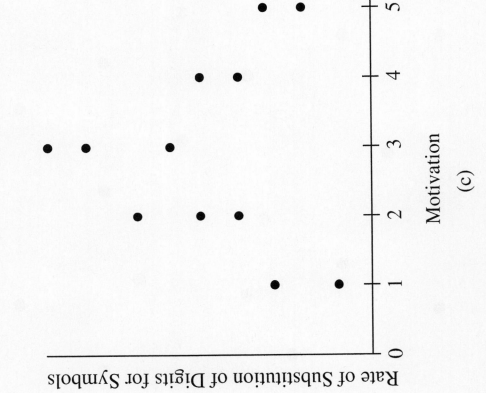

© 1994 by Prentice-Hall, Inc.
A Paramount Communications Company
Englewood Cliffs, New Jersey 07632

Aron/Aron
STATISTICS FOR PSYCHOLOGY

Figure 3.13

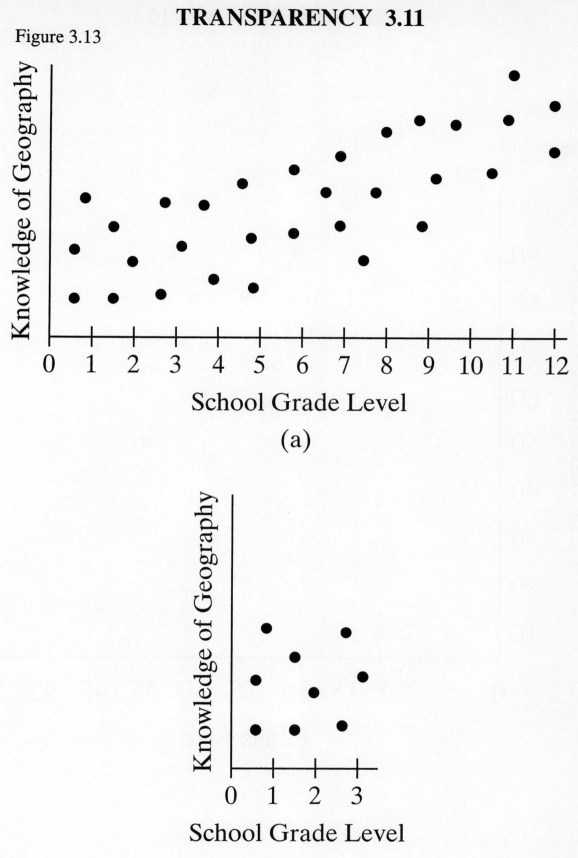

(a)

(b)

Figure 3.14

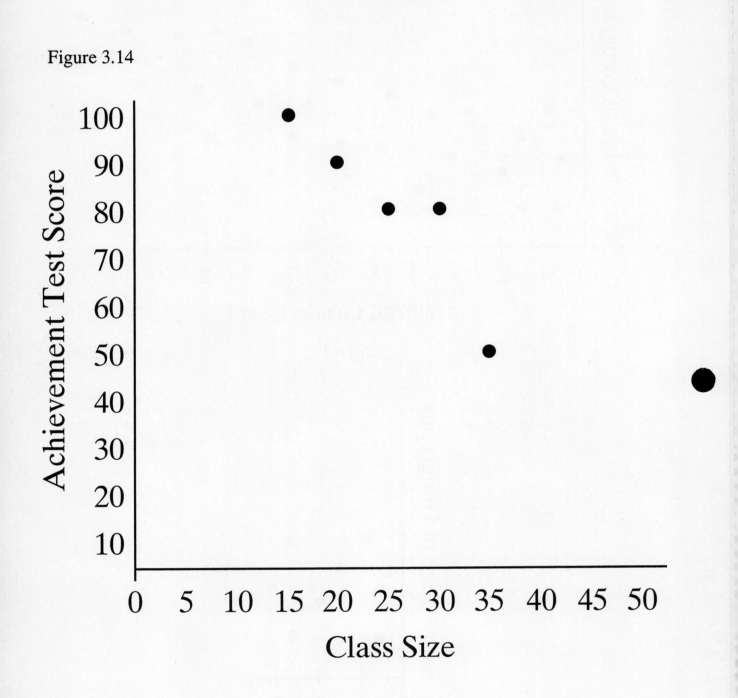

Chapter 4
Prediction

Instructor's Summary of Chapter

Bivariate prediction with Z scores. Bivariate prediction (or regression) makes predictions about a dependent variable based on knowledge of a person's score on a predictor variable. The best model for predicting a person's Z score on the dependent variable is to multiply the standardized regression coefficient (beta) times the person's Z score on the predictor variable. In bivariate prediction, beta is r.

Bivariate prediction with raw scores. Predictions with raw scores can be made by converting the person's score on the predictor variable to a Z score, multiplying it by beta, and then converting the resulting predicted Z score on the dependent variable to a raw score. Direct raw-score-to-raw-score prediction requires computing the raw-score regression coefficient (b) and the regression constant (a): predicted $Y = a + (b)(X)$. Predicted values using this formula form the regression line on the scatter diagram. Its slope is b and its Y-intercept is a.

Proportionate reduction in error. The accuracy of prediction can be estimated by applying the prediction model to the data on which the original correlation was based. The difference between each actual score and what would have been predicted for that subject using the prediction model is called error. Squaring these errors and summing them gives the sum of squared errors (SS_E). One then compares SS_E to the sum of squared error when using just the mean of the dependent variable as your predicted score (SS_T). Proportionate reduction in error (r^2) = $(SS_T - SS_E)/SS_T$.

Multiple regression. In multiple regression a dependent variable is predicted using two or more predictor variables. In a multiple regression model, each score on the predictor variable is multiplied by its own regression coefficient and the products are summed to make the prediction (a single regression constant is added when using raw scores). Each regression coefficient indicates the relation of the predictor to the dependent variable in the context of the other predictor variables. The multiple correlation coefficient describes the overall degree of association between the dependent variable and the predictor variables taken together.

Controversies and limitations. Bivariate and multiple regression have all the same limitations as bivariate correlation. In addition, in multiple regression there is ordinarily considerable ambiguity in interpreting the relative importance of the different predictor variables.

Box 4.1. Clinical versus Statistical Prediction. Some evidence indicates that predictions based on multiple regression rules are more accurate than those made by clinicians and other experts using their experience and intuition. Research has focused on the reasons human information processing does less well, how people can be trained to make more accurate decisions, why people often prefer human decision makers in spite of their limitations, and systems for integrating human and statistical prediction approaches.

Lecture 4.1: Bivariate Regression

Materials

Lecture outline
Transparencies 3.1 and 4.1 through 4.13
    (If using transparencies based on your class's questionnaires, replace 4.4, 4.5, 4.9 and 4.12 with
    4.4R, 4.5R, 4.9R, and 4.12R.)

Outline for Blackboard

    I.   Review
    II. Prediction
    III. Bivariate Prediction with Z Scores
    IV. Bivariate Prediction with Raw Scores
    V. The Regression Line
    VI. Review this Class

Instructor's Lecture Outline

NOTE: Warn class that this chapter's material is a jump in difficulty, and will require extra
        effort on their part.

I.   Review
    A. Idea of descriptive statistics and importance for their own right and as a foundation for the rest of
       the course.
    B. Description of a single distribution in terms of frequency tables, graphs, mean, and standard
       deviation.
    C. Describing a particular score's relation to the distribution using Z scores.
    D. Describing the relation between two distributions using a scatter diagram.
    E. Describing the relation between two distributions by computing the correlation coefficient (r).
        1. r is the average of the cross-products of Z scores.
        2. r ranges from +1 for a perfect positive linear correlation, to 0 for no correlation, to -1 for a perfect negative
          linear correlation.

II. Prediction
    A. Psychologists construct mathematical rules to predict a person's scores on one variable from
       knowledge of their score on another variable. For example, predicting college grades from SAT
       scores.
    B. Prediction is also called "regression."
    C. Show TRANSPARENCY 4.1 (table from text of names for predictor and dependent variables)
       and discuss.
    D. In bivariate prediction we predict a person's score on a dependent variable using the person's
       score on a single predictor. Example: Predicting a person's score on college grades from
       knowledge of their SAT score.
    E. In multiple prediction (called "multiple regression") we predict a person's score on a dependent
       variable using the person's scores on each of two or more predictor variables. Example:

Predicting a persons's score on college grades from knowledge of their SAT score and their high school GPA.

III. Bivariate Prediction with *Z* Scores

A. Show TRANSPARENCY 4.2 (about bivariate prediction with Z scores) for discussion of points B through H below.

B. The basic approach to predicting a person's score on the dependent variable is to multiply a person's *Z* score on the predictor variable times the *regression coefficient* (beta).

C. Thus a bivariate prediction rule is of the form: predicted $Z_Y = (\beta)(Z_X)$.

D. Note that instead of writing "predicted *Z*" we can put a "hat" over the Z.

E. If there is a perfect positive correlation, the *Z* score of the dependent variable will always be the same as the *Z* score on the predictor variable.
   1. For example, if a person's score on the predictor variable is 1.8 standard deviations above the mean, the person's score on the dependent variable will be 1.8 standard deviations above the mean.
   2. Thus, the prediction rule is just predicted $Z_Y = Z_X$.
   3. This is the same as saying that ß = 1.

F. If there is a perfect negative correlation, the *Z* score of the dependent variable will always be the same as the *Z* score on the predictor variable except for having opposite sign.
   1. For example, if a person's score on the predictor variable is 1.8 standard deviations above the mean, the person's score on the dependent variable will be 1.8 standard deviations below the mean.
   2. Thus, the prediction rule is just predicted $Z_Y = -Z_X$.
   3. This is the same as saying that ß = -1.

G. If there is no correlation, the best prediction for the *Z* score of the dependent variable will be its mean (a *Z* score of 0), since the person's score on the predictor variable is irrelevant to their score on the dependent variable.
   1. For example, if a person's score on the predictor variable is 1.8 standard deviations above the mean, your best prediction for the person's score on the dependent variable (assuming you have no other information) is that it will be the mean (a *Z* score of 0) of the dependent variable.
   2. Thus, the prediction rule is just predicted $Z_Y = 0$.
   3. This is the same as saying that ß = 0.

H. In each of these cases, $\beta = r$. This is the general rule for bivariate prediction.

I. For example, if the correlation between the predictor and dependent variable is $r = .4$ and a person's score on the predictor variable is 1.8 standard deviations above the mean, then your best prediction for the person's score on the dependent variable is a *Z* score of .4 times 1.8, which comes out to .72. That is, you would predict their score to be .72 standard deviations above the mean.

J. Show TRANSPARENCY 4.3 (manager's stress example from text) and discuss.

K. Show TRANSPARENCY 4.4 or 4.4R (prone to fears predicting being highly sensitive, from class questionnaire) and discuss *Z* score predictions.

L. Show TRANSPARENCY 4.5 or 4.5R (fond of mother predicting fall in love hard, from class questionnaire) and discuss *Z* score predictions.

M. Show TRANSPARENCY 4.6 (EEG coherence predicting creativity test score) and discuss *Z* score predictions.

IV. Bivariate Prediction with Raw Scores

A. One way to predict from raw scores is to convert the predictor variable to a Z score, make the prediction (of a Z score on the dependent variable), then convert the predicted Z score to a raw score. Show TRANSPARENCY 4.7 (manager's stress example from text) and discuss.

B. Another way is to summarize these steps into a single prediction rule that permits direct raw-score-to-raw-score prediction.

1. The regression equation now uses an a and b, not ß.
2. Show TRANSPARENCY 4.8 (principles of raw-to-raw prediction and manager's stress example) and discuss.
3. Show TRANSPARENCY 4.9 or 4.9R (prone-to-fears predicting Highly Sensitive Scale, from class questionnaire) and discuss.
4. Show TRANSPARENCY 4.10 (EEG coherence predicting creativity) and discuss.

V. The Regression Line

A. Predicted values of Y fall on a straight line called the regression line.

B. Any two points will form a straight line.

C. Show TRANSPARENCY 4.11 (manager's stress example, from text) and discuss, emphasizing a and b.

D. Show TRANSPARENCY 4.12 or 4.12 R (prone to fears and sensitivity, from class questionnaire) and discuss.

E. Show TRANSPARENCY 4.13 (EEG coherence and creativity) and discuss.

VI. Review this Class: Use blackboard outline.

Lecture 4.2: Error and Introduction to Multiple Regression

Materials

Lecture outline

Transparencies 4.4, 4.5, and 4.14 through 4.24

    (If using transparencies based on your class's questionnaires, replace 4.4, 4.5, 4.18, 4.22, and 4.23 with 4.4R, 4.5R, 4.18R, 4.22R, and 4.23R.)

Outline for Blackboard

  I. Review
  II. Error
  III. Proportionate Reduction in Error
  IV. Multiple Regression
  V. Review this Class

Instructor's Lecture Outline

I.   Review

  A. Idea of descriptive statistics and importance for their own right and as a foundation for the rest of the course.

  B. Description of a single distribution in terms of frequency tables, graphs, mean, and standard deviation.

  C. Describing a particular score's relation to the distribution using Z scores.

  D. Describing the relation between two distributions using a scatter diagram and the correlation coefficient.

  E. Predicting a person's score on a dependent variable based on the person's score on a predictor variable.

     1. Z score prediction rule in bivariate prediction:

       a. Predicted $Z_Y = (\beta)(Z_X)$.

       b. $\beta = r$.

     2. Raw score prediction using Z scores: convert X to Z_X, predict Z_Y, and then convert predicted Z_Y to predicted Y.

     3. Direct raw-to-raw prediction:

       a. predicted $Y = a + (b)(X)$.

       b. b is based on β (which you have because it equals r) and the two standard deviations.

       c. a is based on b and the two means.

     4. Regression line is based on the raw score predicted values.

II. Error

  A. Error is difference between predicted and actual score.

     1. Error = Predicted Score - Actual Score.

     2. Example: If for a particular person, the prediction rule predicts a first year GPA of 3.6 and the person's actual GPA is 3.8, error = 3.6 - 3.8 = -.2

  B. Because the concern is with overall amount of error, not the direction of errors, squared error is used.

  C. The accuracy of a prediction rule is related to how much squared error you are likely to make using that rule.

  D. There is no way to know how much squared error you are likely to make using a prediction rule in advance of making the predictions.

E. To estimate the error you would expect to make.
  1. Principle: Determine how much error you would have made using the prediction rule to predict the scores in the sample used to compute the prediction rule.
  2. Example: Show TRANSPARENCY 4.14 (manager's stress example from text) and discuss.
F. Error can be shown graphically.
  1. Principle: Error is the distance between the actual score on the scatter diagram and the regression line (which shows the predicted scores).
  2. Example: Show TRANSPARENCY 4.15 (graph of manager's stress example from text) and discuss.

III. Proportionate Reduction in Error

A. One indication of the accuracy of a prediction rule is how much less squared error you make using it than if you had to make predictions without using the prediction rule.
  1. If predicting a person's score on Y without a prediction rule, the best procedure is to predict the mean of Y.
    a. The error using this procedure is then the actual score minus the mean score.
    b. Example:
      i. You would predict each person to have a first-year GPA equal to the mean GPA, which might be 2.5.
      ii. If actual score was 3.8, error would then be 2.5 - 3.8 = -1.3.
  2. You make more error predicting when using the mean than when using the bivariate prediction rule.
  3. The accuracy of a prediction rule is how much less squared error you expect to make using it than just predicting everyone's score to be the mean.
B. The actual procedure is to find the *proportionate reduction*. This is the proportion of squared error reduced using the bivariate prediction rule (compared to the error you would make predicting from the mean).
C. Show TRANSPARENCY 4.16 (basis of proportionate reduction in error formula) and discuss.
D. Show TRANSPARENCY 4.17 (manager's stress example from text) and discuss.
E. Proportionate reduction in error = r^2.
F. When computing it may be easier to use all Z scores (particularly if you have already converted the scores to Z scores to compute the correlations).
G. Show TRANSPARENCY 4.18 or 4.18R (being prone to fears predicting being highly sensitive, from class questionnaire) and discuss Z score error and squared error and computation of proportionate reduction in error.
H. Show TRANSPARENCY 4.19 (EEG coherence predicting creativity test score) and discuss.
I. Show TRANSPARENCY 4.20 (EEG coherence predicting creativity test score graph) and discuss graphic display of error in light of proportionate reduction in error.

IV. Multiple Regression

A. Principle: Extension of bivariate regression to more than one predictor variable.

B. Regression coefficients.
   1. Raw score (*b*s).
   2. Z score (ßs, which are *not* equal to *r*s).

C. With *Z* scores, predicted value of dependent variable for a particular individual is the sum of each regression coefficient times that person's score on the corresponding independent variable.

D. Show TRANSPARENCY 4.21 (manager's stress example from text) and discuss.

E. Show TRANSPARENCIES 4.4 and 4.22 or 4.4R and 4.22R (bivariate predictions and multivariate predictions with two predictors, both using Highly Sensitive Scale example from class questionnaire) and discuss.

F. Show TRANSPARENCIES 4.5 and 4.23 or 4.5R and 4.23R (bivariate prediction and multivariate predictions with three predictors, both using fall-in-love-hard example from class questionnaire.)

G. Discuss multiple regression with raw scores—noting that the regression constant must also be included.

H. Show TRANSPARENCY 4.24 (manager's stress example from text) and discuss.

I. Limitations of interpreting betas as relative importance.

J. The multiple correlation coefficient as the square root of R^2.

V. Review this Class: Use blackboard outline.

TABLE 4-1
Terminology for Two Variables in Bivariate Prediction

	Variable Predicted From	Variable Predicted To
Name	Predictor variable	Dependent variable
Alternative name	Independent variable	Criterion variable
Symbol	X	Y
Example	SAT scores	College GPA

TRANSPARENCY 4.2

General prediction rule: $\hat{Z}_Y = (\beta)(Z_X)$

    When $\underline{r} = +1$: $\hat{Z}_Y = Z_X$

                  For example, if $Z_X = 1.8$: $\hat{Z}_Y = Z_X = 1.8$

                  That is, $\beta = 1$

                  Thus: $\hat{Z}_Y = (1)(Z_X) = Z_X$

                  For example, if $Z_X = 1.8$: $\hat{Z}_Y = (1)(1.8) = 1.8$

    When $\underline{r} = -1$: $\hat{Z}_Y = -Z_X$

                  For example, if $Z_X = 1.8$: $\hat{Z}_Y = -Z_X = -1.8$

                  That is, $\beta = -1$

                  Thus: $\hat{Z}_Y = (-1)(Z_X) = Z_X$

                  For example, if $Z_X = 1.8$: $\hat{Z}_Y = (-1)(1.8) = -1.8$

    When $\underline{r} = 0$: $\hat{Z}_Y = -Z_X$

                  For example, if $Z_X = 1.8$: $\hat{Z}_Y = 0$

                  That is, $\beta = 0$

                  Thus: $\hat{Z}_Y = (0)(Z_X) = 0$

                  For example, if $Z_X = 1.8$: $\hat{Z}_Y = (0)(1.8) = 0$

More generally, in bivariate prediction $\beta = \underline{r}$

    Thus the general rule is $\hat{Z}_Y = (\beta)(Z_X) = (\underline{r})(Z_X)$

    For example, if $\underline{r} = .4$ and $Z_X = 1.8$: $\hat{Z}_Y = (.4)(1.8) = .72$

Aron/Aron
STATISTICS FOR PSYCHOLOGY

© 1994 by Prentice-Hall, Inc.
A Paramount Communications Company
Englewood Cliffs, New Jersey 07632

TRANSPARENCY 4.3

Manager's stress example. (From Chapter 3 of the text.)

Number supervised ($\underline{X}$): M = 7   SD=2.37
Stress   ($\underline{Y}$): M = 6   SD=2.61

$$\underline{r} = .88$$

If a manager supervised 10 employees, that manager would have a $\underline{Z}$ score for number supervised of 1.27:

$$\underline{Z}_X = (\underline{X}-\underline{M}_X)/\underline{SD}_X = (10-7)/2.37 = 3/2.37 = 1.27$$

$$\hat{\underline{Z}}_Y = (\beta)(\underline{Z}_X) = (.88)(1.27) = 1.12$$

If a manager supervised 3 employees, that manager has a $\underline{Z}$ score for number supervised of -1.69:

$$\underline{Z}_X = (\underline{X}-\underline{M}_X)/\underline{SD}_X = (3-7)/2.37 = -4/2.37 = -1.69$$

$$\hat{\underline{Z}}_Y = (\beta)(\underline{Z}_X) = (.88)(-1.69) = -1.49$$

Aron/Aron
STATISTICS FOR PSYCHOLOGY

© 1994 by Prentice-Hall, Inc.
A Paramount Communications Company
Englewood Cliffs, New Jersey 07632

TRANSPARENCY 4.4

Examples of bivariate prediction.
(Prediction rule from class questionnaire data.)

Predicted $\underline{Z}_Y = (\underline{r})(\underline{Z}_X)$

Predicted $\underline{Z}$ on Highly Sensitive Scale = $(.33)(\underline{Z}$ on Prone to Fears$)$

Example	Example Scores on Predictor Variable Prone to Fears (M=3.64; SD=1.61)		Correlation	Predicted Value of Dependent Variable Highly Sensitive Scale (M=11.54; SD=3.17)	
	X	Z_X	r	Z_Y	Y
a.	1	-1.64	.33	-.54	9.82
b.	2	-1.02	.33	-.34	10.47
c.	3	-.40	.33	-.13	11.12
d.	4	.22	.33	.07	11.77
e.	5	.84	.33	.28	12.42
f.	6	1.47	.33	.48	13.07
g.	7	2.09	.33	.69	13.72

Aron/Aron
STATISTICS FOR PSYCHOLOGY

TRANSPARENCY 4.5

Examples of bivariate prediction.
(Prediction rule from class questionnaire data.)

Predicted $\underline{Z}_Y$ = $(\underline{r})(\underline{Z}_X)$

Predicted $\underline{Z}$ on fall hard = $(-.29)(\underline{Z}$ on mother fond of infants$)$.

Example	Example Scores on Predictor Variable Was your mother fond of infants? ($\underline{M}$=4.64; $\underline{SD}$=1.92)		Correlation $\underline{r}$ = -.29	Predicted Value of Dependent Variable Do you tend to fall in love very hard? ($\underline{M}$=5.26; $\underline{SD}$=1.84)	
	X	Z_X	r	Z_Y	Y
a.	1	-2.30	-.29	.67	5.92
b.	2	-1.76	-.29	.51	5.62
c.	3	-1.22	-.29	.35	5.32
d.	4	-.68	-.29	.20	5.02
e.	5	-.14	-.29	.04	4.72
f.	6	.40	-.29	-.12	4.42
g.	7	.94	-.29	-.27	4.12

Aron/Aron
STATISTICS FOR PSYCHOLOGY

TRANSPARENCY 4.6

Examples of bivariate prediction.
(Prediction rule from Orme-Johnson & Haynes, 1981.)

Predicted $\underline{Z}_Y = (\underline{r})(\underline{Z}_X)$

Predicted Creativity $\underline{Z} = (.82)(\underline{Z}$ for EEG Coherence)

Example	Example Scores on Predictor Variable EEG Coherence Raw	Z	Correlation r = .82	Predicted Value of Dependent Variable Creativity Test Score Z
a.	.2	-1.45	.82	-1.19
b.	.5	- .36	.82	- .30
c.	1.0	1.45	.82	1.19
d.	.8	.73	.82	.60
e.	.5	- .36	.82	- .30

Aron/Aron
STATISTICS FOR PSYCHOLOGY

TABLE 4-2
Summary, Using Formulas, of Steps for Making Raw Score Predictions With Raw-to-Z and Z-to-Raw Conversions, With an Example

Step	Formula	Example
1	$Z_x = (X - M_x)/SD_x$	$Z_x = (3 - 7)/2.37 = -1.69$
2	$\hat{Z}_Y = (\beta)(Z_x)$	$\hat{Z}_Y = (.88)(-1.69) = -1.49$
3	$\hat{Y} = (SD_Y)(\hat{Z}_Y) + M_Y$	$\hat{Y} = (2.61)(-1.49) + 6 = 2.11$

© 1994 by Prentice-Hall, Inc.
A Paramount Communications Company
Englewood Cliffs, New Jersey 07632

Aron/Aron
STATISTICS FOR PSYCHOLOGY

TRANSPARENCY 4.8

Raw-score-to-raw-score prediction rule:

$$\hat{Y} = \underline{a} + (\underline{b})(\underline{X})$$

Computing $\underline{b}$ (raw score regression coefficient):

$$\underline{b} = (\beta)(SD_Y/SD_X)$$

Computing $\underline{a}$ (raw score regression constant):

$$\underline{a} \text{ (regression constant)} = \underline{M_Y} - (\underline{b})(\underline{M_X})$$

Example from manager's stress study:

$\underline{r}=.88$, $\underline{M_X}=7$, $\underline{SD_X}=2.37$, $\underline{M_Y}=6$, $\underline{SD_Y}=2.61$. $\beta = \underline{r} = .88$.

$\underline{b} = (\beta)(\underline{SD_Y}/\underline{SD_X}) = (.88)(2.61/2.37) = (.88)(1.10) = .97$

$\underline{a} = \underline{M_Y} - (\underline{b})(\underline{M_X}) = 6 - (.97)(7) = 6 - 6.79 = -.79$

$\hat{Y} = \underline{a} + (\underline{b})(\underline{X}) = -.79 + (.97)(\underline{X})$

For a manager supervising 10 employees (that is $\underline{X} = 10$):

$\hat{Y} = -.79 + .(97)(\underline{X}) = -.79 + (.97)(10) = -.79 + 9.7 = 8.91$

For a manager supervising 3 employees (that is $\underline{X} = 3$):

$\hat{Y} = -.79 + .(97)(\underline{X}) = -.79 + (.97)(3) = -.79 + 2.91 = 2.12$

Aron/Aron
STATISTICS FOR PSYCHOLOGY

© 1994 by Prentice-Hall, Inc.
A Paramount Communications Company
Englewood Cliffs, New Jersey 07632

TRANSPARENCY 4.9

Examples of bivariate prediction with raw scores:
Prone to fears ($\underline{X}$) predicting being highly sensitive ($\underline{Y}$).
(Prediction rule from class questionnaire data.)

$\underline{M}_X$=3.64; $\underline{SD}_X$=1.61, $\underline{M}_Y$=11.54, $\underline{SD}_Y$=3.17.   ß = $\underline{r}$ = .33.

$\underline{b}$ = (ß)($\underline{SD}_Y$/$\underline{SD}_X$) = (.33)(3.17/1.61) = (.33)(1.97) = .65

$\underline{a}$ = $\underline{M}_Y$-($\underline{b}$)($\underline{M}_X$) = 11.54-(.65)(3.64) = 11.54-2.37 = 9.17

$\hat{\underline{Y}}$ = $\underline{a}$ + ($\underline{b}$)($\underline{X}$) = 9.17 + (.65)($\underline{X}$)

For a person who gave a rating of 1 on prone to fears ($\underline{X}$=1):

$\hat{\underline{Y}}$ =9.17 + (.65)($\underline{X}$) = 9.17+(.65)(1) = 9.17+.65 = 9.82

For a person who gave a rating of 4 on prone to fears ($\underline{X}$=4):

$\hat{\underline{Y}}$ =9.17 + (.65)($\underline{X}$) = 9.17+(.65)(4) = 9.17+2.6 = 11.77

For a person who gave a rating of 7 on prone to fears ($\underline{X}$=7):

$\hat{\underline{Y}}$ =9.17 + (.65)($\underline{X}$) = 9.17+(.65)(7) = 9.17+4.55 = 13.72

Aron/Aron
STATISTICS FOR PSYCHOLOGY

© 1994 by Prentice-Hall, Inc.
A Paramount Communications Company
Englewood Cliffs, New Jersey 07632

TRANSPARENCY 4.10

Examples of bivariate prediction with raw scores:
EEG coherence predicting creativity.
(Fictional example based on Orme-Johnson & Haynes, 1981.)

EEG Coherence ($\underline{X}$): $\underline{M}_X$ = .6 $\underline{SD}_X$ = .28
Creativity ($\underline{Y}$): $\underline{M}_Y$ = 5 $\underline{SD}_Y$ = 1.67
ß = $\underline{r}$ = .82

$\underline{b}$ = (ß)($\underline{SD}_Y/\underline{SD}_X$) = (.82)(1.67/.28) = (.82)(5.96) = 4.89

$\underline{a}$ = $\underline{M}_Y$-($\underline{b}$)($\underline{M}_X$) = 5-(4.89)(.6) = 5-2.93 = 2.07

   Y = $\underline{a}$ + ($\underline{b}$)($\underline{X}$) = 2.07 + (4.89)($\underline{X}$)

For a person with an EEG coherence of .2 ($\underline{X}$=.2):

   $\underline{\hat{Y}}$ = 2.07 + (4.89)($\underline{X}$) = 2.07+(4.89)(.2) = 2.07+.98 = 3.05

For a person with an EEG coherence of .5 ($\underline{X}$=.5):

   $\underline{\hat{Y}}$ = 2.07 + (4.89)($\underline{X}$) = 2.07+(4.89)(.5) = 2.07+2.45 = 4.52

For a person with an EEG coherence of .8 ($\underline{X}$=.8):

   $\underline{\hat{Y}}$ = 2.07 + (4.89)($\underline{X}$) = 2.07+(4.89)(.8) = 2.07+3.91 = 5.98

Aron/Aron
STATISTICS FOR PSYCHOLOGY

© 1994 by Prentice-Hall, Inc.
A Paramount Communications Company
Englewood Cliffs, New Jersey 07632

TRANSPARENCY 4.11

Figure 4.2

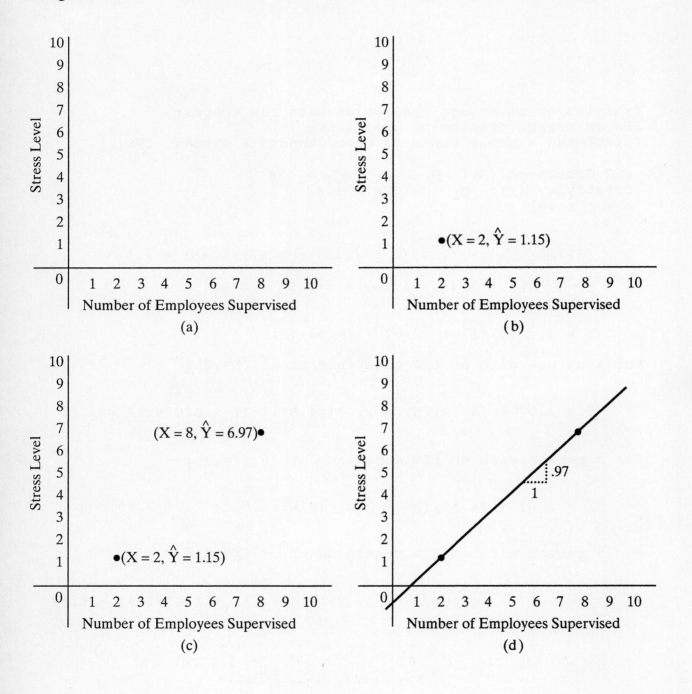

TRANSPARENCY 4.12

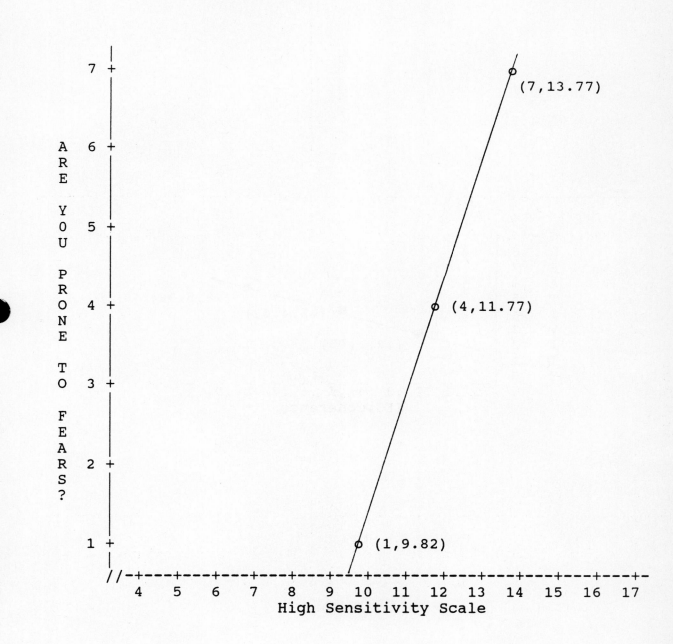

A
R
E

Y
O
U

P
R
O
N
E

T
O

F
E
A
R
S
?

(7,13.77)

(4,11.77)

(1,9.82)

High Sensitivity Scale

Aron/Aron
STATISTICS FOR PSYCHOLOGY

TRANSPARENCY 4.13

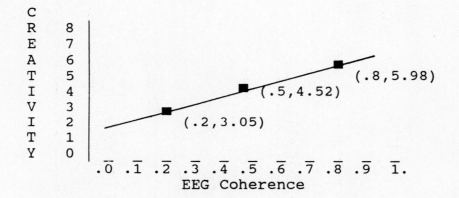

TABLE 4-3
Calculation of Error and Squared Error Using Raw Scores for the Managers' Stress Example (Fictional Data)

Employees Supervised	Stress Level		Error	Error^2
	Actual	Predicted		
X	Y	$\hat{Y}$	$Y - \hat{Y}$	$(Y - \hat{Y})^2$
6	7	5.03	1.97	3.88
8	8	6.97	1.03	1.06
3	1	2.12	-1.12	1.25
10	8	8.91	-.91	.83
8	6	6.97	-.97	.94
				Sum = 7.96

Aron/Aron
STATISTICS FOR PSYCHOLOGY

© 1994 by Prentice-Hall, Inc.
A Paramount Communications Company
Englewood Cliffs, New Jersey 07632

Figure 4.3

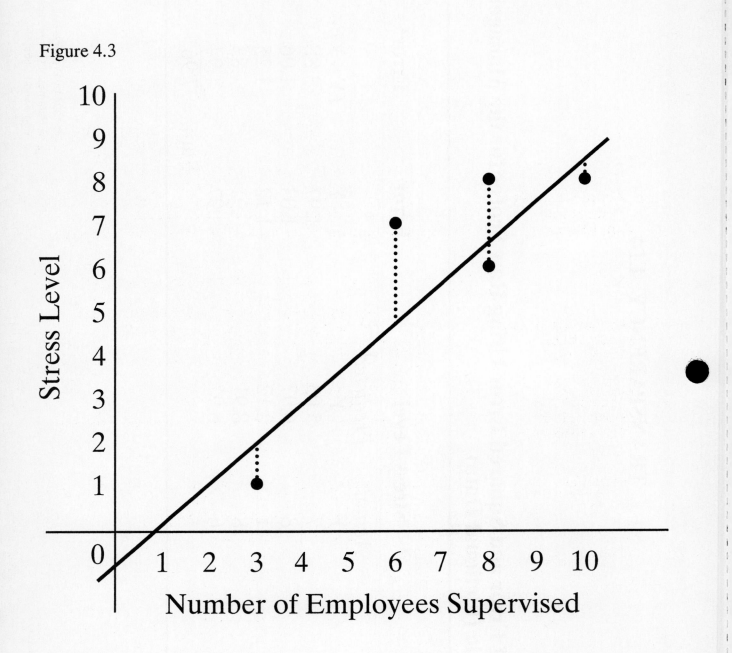

TRANSPARENCY 4.16

$$\text{Proportionate reduction in error} = \frac{\text{Reduction in squared error using bivariate prediction rule over squared error using mean to predict}}{\text{Sum of squared error using mean to predict}}$$

$$\text{Proportionate reduction in error} = \frac{\text{Sum of squared error using mean to predict MINUS Sum of squared error using bivariate prediction rule to predict}}{\text{Sum of squared error using mean to predict}}$$

$$\text{Proportionate reduction in error} = \frac{SS_T - SS_E}{SS_T}$$

Symbols: SS_T = sum of squared error using mean to predict

SS_E = sum of squared error using bivariate prediction rule to predict

Aron/Aron
STATISTICS FOR PSYCHOLOGY

© 1994 by Prentice-Hall, Inc.
A Paramount Communications Company
Englewood Cliffs, New Jersey 07632

TABLE 4-4
Calculation of Proportionate Reduction in Error for the Managers' Stress Example (Fictional Data)

Actual	Predicting Using Mean			Using Prediction Model		
Y	Mean	Error	Error²	Ŷ	Error	Error²
7	6	1	1	5.03	1.97	3.88
8	6	2	4	6.97	1.03	1.06
1	6	−5	25	2.12	−1.12	1.25
8	6	2	4	8.91	−.91	.83
6	6	0	0	6.97	−.97	.94
			$SS_T = 34$			$SS_E = 7.96$

$$\text{Proportionate reduction in error} = \frac{SS_T - SS_E}{SS_T} = \frac{34 - 7.96}{34} = \frac{26.04}{34} = .77$$

Aron/Aron
STATISTICS FOR PSYCHOLOGY

Predictor Prone to Fears		Dependent Highly Sensitive Scale		Dependent Predicted Z Using Mean			Dependent Bivariate Predicted (r=.33)		
Raw	Z	Raw	Z	Z	Error	Error^2	Z	Error	Error^2
1 1	-1.64	14	.78	0	.78	.61	-.54	-1.32	1.74
2 4	.22	15	1.10	0	1.20	1.44	.07	-1.03	1.06
3 2	-1.02	12	.15	0	.15	.02	-.34	-.49	.24
. .	.	.	.	.	.	.	.	.	.
. .	.	.	.	.	.	.	.	.	.
103 2	-1.02	11	-.17	0	-.17	.03	-.34	-.17	.03
104 4	.22	12	.15	0	.15	.02	.07	-.08	.01
105 2	-1.02	6	-1.76	0	-1.76	3.10	-.34	1.42	2.02

$$SS_T = 105.00 \qquad SS_E = 93.52$$

$$\text{Proportionate Reduction in Error} = \frac{SS_T - SS_E}{SS_T} = \frac{105 - 93.52}{105} = \frac{11.48}{105} = .11$$

Another way to get the same result: Since r = .33, r^2 = .33^2 = .11.

Predictor EEG Coherence		Dependent Creativity Test		Prediction from Mean			Bivariate Prediction r = .82		
Raw	Z	Raw	Z	Z	Error	Error²	Z	Error	Error²
.2	-1.45	2	-1.79	0	-1.79	3.20	-1.19	-.60	.36
.5	- .36	6	.60	0	.60	.36	- .30	.90	.81
1.0	1.45	7	1.20	0	1.20	1.44	1.19	.01	.00
.8	.73	5	0	0	0	0	.60	-.60	.36
.5	- .36	5	0	0	0	0	- .30	.30	.09
						====			====
					SS_T =	5.00			1.62

$$\text{Proportionate Reduction in Error} = \frac{SS_T - SS_E}{SS_T} = \frac{5 - 1.62}{5} = \frac{3.38}{5} = .68$$

Another way to get the same result: Since r = .82, r^2 = .82² = .67.

(Note: Difference between .68 and .67 is rounding error.)

Aron/Aron
STATISTICS FOR PSYCHOLOGY

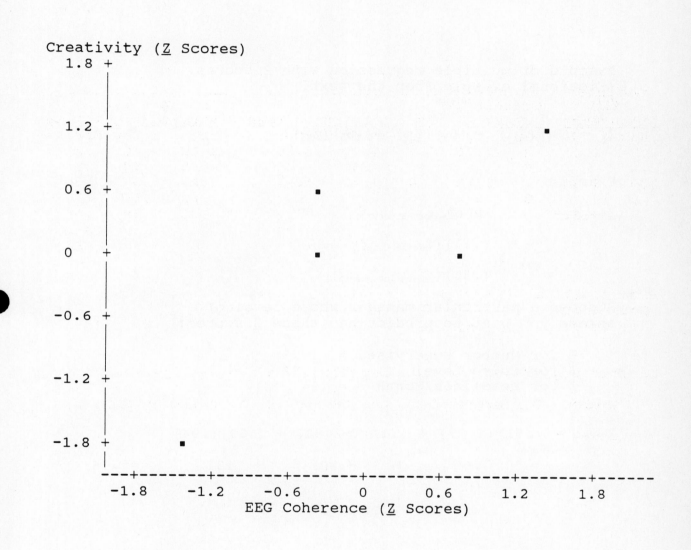

Creativity (Z Scores)

EEG Coherence (Z Scores)

Aron/Aron
STATISTICS FOR PSYCHOLOGY

Example of multiple regression with $\underline{Z}$ scores.
(Fictional example from the text.)

$$\underline{Z}_Y = (\beta_1)(Z_{\underline{X1}}) + (\beta_2)(Z_{\underline{X2}}) + (\beta_3)(Z_{\underline{X3}})$$

Example:

$$\underline{Z}_{\text{Stress}} = \quad (.51)(\underline{Z}_{\text{Number Supervised}})$$

$$+ \quad (.11)(\underline{Z}_{\text{Noise Level}})$$

$$+ \quad (.33)(\underline{Z}_{\text{Deadlines/Month}})$$

Suppose a particular manager whose level of
stress you want to predict had these $\underline{Z}$ scores:

$\underline{Z}$ for Number Supervised = 1.27
$\underline{Z}$ for Noise Level = -1.82
$\underline{Z}$ for Deadlines/Month = .94

$$\hat{\underline{Z}}_{\text{Stress}} = (.51)(1.27) + (.11)(-1.81) + (.33)(.94)$$

$$= \quad .65 \quad + \quad -.20 \quad + \quad .31 \quad = .76$$

Aron/Aron
STATISTICS FOR PSYCHOLOGY

© 1994 by Prentice-Hall, Inc.
A Paramount Communications Company
Englewood Cliffs, New Jersey 07632

TRANSPARENCY 4.22

Examples of prediction in multiple regression.
(Prediction rule from class questionnaire data.)

Predicted $\underline{Z}_Y$ = (ß1)($\underline{Z}_x$1) + (ß2)($\underline{Z}_x$2) + ...

Predicted $\underline{Z}$ for Sensitivity = (.24)($\underline{Z}$ for Prone to Fears)
+ (.29)($\underline{Z}$ for Avoid Upsetting...)

| | Example Combinations of Predictor Variable Scores | | | | Predicted Value of Dependent Variable | |
| | Prone to Fears $\underline{M}$=3.64; $\underline{SD}$=1.61) ß = .24 | | Avoid Upsetting or Overwhelming Situations ($\underline{M}$=3.74; $\underline{SD}$=1.77) ß = .29 | | Highly Sensitive Scale ($\underline{M}$=11.54; $\underline{SD}$=3.17) | |
Examples	X_1	Z_{X1}	X_2	Z_{X2}	$\underline{Z}_Y$	$\underline{Y}$
a.	1	-1.64	1	-1.55		
	(ß$_1$) (Z_{X2}) (.24)(-1.64) -.39		+ (ß$_2$) (Z_{X2}) + (.29)(-1.55) + -.45		= -.84	8.87
b.	2	-1.02 (.24)(-1.02) -.24	1	-1.55 + (.29)(-1.55) + -.45	= -.69	9.34
c.	3	-.40	1	-1.55	-.54	9.81
d.	4	.22	1	-1.55	-.40	10.29
e.	5	.84	1	-1.55	-.25	10.76
f.	6	1.47	1	-1.55	-.10	11.23
g.	7	2.09	1	-1.55	.05	11.70
h.	1	-1.64	2	-.98	-.68	9.39
i.	1	-1.64	6	1.28	-.02	11.47
j.	3	-0.40	1	-1.55	-.54	9.81
k.	2	-1.02	7	1.84	.29	12.46
l.	3	-.40	5	.71	.11	11.89
m.	2	-1.02	2	-.98	-.53	9.86

Aron/Aron
STATISTICS FOR PSYCHOLOGY

© 1994 by Prentice-Hall, Inc.
A Paramount Communications Company
Englewood Cliffs, New Jersey 07632

TRANSPARENCY 4.23

Examples of prediction in multiple regression.
(Prediction rule from class questionnaire data.)

Predicted $\underline{Z}_Y = (\beta_1)(\underline{Z}_{x1}) + (\beta_2)(\underline{Z}_{x2}) + (\beta_3)(\underline{Z}_{x3})$

Predicted $\underline{Z}$ for Falling Hard $= \;(-.26)(\underline{Z}$ for Mother Fond of Infants$)$
$+ (.15)(\underline{Z}$ for Cry Easily$)$
$+ (.27)(\underline{Z}$ for Think Movies Next Day$)$

	Was your mother mother fond of infants? ($\underline{M}=4.64$; $\underline{SD}=1.92$) $\beta = -.26$		Do you cry easily? ($\underline{M}=3.59$; $\underline{SD}=1.94$) $\beta = .15$		Do you think about movies the next Day? ($\underline{M}=4.58$; $\underline{SD}=1.62$) $\beta = .27$		Predicted Do you tend to fall in love very hard? ($\underline{M}=5.26$; $\underline{SD}=1.84$)	
Ex	$\underline{X}_1$	$\underline{Z}_{x1}$	$\underline{X}_2$	$\underline{Z}_{x2}$	$\underline{X}_3$	$\underline{Z}_{x3}$	$\underline{Z}_Y$	$\underline{Y}$
a.	1	-2.30	1	-1.34	1	-2.21		

$$(\beta_1)\;(\;\underline{Z}_{x1})\;+\;(\beta_2)\;(\underline{Z}_{x2})\;+\;(\beta_3)\;(\;\underline{Z}_{x3})$$
$$(-.26)(-2.30)\quad(.15)(-1.34)\quad(.27)(-2.21)$$
$$.60\quad+\quad -.20\quad+\quad -.60\quad=\quad -.20\quad 4.26$$

Ex	$\underline{X}_1$	$\underline{Z}_{x1}$	$\underline{X}_2$	$\underline{Z}_{x2}$	$\underline{X}_3$	$\underline{Z}_{x3}$	$\underline{Z}_Y$	$\underline{Y}$
b.	2	-1.76	1	-1.34	2	-1.59	-.17	4.31
c.	3	-1.22	1	-1.34	1	-2.21	-.48	3.72
d.	4	-.68	1	-1.34	1	-2.21	-.62	3.45
e.	5	-.14	1	-1.34	1	-2.21	-.76	3.18
f.	6	.40	1	-1.34	1	-2.21	-.90	2.91
g.	7	.94	1	-1.34	1	-2.21	-1.04	2.64
h.	1	-2.30	2	-.82	1	-2.21	-.12	4.41
i.	1	-2.30	3	-.30	1	-2.21	-.04	4.56
j.	1	-2.30	2	-.82	2	-1.59	.05	4.73
k.	1	-2.30	6	1.24	3	-.98	.52	5.64
l.	3	-1.22	1	-1.34	7	1.49	.52	5.64
m.	2	-1.76	7	1.76	3	-.98	.46	5.52
n.	3	-1.22	5	.73	6	.88	.66	5.91

Aron/Aron
STATISTICS FOR PSYCHOLOGY

© 1994 by Prentice-Hall, Inc.
A Paramount Communications Company
Englewood Cliffs, New Jersey 07632

TRANSPARENCY 4.24

Example of multiple regression with raw scores.
(Fictional example from the text.)

$$\underline{Y} = \underline{a} + (\underline{b}_1)(\underline{X}_1) + (\underline{b}_2)(\underline{X}_2) + (\underline{b}_3)(\underline{X}_3)$$

$\hat{\text{Stress}}$ =   -4.70

+   (.56)(Number Supervised)

+   (.06)(Noise Level in Decibels)

+   (.86)(Deadlines/Month)

Suppose for a particular manager whose
level of stress you want to predict:

    Number Supervised = 8
    Noise Level   = 85
    Deadlines/Month   = 4

$\hat{\text{Stress}}$ = -4.70 + (.56)(8) + (.06)(85) + (.86)(4)

   = -4.70 +   4.48   +   5.1   +   3.44 = 8.32

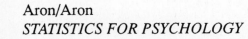

Aron/Aron
STATISTICS FOR PSYCHOLOGY

Chapter 5
Some Ingredients for Inferential Statistics:
The Normal Curve, Probability, and Population versus Sample

Instructor's Summary of Chapter

The normal curve. Many distributions of variables in psychology research approximately follow a bell-shaped, symmetrical, unimodal distribution called the normal curve. Because it is precisely mathematically defined, there are a known percentage of cases between any two points on a normal curve.

50%-34%-14%-2% rule of thumb. When working with normal curves, 50% of the cases fall above the mean, about 34% between the mean and one standard deviation above the mean, 14% between one and two standard deviations, and 2% beyond two standard deviations.

Normal curve tables give the percentage of cases between the mean and any particular positive Z score. Using such a table, and knowing that the curve is symmetrical and that 50% of the cases fall above the mean, it is possible to determine the percentage of cases above or below any particular Z score, and also the Z score corresponding to the point at which a particular percentage of cases begins.

Probability. Most psychologists consider the probability of an event to be its expected relative frequency, though some treat it as the subjective degree of belief that the event will happen. Probability is usually calculated as the proportion of successful outcomes to the total possible outcomes. It is symbolized by p and has a range from 0 (event is impossible) to 1 (event is certain). The normal distribution can be thought of as providing a way to know the probabilities of scores being within particular ranges of values. (That is, it can be thought of as a probability distribution.)

Sample and population. A sample is an individual or group which is studied—usually as representative of a larger group, the population, which is not practical or possible to study in its entirety. Ideally, the sample is selected from the population using a strictly random procedure.

Sample statistics and population parameters. The mean, variance, and so forth of a sample are called sample statistics; when of a population, they are called population parameters and are symbolized by Greek letters—μ for mean and σ^2 for variance. Most of the techniques in the rest of this book make probabilistic inferences in order to draw conclusions about populations based on information from samples. In this process, populations are usually assumed to be normally distributed.

Controversies. One question is about whether normal distributions are really that typical of the populations of scores for the variables we study in psychology. Another debate, raised by those who follow a "Bayesian" approach to statistics, is whether we should explicitly construct our statistical procedures to take into account the researcher's initial subjective probabilities. Finally, the representativeness of the samples that psychologists use, which are typically not obtained through strict random selection, has been contested—though there are also reasons to think that for the topics most psychologists study, this may not matter very much.

How the procedures of this chapter are described in research articles. Research articles rarely discuss normal curves (except briefly when the distribution involved seems not to be normal) or probability (except in the context of significance testing, to be covered starting in Chapter 6). However, procedures of sampling, particularly when the study is a survey, are usually described, and the representativeness of a sample when random sampling could not be used may be discussed.

Box 5.1. De Moivre, the Eccentric Stranger Who Invented the Normal Curve. Briefly describes the life of Abraham de Moivre, who in a 1733 paper was the first to describe the principles of the normal curve (without actually drawing it).

Box 5.2. Pascal Begins Probability Theory at the Gambling Table, Then Learns to Bet on God. Briefly describes the life of Blaise Pascal, who in 1654 solved the "problem of points" (a gambling puzzle) and in so doing began probability theory.

Box 5.3. Surveys, Polls, and 1948's Costly "Free Sample." Discusses the mistaken predictions of the major polling organizations about the 1948 U.S. presidential election and how polling organizations today approximate probability sampling procedures.

Lecture 5.1: The Normal Curve

Materials

Lecture outline

Deck of 90 large index cards made up so that on each is
written a single numeral from one to nine—ten with 1s, ten with 2s, etc.

Transparencies 1.6, 1.7, 1.9, 2.11, and 2.12, and 5.1 through 5.3.
(If using transparencies based on your class's questionnaires, replace 1.7 and 2.12 with 1.7R and 2.12R.)

Outline for Blackboard

I. Review

II. Characteristics of the Normal Curve

III. Why Approximations to the Normal Curve Are So Common

IV. Why Some Distributions Are Not Near Normal

V. Normal Curve and Z scores

VI. Review this Class

Instructor's Lecture Outline

I.   Review
  A. Idea of descriptive statistics and brief overview of major types:
      1. Univariate: tables, graphs, mean, standard deviation, variance, and Z scores.
      2. Bivariate: correlation coefficient and proportionate reduction in error.
  B. Review principle of the histogram and frequency polygon for summarizing a group of data.
      Show TRANSPARENCIES 1.6 and 1.9 (histogram and frequency polygon of student's stress
      ratings from text) and discuss.
  C. Review normal curve as introduced in Chapter 1 as a symmetrical, unimodal distribution. Show
      TRANSPARENCY 1.7 or 1.7R (histograms approximating normal curves for Highly Sensitive
      Scale, from class questionnaire) and discuss.
  D. Review Z scores.
      1. Principle: Number of standard deviations above or below the mean.
      2. Raw to Z score: $Z = (X-M)/SD$.
      3. Z to Raw score: $X = M + (Z)(SD)$.
      4. Show TRANSPARENCY 2.13 (Z score examples from horn honking study) and discuss.
      5. Z score as distance in a histogram or frequency polygon.
      6. Show TRANSPARENCY 2.12 or 2.12R (Z score examples, with histogram, for Highly Sensitive Scale, from
         class questionnaire) and discuss.

II.   Characteristics of the Normal Curve
  A. Show TRANSPARENCY 5.1 (normal curve) when discussing points A through C below.
  B. Most cases fall near the center, fewer at extremes (that is, it is unimodal).
  C. Symmetrical.
  D. It is a theoretical distribution.
      1. Precisely mathematically defined by a formula with two variables:
          a. mean—where its center is located.
          b. standard deviation—how spread out the numbers along its base are.
      2. Thus, when using Z scores (where M is always 0 and SD is always 1), the normal curve is always the same.

3. Does not *exactly* correspond to any distribution in nature (though it certainly does approximately).

4. The scale has no limit below or above.

5. It is a smooth curve because, as either a histogram or frequency polygon, the "intervals" of the scale are infinitesimal.

III. Why Approximations to the Normal Curve Are So Common

A. Approximations are extremely common (for each example explain that most cases would fall towards the center with decreasing numbers of cases going out from the center at about equal rates).

1. Examples:

a. Scores of a group of people on a test of self-esteem.

b. Weights of all newborn foals in Kentucky in 1994.

c. Length of all adult sardines in a particular school of them.

d. Number of letters in the lines in a particular book.

2. Ask students to come up with some examples.

B. An intuitive understanding of why it is so common.

1. Each actual case is influenced by many things, each of which is essentially random.

2. The combination of these random offsetting events is likely to be a middle score in most cases (making the distribution unimodal).

3. When it is not a middle score, there are equal chances of an imbalance of the random influences being in either direction (making the distribution symmetrical).

C. The Central Limit Theorem.

1. Each case representing a combination of random influences producing a normal-shaped curve is called the Central Limit Theorem.

2. The intuitive understanding given above (in point B) explains only why the normal curve is unimodal and symmetrical, but not why it has its precise shape. However, it can be proven mathematically that taking cases representing a combination of random influences will in the long run produce an exact normal curve.

D. Class demonstration (if time does not permit the entire exercise, omit the version using two cards each).

1. On the blackboard draw a base for a histogram with nine intervals, labeled 1, 2, 3, 4, 5, 6, 7, 8, and 9.

2. Ask three students to volunteer to come to the front of the class. Make one a Shuffler, one a Reader, and one a Grapher.

3. Show the class the deck of cards and how it consists of 10 ones, 10 twos, etc., through 10 nines.

4. Have the Shuffler shuffle the deck very thoroughly.

5. Have the Reader take the top card from the deck, show it and read it to the class, then put the card back in the deck.

6. Have the Grapher put a box on the histogram for that number.

7. Repeat this process (Steps 4 through 6) about 20 times and note how the distribution is approximately rectangular and discuss.

8. Select three new volunteers—this time be sure that the Reader is someone who feels they can figure numbers in their head fairly easily.

9. Draw a new baseline for a histogram with intervals of 1, 1.5, 2, 2.5, etc., through 9.

10. Have the Shuffler thoroughly shuffle the cards.

11. Have the Reader take the top TWO cards from the deck, show them to the class, and speak out their AVERAGE, then put the two cards back in the deck.

12. Have the Grapher put a box on the histogram for that number.

13. Repeat this process (Steps 10 through 12) 15-20 times and note how the distribution is now approximately unimodal and symmetrical. (You may have to judge a good point to stop before or after 20 to make this point seem clear.)

14. Discuss how the result of using two cards demonstrates the Central Limit Theorem and why normal curves are so common in nature. (You might also mention that, when drawing two cards, strictly speaking they should have taken each card out one at a time and put it back before the second—but that for this demonstration it was good enough to approximate equal chances for each possible pair of numbers to arise.)

15. Repeat whole process again with new volunteers and using three cards each time.

16. Note how curve is more clearly normal—and how it is narrower than when only two were used.

a. Explain that this principle (that the more cases in each group that are combined, the narrower the curve) will be considered in great detail later in the course.

b. Note that the idea is that it is unlikely that two cards will BOTH be nines or BOTH ones, making these extremes unlikely. But it is even less likely that THREE cards would all be ones or nines, making these extremes in this case VERY unlikely. But middle scores can be made up of many more combinations—more when there are three cards than when there are two each.

IV. Why Some Distributions Are Not Near Normal

A. Limits on maximum or minimum score possible.

    1. Ceiling effects. Example: Percent correct on an essay test (maximum = 100).

    2. Floor effects. Example: Distribution of SAT scores at college (only those with adequate levels are accepted, so there is a floor effect compared to the whole population).

B. Limited number of categories, making a noncontinuous distribution (a nonsmooth curve). Example: Number of children in a family.

C. Growth and time effects. Example: Height of children (will be rectangular if not all at same age).

D. Direct random effects. Example: Distribution of roulette wheel numbers.

V. Normal Curve and Z Scores

A. Because the normal curve is an exactly defined distribution, if a distribution is normal then there is an exact relation between any of its Z scores and the percent of cases above and below it. Thus:

    1. If you know the Z score, you can determine the percent of cases above or below that Z score. Example: If a Z score is 2.5, what percent of people have higher scores?

    2. If you know your relative position to others in a distribution (the proportion above or below you), you can determine your Z score. Example: If you are in the top 10% on a test, what is your Z score?

B. One aspect of the normal curve is that it is perfectly symmetrical around the mean. Thus,

    1. If a person has a Z score of 0, there are exactly 50% of the cases above that person and 50% below.

    2. If a person has 50% of the people below them and 50% above them, there Z score is 0.

C. 50%-34%-14% Approximations. Show TRANSPARENCY 5.2 (normal curve figure showing percentage approximations from text) and discuss.

    1. Approximately 34% of cases fall between the mean and one standard deviation. Thus,

        a. If a person has a Z score of 1:

            i. 34% of the people have Z scores between that person and the mean.

            ii. Because 50% of the cases are below the mean, 84% of the people fall below that person.

            iii. Because there is total of 50% of the cases above the mean, 16% of the people fall above that person.

        b. If a person has a Z score of -1:

            i. 34% of the people have Z scores between that person and the mean.

            ii. Because 50% of the cases are below the mean, 16% of the people fall below that person.

            iii. Because there are 50% of the cases above the mean, 84% of the people are above that person.

        c. 84% of the people have Z scores between +1 and -1.

    2. Approximately 14% of cases fall between one and two standard deviations from the mean. Thus,

        a. If a person has a Z score of 2:

            i. 14% of the people have Z scores between that person and someone only one standard deviation above the mean.

            ii. 98% of the cases are below that person (14%+34%+50%)—that is, the person is in the top 2% of the cases.

        b. If a person has a Z score of -2:

            i. 14% of the people have Z scores between that person and someone only one standard deviation below the mean.

            ii. 2% of the cases are below that person (50%-14%-34%)—that is, 98% of the cases are above that person on this measure.

        c. 98% of the people have Z scores between +2 and -2.

3. Ask class:
   a. If a person has 50% of the people above him or her, what is that person's Z score? [0]
   b. If a person has 50% of the people below him or her, what is that person's Z score? [0]
   c. If a person has 2% of the people above him or her, what is that person's Z score? [+2]
   d. If a person has 2% of the people below him or her, what is that person's Z score? [-2]
   e. If a person has 16% of the people above him or her, what is that person's Z score? [1]
   f. What percent of the people fall between the mean and a Z score of +2? [48%]
   g. What percent of the people fall between a Z score of +1 and a Z score of 0? [34%]
   h. What percent of the people fall between a Z score of -1 and a Z score of 0? [34%]
   i. What percent of the people fall between a Z score of +1 and a Z score of -2? [82%]
4. This procedure can be done with raw scores when you know a distribution's M and SD, because then raw scores can be connected to Z scores and vice versa. Example: If a person has a score of 36 on a test and the mean is 30 and the standard deviation is 6, then that person's Z score is +2, and—assuming the distribution of scores on this test is normal—that person is in the top 16% of all scorers.

D. Using normal curve tables.
   1. The percent of cases between the mean and any particular Z score can be computed mathematically from the equation for the normal curve.
   2. Because this is quite tedious to do, there are tables.
   3. Show TRANSPARENCY 5.3 (normal curve table) and discuss, giving several examples.

VI. Review this Class: Use blackboard outline.

Lecture 5.2: Probability and Sample and Population

Materials

Lecture outline
Transparencies 5.1 through 5.8

Outline for Blackboard

I.   **Review**
II. **Probability**
III. **Sample and Population**
IV. **Relation of Normal Curve, Probability, and Sample versus Population**
V.   **Review this Class**

Instructor's Lecture Outline

I.   Review
   A. Idea of descriptive statistics.
   B. Show TRANSPARENCY 5.1 (normal curve) and discuss key characteristics.
   C. Normal curve and Z scores.
      1. Show TRANSPARENCY 5.2 (normal curve showing approximation rule percentages from text) and discuss.
      2. Show TRANSPARENCY 5.3 (normal curve table) and discuss.

II. Probability
   A. Scientific research does not permit determining the definite truth or falsity of theories or applied procedures.
   B. But inferential statistics are applied to results of research to make probabilistic conclusions about theories or applied procedures.
   C. Probability is a large and controversial topic, but there are only a few key ideas you need to know to understand basic inferential statistical procedures.
   D. Show TRANSPARENCY 5.4 (interpretations of probability) and discuss.
   E. Show TRANSPARENCY 5.5 (computing probability) and discuss.
   F. The normal distribution can also be thought of as a probability distribution. The proportion of cases between any two Z scores is the same as the probability of selecting a case between those two Z scores. Show TRANSPARENCY 5.2 again and discuss.

III. Sample and Population
   A. Show TRANSPARENCY 5.6 (pot of beans and spoonful from text) and discuss.
      1. A *population* (like the pot of beans) is the entire set of things of interest.
      2. A *sample* (like the spoonful) is the subset of the population about which you actually have information.
   B. Why samples are studied (instead of populations).
      1. Usually more practical.
      2. Goal of science is to make generalizations or predictions about events beyond our reach.
   C. General strategy of psychology research is to study a sample of individuals who are believed to be representative of the general population (or of some particular population of interest).
   D. At the minimum, researchers try to study people who at least do not differ from the general population in any systematic way which would be expected to matter for that topic of research.
   E. Methods of sampling.

1. *Random selection*: The researcher obtains a complete list of all the members of a population and randomly selects some number of them to study.
   a. Example: Telephone survey using a random numbers table to select from a listing of all residential phone numbers (presuming all people with phones is the population).
   b. Example: Putting all the names of the students in a large class on equal size slips in a hat, shaking up the hat, and taking out 40 names (students in the class are the population).
2. *Haphazard selection*: Selecting whoever is available without any systematic plan. This is likely to yield a biased sample.
   a. Example: Surveying each person you run into in the street. (Ask class and then discuss why not random.)
   b. Example: Selecting students from a class who happen to be sitting in the front row. (Ask class and then discuss why not random.)
3. Ask class for some ideas for doing random sampling when planning research involving each of the following populations. Discuss strengths and weaknesses of each idea:
   a. Presidents of top corporations.
   b. U.S. voters.
   c. Chimpanzees, to be used in a study of language acquisition.
4. In psychology research it is rarely possible to employ true random sampling.
   a. Researchers try to study a sample not systematically unrepresentative of the population in any relevant way.
   b. Researchers often assume that the pattern of relationships will be reasonably unaffected by the particular sample, even though means may differ from sample to sample.

F. Statistical terminology for samples and populations. Show TRANSPARENCY 5.7 (symbols for population parameters and sample statistics for mean, variance, and standard deviation, from text) and discuss.

IV. **Relation of Normal Curve, Probability, and Sample versus Population.** Show TRANSPARENCY 5.8 (relation as expressed in a typical psychology experiment) and discuss. Tell them the awkward wording of the research questions has a purpose they will soon learn.

V. **Review this Class:** Use blackboard outline.

TRANSPARENCY 5.1

Figure 5.1

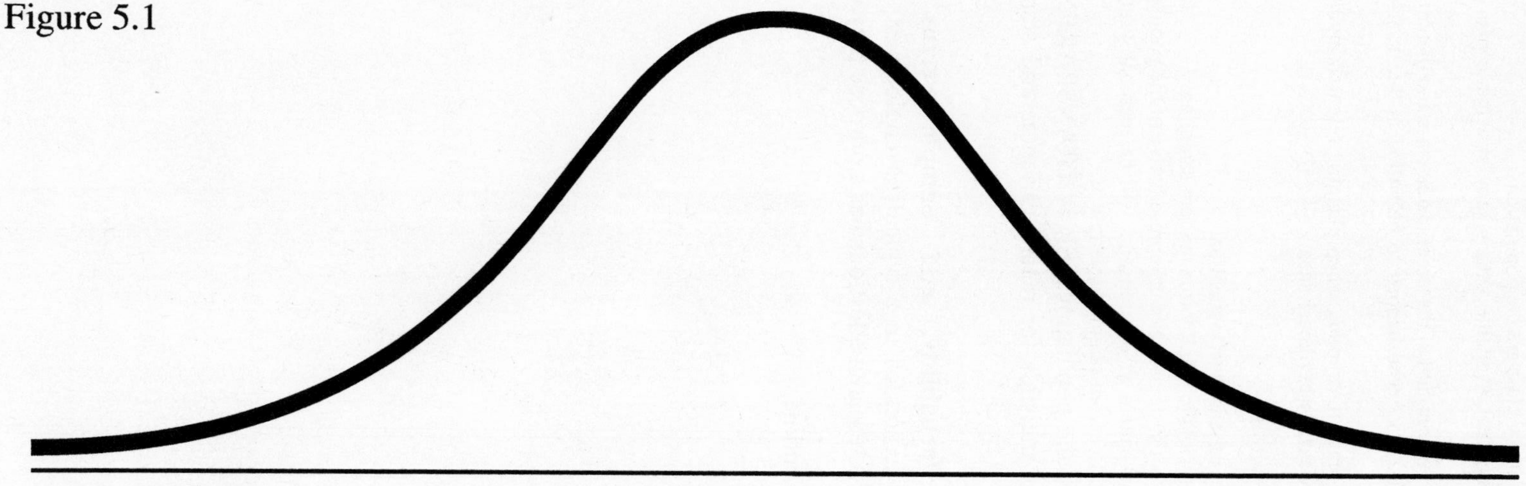

Aron/Aron
STATISTICS FOR PSYCHOLOGY

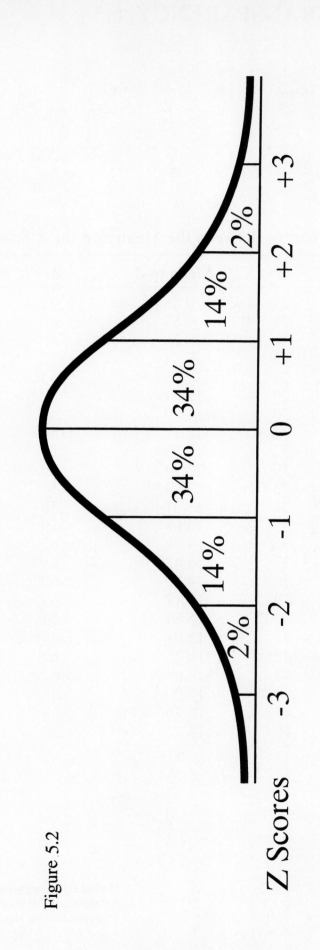

Figure 5.2

© 1994 by Prentice-Hall, Inc.
A Paramount Communications Company
Englewood Cliffs, New Jersey 07632

TRANSPARENCY 5.3

TABLE B-1
Normal Curve Areas:
Percentage of the Normal Curve Between the Mean and the Z Scores Shown

Z	% Mean to Z	Z	% Mean to Z	Z	% Mean to Z
.00	.00	.24	9.48	.47	18.08
.01	.40	.25	9.87	.48	18.44
.02	.80	.26	10.26	.49	18.79
.03	1.20	.27	10.64	.50	19.15
.04	1.60	.28	11.03	.51	19.50
.05	1.99	.29	11.41	.52	19.85
.06	2.39	.30	11.79	.53	20.19
.07	2.79	.31	12.17	.54	20.54
.08	3.19	.32	12.55	.55	20.88
.09	3.59	.59	22.24	.56	21.23
.10	3.98	.33	12.93	.57	21.57
.11	4.38	.34	13.31	.58	21.90
.12	4.78	.35	13.68	.59	22.24
.13	5.17	.36	14.06	.60	22.57
.14	5.57	.37	14.43	.61	22.91
.15	5.96	.38	14.80	.62	23.24
.16	6.36	.39	15.17	.63	23.57
.17	6.75	.40	15.54	.64	23.89
.18	7.14	.41	15.91	.65	24.22
.19	7.53	.42	16.28	.66	24.54
.20	7.93	.43	16.64	.67	24.86
.21	8.32	.44	17.00	.68	25.17
.22	8.71	.45	17.36	.69	25.49
.23	9.10	.46	17.72	.70	25.80

Aron/Aron
STATISTICS FOR PSYCHOLOGY

© 1994 by Prentice-Hall, Inc.
A Paramount Communications Company
Englewood Cliffs, New Jersey 07632

TRANSPARENCY 5.4

Two interpretations of probability:

1. Long-run relative-frequency interpretation:

    Probability is the long run, expected relative frequency of a
    particular outcome.

> <u>Outcome</u> = result of an experiment or event.
>
> <u>Frequency</u> = how many times something occurs.
>
> <u>Relative frequency</u> = number of times something occurs
> relative to the number of times it
> could have occurred.
>
> <u>Long run relative frequency</u> = what you would expect to get,
> in the long run, if you were
> to repeat the experiment many
> times.

2. Subjective interpretation:

    Probability is how certain one is that a particular outcome will
    occur.

Aron/Aron
STATISTICS FOR PSYCHOLOGY

© 1994 by Prentice-Hall, Inc.
A Paramount Communications Company
Englewood Cliffs, New Jersey 07632

TRANSPARENCY 5.5

Computing probability.

$$p = \frac{\text{Number of possible successful outcomes}}{\text{Number of all possible outcomes}}$$

Examples

Probability of a head (one possible outcome)
on a single coin flip (two possible outcomes)

$$p = \frac{\text{Number of possible successful outcomes}}{\text{Number of all possible outcomes}} = \frac{1}{2} = .5$$

Probability buying a winning ticket for a lottery
(800 winners) in which there are 1,000,000 tickets sold.

$$p = \frac{\text{Number of possible successful outcomes}}{\text{Number of all possible outcomes}} = \frac{800}{1,000,000} = .0008$$

Probability of getting accepted into a class (140 places) in which
400 have applied (assuming acceptance is by a random draw).

$$p = \frac{\text{Number of possible successful outcomes}}{\text{Number of all possible outcomes}} = \frac{140}{400} = .35$$

Aron/Aron
STATISTICS FOR PSYCHOLOGY

© 1994 by Prentice-Hall, Inc.
A Paramount Communications Company
Englewood Cliffs, New Jersey 07632

TRANSPARENCY 5.6

Figure 5.9

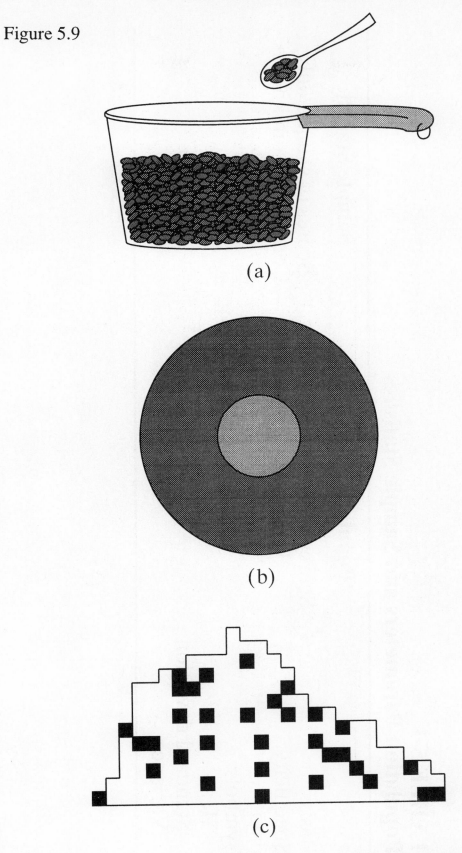

(a)

(b)

(c)

TABLE 5-1
Population Parameters and Sample Statistics

	Population Parameter	Sample Statistic
Basis:	Scores of entire population	Scores of sample only
Specificity:	Usually unknown	Computed from known
data		
Symbols:		
Mean	μ	M
Standard deviation	σ	SD
Variance	σ^2	SD^2

Aron/Aron
STATISTICS FOR PSYCHOLOGY

TRANSPARENCY 5.8

Example of relation of normal curve, probability, and sample versus population as expressed in the typical psychology experiment.

Experiment comparing words recalled by 20 subjects randomly assigned to learn words one at a time versus words recalled by another 20 subjects randomly assigned to learn words all at once.

Hypothetical Populations

(Each Assumed to be Normally Distributed on Number of Words Recalled)

Population of people Who Learn Words One at a Time	Population of People Who Learn Words All at Once

Samples Intended to Represent the Above Populations

20 Subjects Randomly Assigned to Learn Words One at a Time	20 Subjects Randomly Assigned to Learn Words All at Once

Research question: Given the results of the study (some difference between the means of the two groups in number of words recalled), what is the probability of getting such a large difference in these two samples if in fact the means of the two populations were not different.

Aron/Aron
STATISTICS FOR PSYCHOLOGY

Chapter 6
Introduction to Hypothesis Testing

Instructor's Summary of Chapter

This chapter introduces the logic and key terminology of hypothesis testing, illustrating it with the rare-in-practice research situation in which a sample of one case is being compared to a known population.

The basic principle of hypothesis testing is to examine the probability that the outcome of a study could have arisen even if the true situation was that the experimental treatment made no difference. If this probability is low, this scenario is rejected and the theory from which the treatment was proposed is supported.

Research hypothesis and null hypothesis. The expectation of a difference or effect is the research hypothesis (or "alternative hypothesis"). The hypothetical situation in which there is no difference or effect is the null hypothesis.

Possible conclusions. When an obtained result would be extremely unlikely if the null hypothesis were true, the null hypothesis is rejected and the research hypothesis supported. If the obtained results are not very extreme, the study is inconclusive.

Conventional significance levels and directional and nondirectional tests. Psychologists usually consider a result too extreme if it is less than 5% likely, though a more stringent 1% cutoff is sometimes used. These percentages may apply to the probability of the result being extreme in a single predicted direction, a directional or one-tailed test, or to the probability of it being extreme in either direction, a nondirectional or two-tailed test.

Five steps of hypothesis testing: (1) Reframe the question into a research hypothesis and a null hypothesis about populations. (2) Determine the characteristics of the comparison distribution. (3) Determine the cutoff sample score on the comparison distribution at which the null hypothesis should be rejected. (4) Determine the score of your sample on the comparison distribution. (5) Compare the scores obtained in Steps 3 and 4 to decide whether to reject the null hypothesis.

Controversy. A continuing controversy is about when it is appropriate to use a one-tailed test, a two-tailed test, or possibly a combination in which there is more of the percentage at one tail.

How the procedures of this chapter are described in research articles. Research articles typically report the results of hypothesis testing by noting that a result was or was not significant, and giving the probability level cutoff (usually 5% or 1%) at which the decision was made.

Box 6.1. Nothing Happened—It Worked! Reviews a classic article by Greenwald (1975) which argues that there is an unfair prejudice against null hypothesis results.

Lecture 6.1: Introduction to Hypothesis Testing

Materials

Lecture outline
Transparencies 5.2, 5.3, 5.5, 5.7, 5.8, and 6.1 through 6.6

Outline for Blackboard

I.   **Review**
II. **Inferential Statistics**
III. **Example of Hypothesis Testing**
IV. **Language and Steps of Hypothesis Testing**
V.   **Additional Examples**
VI. **Review this Class**

Instructor's Lecture Outline

I.   Review
A. Idea of descriptive statistics and importance in their own right.
B. The normal curve.
  1. Approximations to it are common in distributions used in psychology.
  2. Show TRANSPARENCY 5.2 (normal curve showing 50%-34%-14% rules of thumb) and discuss relation of Z scores and percentages of cases when using a normal curve.
  3. Show TRANSPARENCY 5.3 (normal curve table) and discuss using normal curve tables.
C. Probability. Show TRANSPARENCY 5.5 (calculation of probability) and discuss.
D. Sample and population. Show TRANSPARENCY 5.7 (symbols for population parameters and sample statistics, from text) and discuss.
E. Relation of normal curve, probability, and sample and population. Show TRANSPARENCY 5.8 (relation as expressed in a typical psychology experiment) and discuss.

II.   Inferential Statistics
A. Involve making inferences about populations based on information in samples (as compared to descriptive statistics, which merely summarize attributes of known data).
B. Especially important because we use them to draw conclusions about the world in general (i.e., populations we can not measure as a whole) based on results obtained from a particular experiment or study (i.e., a sample).
C. This and the next few classes use research examples that would not usually occur in order to introduce the key logic in the simplest possible context. (Later you learn to apply the logic in more realistic research situations.)

III. Example of Hypothesis Testing

A. A person says she can identify people of above average intelligence with her eyes closed.

B. The plan is then made to take her to a stadium full of randomly selected people from the general public and ask her to pick someone with her eyes closed who is of above average intelligence.

C. It is known in advance that the distribution of intelligence is normal with $M = 100$ and $SD = 15$.

D. Therefore, we know in advance that if she picks someone with an IQ of, say, 145, that is extremely unlikely to have been by chance. (That is, $Z = +3$, from normal curve table, probability is .13%.)

E. In fact, if she picks someone with an IQ of even 130 ($Z = 2.0$), there is only about a 2% chance and we would probably be convinced.

F. But if she picks someone with an IQ of 115 ($Z = 1$), there is a 16% chance she could have done this by chance and we would probably not be convinced.

G. Thus, we set *in advance* a score by which we will be convinced. In this case, maybe 130 (for a 2% chance).

H. Or we might pick, say, a 1% chance, corresponding to a Z score of 2.33 or IQ = about 135.

I. Or for a 5% chance, for a Z score of about 1.64, for an IQ of $(15)(1.64) + 100 = 24.6 + 100 = 124.6$.

J. We then conduct the experiment.

    1. Suppose she picks someone with an IQ of 140? This should convince us that she has done something that would not just result from chance. (That is, she has some special ability, either employing intuition, a hoax, or whatever.)

    2. Suppose she picks someone with an IQ of 90? We would clearly not be convinced.

    3. Suppose she picks someone with an IQ of 115? This result is inconclusive—she did what she said she would do, but it might have just happened by chance.

IV. Language and Steps of Hypothesis Testing

A. Show TRANSPARENCY 6.1 (steps of hypothesis testing for IQ example) and at each step discuss language and logic of this step.

B. Show TRANSPARENCY 6.2 (normal curve illustration of IQ example) and discuss.

V. Additional Examples

A. Show TRANSPARENCIES 6.3 and 6.4 (millionaire's happiness example from text) and discuss.

B. Show TRANSPARENCIES 6.5 and 6.6 (stress-reduction method example) and discuss.

VI. Review this Class: Use blackboard outline.

Lecture 6.2: Significance Levels and Directional Tests

Materials

Lecture outline
Transparencies 6.3, 6.4, and 6.6 through 6.12

Outline for Blackboard

I. **Review**
II. **Significance Levels**
III. **One- and Two-Tailed Tests**
IV. **Additional Examples**
V. **Review this Class**

Instructor's Lecture Outline

I. **Review**
   A. Descriptive and inferential statistics.
   B. Basic language and steps of hypothesis testing. Show TRANSPARENCIES 6.3 and 6.4 (millionaire's happiness example from text) and discuss each step.

II. **Significance Levels**
   A. Significance level as percent cutoff in Step 3 of hypothesis testing process.
   B. The lower the significance level (smaller percentage), the more sure we are if result is that extreme. Example: If .0001% was used, we would be very sure that if a result was that extreme we are correct in rejecting the null hypothesis.
   C. The higher the level (larger percentage), the less likely we are to have to deal with inconclusive results. Example: If 25% was used, almost any result in the direction we predicted would permit us to reject the null hypothesis.
   D. Psychologists usually use 5% as a conventional compromise.
   E. Some psychologists prefer to use 1%.
   F. We consider these issues in much more detail in Chapters 7 and 8.
   G. How results are expressed.
      1. A result which is more extreme than the set significance level is said to be "statistically significant."
      2. The 5% level is often expressed as .05 and 1% as .01.
      3. A significant result may be described as $p < .05$.
         a. p is for probability of getting this result if the null hypothesis is true.
         b. $p < .05$ means that the probability of getting this result if the null hypothesis is true is less than .05 (5%).
      4. Sometimes researchers give more extreme (lower percentage) p levels to describe results that are very extreme, even though the original cutoff set was only 5% or 1%.

III. One- and Two-Tailed Tests

A. A researcher's prediction is often that those receiving some experimental treatment will score higher than those not receiving it.

1. For example, the prediction was that those receiving $1,000,000 would be happier than the general public.
2. This is called a *directional hypothesis* because it predicts a direction of result.
3. It is also called a *one-tailed test* because the hypothesis test looks for an extreme result at just one end (in this case, the high end) of the curve.
4. Example: Show TRANSPARENCY 6.4 (millionaire example figure) and discuss.
5. A study could predict a directional result (that is, involve a one-tailed test) in which the prediction was that the score of the sample receiving the experimental treatment would be *lower*.
6. Example: Show TRANSPARENCY 6.6 (stress-reduction-method example figure) and discuss.

B. Sometimes, however, a researcher predicts that an experimental treatment will make a difference, but does not know whether it will create higher or lower scores.

1. Example: A polluting substance accidentally put in the water in a particular region may be suspected to affect the part of the brain that controls sleep, but it is not known whether it will increase or decrease amount of sleep (or have no effect).
2. This is an example of a *nondirectional hypothesis* because the research hypothesis is that the polluting substance will affect sleep, but no direction of effect is specified.
3. This is called a *two-tailed test* because extreme results at *either* extreme or tail (a lot or a little sleep) would make us want to reject the null hypothesis that the substance did not affect sleep.
4. In setting the cutoff for rejecting the null hypothesis, the overall probability (say 5%) must be divided between the two tails (say 2.5% at each).
5. This dividing means that, to be significant, a score must be more extreme than if a one-tailed test were used.
6. Show TRANSPARENCY 6.7 (illustration of one- and two-tailed tests from text) and discuss.
7. Show TRANSPARENCIES 6.8 and 6.9 (pollution and sleep example) and discuss.
8. Show TRANSPARENCIES 6.10 and 6.11 (therapy and depression level example from text) and discuss.

C. When to use one-tailed and two-tailed tests.

1. If you use a one-tailed test and result comes out in unexpected direction, no matter how extreme, it can not be considered significant.
2. To avoid this problem, researchers generally prefer to use two-tailed tests except where it is very clear that only one direction of outcome would be of interest. Example: Test of new procedure in which if it either did not work or made things worse, it would not be used.
3. Some researchers, however, use one-tailed tests whenever there is any basis for making the prediction.
4. The whole topic is very controversial.

IV. Additional Examples: Show TRANSPARENCY 6.12 (worked out numerical examples) and discuss each problem.

V. Review this Class: Use blackboard outline.

TRANSPARENCY 6.1

Steps of hypothesis testing:

1. Reframe the question into a research hypothesis and a null
   hypothesis about populations.

   Population 1: People chosen by woman with her eyes closed.
   Population 2: People in general.

   Research Hypothesis: Those chosen are more intelligent (Population 1
                        has a higher mean intelligence than Population
                        2).
   Null Hypothesis: Those chosen are not more intelligent (Population 1
                    does not have a higher mean intelligence than
                    Population 2).

2. Determine the characteristics of the comparison distribution.

   Known normal distribution, with μ = 100 and σ = 15.

3. Determine the cutoff sample score on the comparison
   distribution at which the null hypothesis should be rejected.

   For 5% probability (top 5% of comparison distribution),
   $\underline{Z}$ needed is 1.64.

4. Determine the score of your sample on the comparison
   distribution.

   Person picks out individual with IQ=140.

   $\underline{Z} = (\underline{X}-\mu)/\sigma = (140-100)/15 = 40/15 = 2.67.$

5. Compare the scores obtained in Steps 3 and 4 to decide
   whether to reject the null hypothesis.

   Score on 4 (2.67) is higher than score on 3 (1.64).

   Therefore, <u>reject</u> the null hypothesis; the research hypothesis is
   supported.

Aron/Aron
STATISTICS FOR PSYCHOLOGY

© 1994 by Prentice-Hall, Inc.
A Paramount Communications Company
Englewood Cliffs, New Jersey 07632

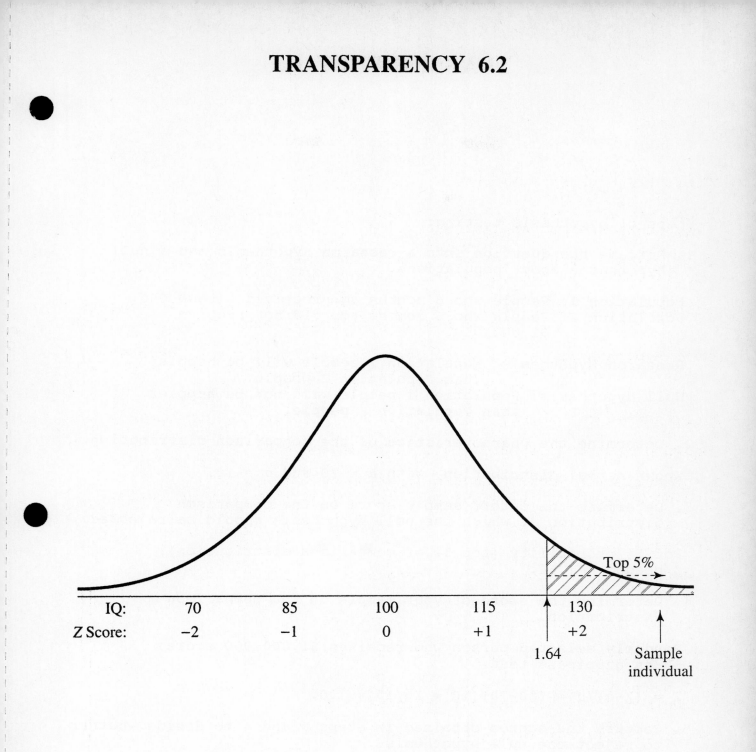

| IQ: | 70 | 85 | 100 | 115 | | 130 |
| Z Score: | −2 | −1 | 0 | +1 | | +2 |

1.64

Top 5%

Sample individual

TRANSPARENCY 6.3

Steps of hypothesis testing:

1. Reframe the question into a research hypothesis and a null hypothesis about populations.

   Population 1: People who 6 months ago received $1,000,000.
   Population 2: People who 6 months ago did not receive $1,000,000.

   Research Hypothesis: Population 1 people will be happier than Population 2 people.
   Null Hypothesis: Population 1 people will not be happier than Population 2 people.

2. Determine the characteristics of the comparison distribution.

   Known normal distribution, with $\mu = 70$ and $\sigma = 10$.

3. Determine the cutoff sample score on the comparison distribution at which the null hypothesis should be rejected.

   For 5% probability (top 5% of comparison distribution), $\underline{Z}$ needed = 1.64.

4. Determine the score of your sample on the comparison distribution.

   Randomly selected person who receives $1,000,000 scores 80 on happiness test.

   $\underline{Z} = (\underline{X}-\mu)/\sigma = (80-70)/10 = 10/10 = 1.0$.

5. Compare the scores obtained in Steps 3 and 4 to decide whether to reject the null hypothesis.

   Score on 4 (1.0) is <u>not</u> higher than score on 3 (1.64).

   Therefore, <u>do not</u> reject the null hypothesis; the result is inconclusive.

Aron/Aron
STATISTICS FOR PSYCHOLOGY

© 1994 by Prentice-Hall, Inc.
A Paramount Communications Company
Englewood Cliffs, New Jersey 07632

TRANSPARENCY 6.4

Figure 6.5

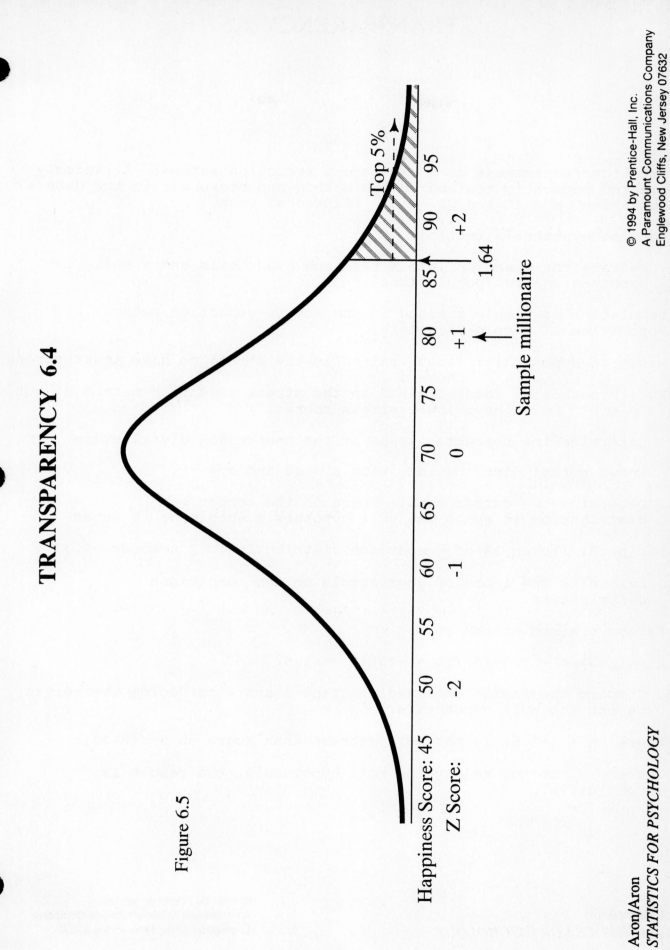

Happiness Score: 45   50   55   60   65   70   75   80   85   90   95

Z Score: -2   -1   0   +1   +2

Top 5%

1.64

Sample millionaire

Aron/Aron
STATISTICS FOR PSYCHOLOGY

TRANSPARENCY 6.5

Test of effectiveness of a new stress reduction method: A randomly selected person is trained in the method and measured; in the general population, $\underline{M}$ = 40 and $\underline{SD}$ = 10. (Fictional study.)

Steps of hypothesis testing:

1. Reframe the question into a research hypothesis and a null hypothesis about populations.

   Population 1: People trained in the stress-reduction method.
   Population 2: People in general.

   Research Hypothesis: Those trained in the method do have lower stress scores.
   Null Hypothesis: Those trained in the stress reduction method do not have lower stress scores.

2. Determine the characteristics of the comparison distribution.

   Known normal distribution, with μ = 40 and σ = 10.

3. Determine the cutoff sample score on the comparison distribution at which the null hypothesis should be rejected.

   For $\underline{p}$<.01 (lower 1% of comparison distribution), $\underline{Z}$ needed= -2.33.

4. Determine the score of your sample on the comparison distribution.

   Person trained scores 25.

   $\underline{Z}$ = $(\underline{X}-\mu)/\sigma$ = (25-40)/10 = -15/10 = -1.5.

5. Compare the scores obtained in Steps 3 and 4 to decide whether to reject the null hypothesis.

   Score on 4 (-1.5) is not more extreme than score on 3 (-2.33).

   Therefore, do not reject the null hypothesis; the result is inconclusive.

Aron/Aron
STATISTICS FOR PSYCHOLOGY

© 1994 by Prentice-Hall, Inc.
A Paramount Communications Company
Englewood Cliffs, New Jersey 07632

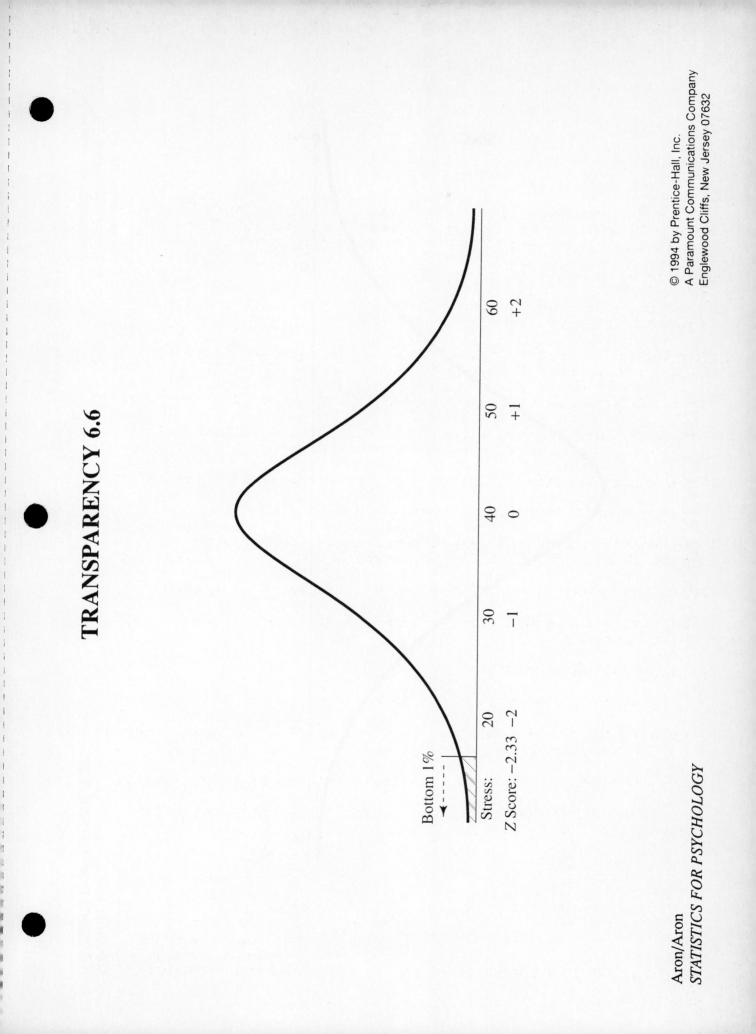

TRANSPARENCY 6.6

Bottom 1%

Stress: 20 30 40 50 60
Z Score: −2.33 −2 −1 0 +1 +2

Aron/Aron
STATISTICS FOR PSYCHOLOGY

TRANSPARENCY 6.7

Figure 6.6

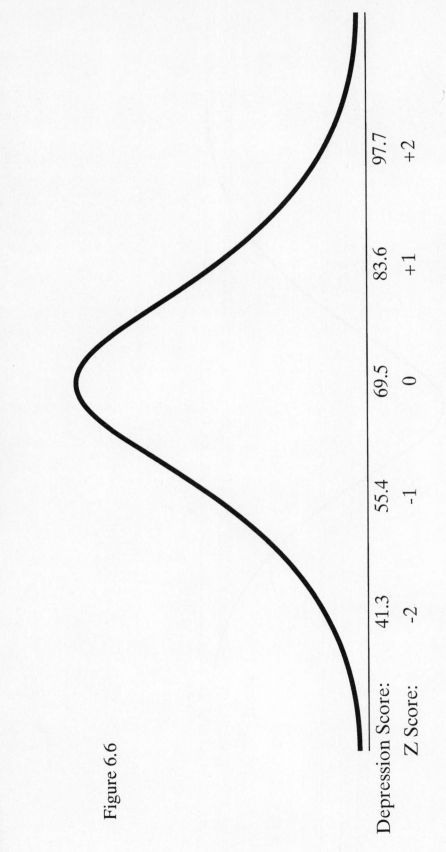

Depression Score:	41.3	55.4	69.5	83.6	97.7
Z Score:	-2	-1	0	+1	+2

© 1994 by Prentice-Hall, Inc.
A Paramount Communications Company
Englewood Cliffs, New Jersey 07632

Aron/Aron
STATISTICS FOR PSYCHOLOGY

TRANSPARENCY 6.8

Study of amount of sleep of person living in a polluted region. (Fictional data.)

Steps of hypothesis testing:

1. Reframe the question into a research hypothesis and a null hypothesis about populations.

   Population 1: People who live in the polluted region.
   Population 2: People in general.

   Research Hypothesis: Those from the region sleep a different amount.
   Null Hypothesis: Two populations are the same on the amount of sleep.

2. Determine the characteristics of the comparison distribution.

   Known normal distribution, with $\mu = 8$ and $\sigma = 1$.

3. Determine the cutoff sample score on the comparison distribution at which the null hypothesis should be rejected.

   For 5% probability (bottom 2.5% and top 2.5% of comparison distribution), $\underline{Z}$ needed is above +1.96 or below -1.96.

4. Determine the score of your sample on the comparison distribution.

   Person from polluted region sleeps 10.2 hours.

   $\underline{Z} = (\underline{X}-\mu)/\sigma = (10.2-8)/1 = 2.2/1 = 2.2$.

5. Compare the scores obtained in Steps 3 and 4 to decide whether to reject the null hypothesis.

   Score on 4 (2.2) is more extreme (higher) than high cutoff on 3 (±1.96).

   Therefore, reject the null hypothesis; the research hypothesis is supported.

Aron/Aron
STATISTICS FOR PSYCHOLOGY

© 1994 by Prentice-Hall, Inc.
A Paramount Communications Company
Englewood Cliffs, New Jersey 07632

TRANSPARENCY 6.9

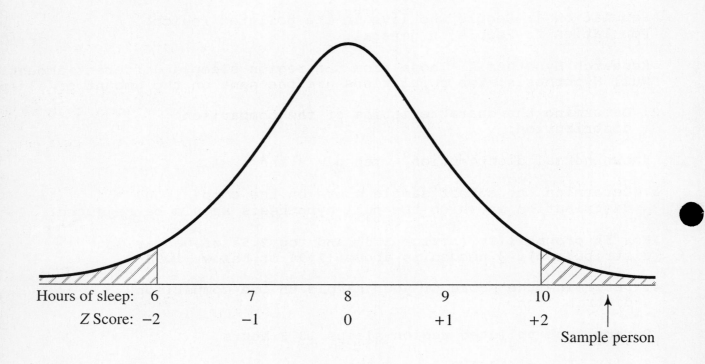

Hours of sleep: 6   7   8   9   10

Z Score: −2   −1   0   +1   +2

Sample person

Aron/Aron
STATISTICS FOR PSYCHOLOGY

TRANSPARENCY 6.10

Study of depressed patient exposed to a new form of therapy. (Fictional data.)

Steps of hypothesis testing:

1. Reframe the question into a research hypothesis and a null hypothesis about populations.

   Population 1: Patients diagnosed as depressed who receive the new therapy.
   Population 2: Patients diagnosed as depressed who receive the standard therapy.

   Research Hypothesis: After 4 weeks those receiving the new therapy have a different amount of depression than those receiving standard.
   Null Hypothesis: After 4 weeks the two populations are the same on depression.

2. Determine the characteristics of the comparison distribution.

   Known approximately normal distribution, μ = 69.5 and σ= 14.1.

3. Determine the cutoff sample score on the comparison distribution at which the null hypothesis should be rejected.

   For 5% probability (bottom 2.5% and top 2.5% of comparison distribution), $\underline{Z}$ needed is above +1.96 or below -1.96.

4. Determine the score of your sample on the comparison distribution.

   After 4 weeks, patient receiving new therapy scored 41 on depression.

   $\underline{Z}$ = $(\underline{X}-\mu)/\sigma$ = (41-69.5)/14.1 = -28.5/14.1 = -2.02.

5. Compare the scores obtained in Steps 3 and 4 to decide whether reject the null hypothesis.

   Score on 4 (-2.02) is lower (more negative) than score on 3 (-1.96).

   Therefore reject the null hypothesis; the research hypothesis is supported.

Aron/Aron
STATISTICS FOR PSYCHOLOGY

TRANSPARENCY 6.11

Figure 6.8

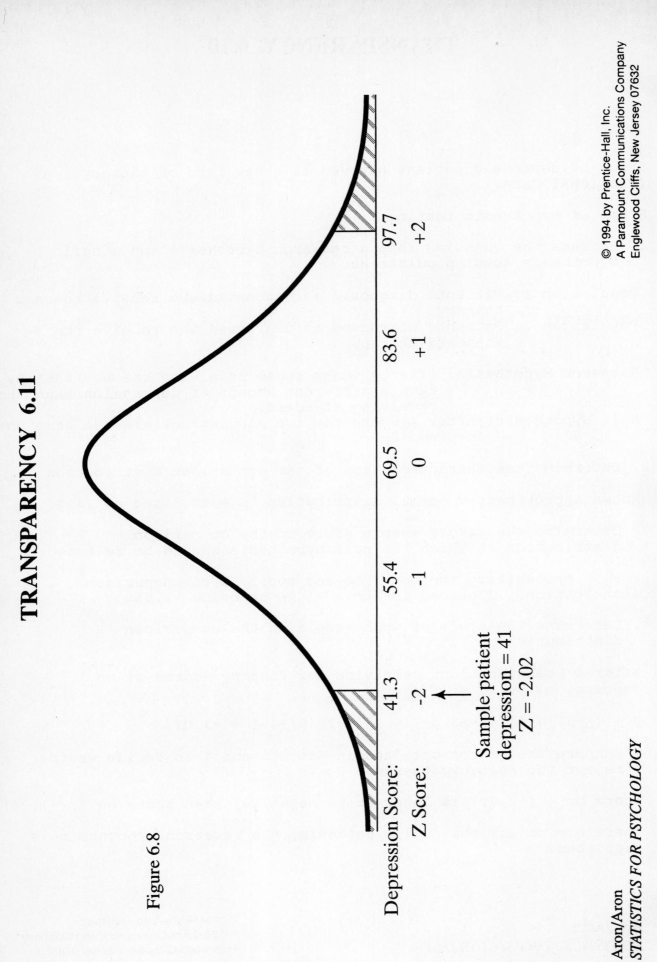

Depression Score:	41.3	55.4	69.5	83.6	97.7
Z Score:	-2	-1	0	+1	+2

Sample patient
depression = 41
Z = -2.02

© 1994 by Prentice-Hall, Inc.
A Paramount Communications Company
Englewood Cliffs, New Jersey 07632

Aron/Aron
STATISTICS FOR PSYCHOLOGY

TRANSPARENCY 6.12

Examples (using an abbreviated hypothesis testing format).

Study	Population μ	σ	Sample Score	Signif Level	Tails	
A	5	2	8	.05	1	(high predicted)

$\underline{Z}$ needed = 1.64   Obtained $\underline{Z}$ = (8-5)/2 = 3/2 = 1.5

Do NOT reject null hypothesis.

+++

Study	Population μ	σ	Sample Score	Signif Level	Tails
B	80	6	60	.05	2

$\underline{Z}$ needed = ±1.96   Obtained $\underline{Z}$ = (60-80)/6 = -20/6 = -3.33

Reject null hypothesis.

+++

Study	Population μ	σ	Sample Score	Signif Level	Tails	
C	8.5	2.1	5.3	.01	1	(low predicted)

$\underline{Z}$ needed = -2.33   Obtained $\underline{Z}$ = (5.3-8.5)/2.1 = -3.2/2.1 = -1.52

Do NOT reject null hypothesis.

Aron/Aron
STATISTICS FOR PSYCHOLOGY

© 1994 by Prentice-Hall, Inc.
A Paramount Communications Company
Englewood Cliffs, New Jersey 07632

Chapter 7
Hypothesis Tests with Means of Samples

Instructor's Summary of Chapter

The distribution of means. The distribution of means of *all* possible samples (from a particular population) of a given size can be thought of as the result of taking a very large number of samples, each of the same size and each drawn randomly from the population of individual cases, and making a distribution of the means of these samples.

Characteristics of a distribution of means. The distribution of means has the same mean as the population of individual cases. It has a smaller variance because the means of samples are less likely to be extreme than are individual cases—specifically, its variance is the variance of the population of individual cases divided by the size of the sample. Its shape approximates a normal curve if either (a) the population of individual cases is normally distributed or (b) the sample size being considered is 30 or more.

Hypothesis testing with a single sample with N > 1. Hypothesis tests involving a single sample of more than one individual and a known population are conducted in exactly the same way as the hypothesis tests of Chapter 6 (where the single sample was of a single individual), except that the comparison distribution is a distribution of means of samples of that size.

Type I and Type II error. In hypothesis testing in general, there are two kinds of correct outcomes: (a) the null hypothesis is rejected and the research hypothesis is actually true, and (b) the null hypothesis is not rejected and the research hypothesis is actually false. There are also two kinds of errors. A Type I error is when the null hypothesis is rejected but the research hypothesis is actually false. A Type II error is when the null hypothesis is not rejected but the research hypothesis is actually true.

Controversy. A controversy that relates to "known populations" is the use of norms based on large-scale administrations of standardized tests to assess individuals who may come from a group not represented by the population studied (that is, a group systematically excluded from the large-scale administrations).

How the procedures of this chapter are described in research articles. The standard deviation of the distribution of means—the *standard error* or *SE* is sometimes used to describe the expected variability of means, particularly in bar graphs in which the standard error may be shown as the length of a line above and below the top of each bar (instead of the more common use of the standard deviation for this purpose).

Box 7.1. More About Polls: Sampling Errors and Errors in Thinking about Samples. Error in an opinion poll is measured with a procedure similar to finding the standard deviation of the distribution of means—the proportion of the population the sample represents is not usually relevant.

Lecture 7.1: The Distribution of Means

Materials

Lecture outline
Transparencies 6.5, 6.6, and 7.1 through 7.5

Outline for Blackboard

I.   **Review**
II. **Introduction to the Distribution of Means**
III. **The Mean of the Distribution of Means**
IV. **The Variance and Standard Deviation of the Distribution of Means**
V.   **The Shape of the Distribution of Means**
VI. **Summary of How to Determine the Characteristics of the Distribution of Means**
VII.**Review this Class**

Note to the Instructor

The demonstration described in Lecture 5.1, creating a normal curve, can be used with this lecture instead, or in addition. Just emphasize how the variance of the distribution of means decreases with increasing sample size.

If this demonstration is used here, you will need the appropriate materials (the deck of index cards), as described for that lecture, and the outline should be amended accordingly.

Instructor's Lecture Outline

I.   Review
    A. Idea of descriptive and inferential statistics.
    B. Basic logic of hypothesis testing.
        1. Show TRANSPARENCIES 6.5 and 6.6 (stress-reduction-method example) and discuss the logic and language.
        2. Highlight role of comparison distribution in this process, noting that *in this case* it is the same as the population distribution, but in material covered later in the course, it usually will not be.

II. Introduction to the Distribution of Means
    A. Refer to stress-reduction example.
        1. What if *two* persons were trained in this procedure and *both* scored 25?
        2. Or *twenty* persons were tested and all scored about 25, or averaged 25?
        3. These examples are about hypothesis testing when the sample is greater than one (the situation considered up to now involved samples of one).
    B. The question this situation raises is, "If the null hypothesis is true, what is the probability of randomly selecting a sample of two or more scores with an *average* of some particular number?"
    C. The probability of a single individual having a score of 25 is higher than the probability of two individuals both having 25 (or of the two averaging 25).
        1. That is, the probability of a group's average being any particular extreme score is smaller than the probability of an individual's score being that extreme.

160

2. This is because if there is, say, a 10% chance of one person getting a score that extreme, for two to get a score that extreme, this event with a 10% chance would have to happen twice, and the chances of that are certainly less than 10%.

3. Another example: One is pretty unlikely to win a lottery once, but even more unlikely to win it twice in a row.

D. To determine just what the probability is of a group of a particular size having average scores at various extremes, we need a distribution of averages of groups of this size.

E. Such a distribution is a distribution of all possible means of samples of this size drawn randomly from the population.

F. For short, we call this a *distribution of means*.

G. Statisticians usually call this a *sampling distribution of the mean*. (We avoid that language because students get it mixed up with the distribution of the sample itself.)

H. Developing an intuitive understanding of this distribution, using as an example a distribution of means of all possible samples of 5 cases each.

1. Start with a population of individual cases.
2. Select a sample of five cases and compute its mean.
3. Select another sample of five cases and compute its mean.
4. Continue this process a very large number of times.
5. Create a histogram of these means.
6. This histogram is a distribution of means.

I. At this point remind students of demonstration in Lecture 5.1 (or carry it out here if it was not done then). Note that in that demonstration they also created a distribution of means, of samples of 1, 2, and 3.

J. An actual distribution of means as used in statistics is not constructed in this way—it would be impractical. But fortunately its characteristics (mean, variance, shape) can be determined mathematically based on the characteristics of the population and the size of the sample.

K. Show TRANSPARENCY 7.1 (three kinds of distributions) and discuss.

III. The Mean of the Distribution of Means

A. The mean of the distribution of means is the same as the mean of the population.

B. This is because there is no reason all these randomly selected sample means should be systematically higher or lower than the population mean.

IV. The Variance and Standard Deviation of the Distribution of Means

A. If each sample has only one case, the distribution of means is the same as the population (in every respect, because it is the same thing).

B. When the sample size is two or more, the variance of the distribution of means is always smaller than the variance of the population.

1. This is because the probability of getting two or more extreme cases in the same direction is less than getting just one.
2. This is the same principle we saw in the demonstration in Lecture 5.1 (or in this lecture).
3. This is the reason so many distributions in nature are normally distributed:
   a. The offsetting influences that combine to form each case represent, in effect, an average of a random sample of scores.
   b. That is, most normal curves, in a deep sense, are really distributions of means themselves.
   c. But do not confuse this general sense in which all normal curves can be thought of as distributions of means with the specific distinction between a population as a distribution of single cases and a distribution of means.

C. The more cases in each sample, the smaller the variation.

D. Exact rule: The variance of the distribution of means is the population variance divided by the number of cases in each of the samples.

E. Examples:
   1. Population variance is 12:
      a. Distribution of means of samples of 2 cases each has a variance of 6.
      b. Distribution of means of samples of 3 cases each has a variance of 4.
      c. Distribution of means of samples of 6 cases each has a variance of 2.
      d. Distribution of means of samples of 12 cases each has a variance of 1.
      e. Distribution of means of samples of 24 cases each has a variance of 1/2.
      f. Ask class: What if samples are 4 cases each? [3]
      g. Ask class: What if samples are 36 cases each? [1/3]
   2. Population variance is 50. Ask class, what is variance of distribution of means of samples of the following sizes:
      a. 2? [25]
      b. 5? [10]
      c. 10? [5]
      d. 25? [2]
      e. 50? [1]
      f. 100? [.5]
      g. 500? [.1]
F. Formula: Show TRANSPARENCY 7.2 (formula and above examples) and discuss.
G. Standard deviation of distribution of means is the square root of its variance: Show TRANSPARENCY 7.3 (σ_M formula and examples) and discuss.

V. The Shape of the Distribution of Means
A. If population is normal, distribution of means is normal.
B. If population is not normal (and $N > 1$).
   1. Distribution of means is still unimodal and symmetrical.
   2. The larger the sample size, the closer the distribution of means is to being normal.
   3. For practical purposes, if $N = 30$ or more, distribution of means can be considered normal.

VI. Summary of How to Determine the Characteristics of the Distribution of Means
A. Show TRANSPARENCY 7.4 (graphic of characteristics of distribution of means from text) and discuss.
B. Show TRANSPARENCY 7.5 (examples of determining characteristics of distribution of means) and discuss.

VII. Review this Class: Use blackboard outline.

Lecture 7.2: Hypothesis Testing with a Sample of More than One

Materials

Lecture outline
Transparencies 6.5, 6.6, 7.4, and 7.6 through 7.13

Outline for Blackboard

I.   **Review**
II. **The Distribution of Means as a Comparison Distribution**
III. **Hypothesis Testing with a Distribution of Means**
IV. **Additional Examples**
V.   **Possible Outcomes of Hypothesis Testing**
VI. **Review this Class**

Instructor's Lecture Outline

I.   Review
A. Idea of descriptive statistics and inferential statistics.
B. Basic logic of hypothesis testing.
   1. Show TRANSPARENCIES 6.5 and 6.6 (stress-reduction-method example) and discuss logic and language.
   2. Highlight role of comparison distribution in this process, noting that *in this case* it is the same as the population distribution, but it is not always.
C. Distribution of means.
   1. Principle: Distribution of means of all possible samples of a given size drawn randomly from the population.
   2. Intuitive understanding as constructed by taking a sample, computing its mean, repeating this many times and then making a histogram of these means.
   3. Show TRANSPARENCY 7.4 (graph of characteristics of distribution of means) and discuss.
   4. Specific rules: Show TRANSPARENCY 7.6 (table comparing three types of distributions, from text) and discuss.

II. The Distribution of Means as a Comparison Distribution
A. The main importance of the distribution of means is that it is the comparison distribution when the sample is greater than one.
B. If you were to use the population distribution for the comparison distribution, you would be seeing where a number representing an average of several cases falls along a distribution of individual cases.
C. The distribution of means is the appropriate comparison distribution because it is the same as what is being compared—it is a particular mean being compared by seeing where it falls along a distribution of means.

III. Hypothesis Testing with a Distribution of Means
A. Procedure is the same as before, except for using this new comparison distribution.
B. Show TRANSPARENCIES 7.7 and 7.8 (stress-reduction-method example with $N=20$) and discuss.

IV. Additional Examples
A. Show TRANSPARENCIES 7.9 and 7.10 (achievement test example from text) and discuss.

B. Show TRANSPARENCIES 7.11 and 7.12 (pollution-affecting-sleep example from Chapter 6 lectures with $N = 50$) and discuss.

V. **Possible Outcomes of Hypothesis Testing:** Show TRANSPARENCY 7.13 (table showing four possible outcomes from text) and discuss using examples from this lecture.

VII. **Review this Class:** Use blackboard outline.

NOTE: Warn students that Chapter 8 will take more than the usual attention to master.

TRANSPARENCY 7.1

Figure 7.5

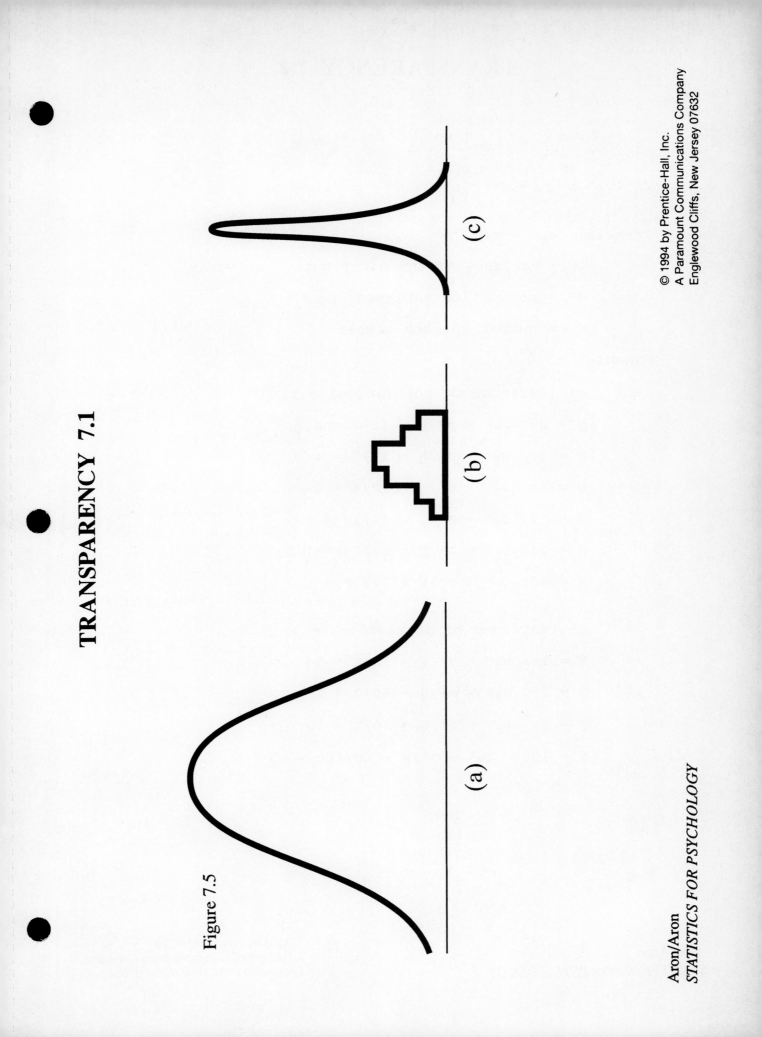

(a)

(b)

(c)

Aron/Aron
STATISTICS FOR PSYCHOLOGY

TRANSPARENCY 7.2

Formula: $\sigma_M^2 = \sigma/\underline{N}$

σ_M^2 = variance of the distribution of means

σ^2 = population variance

$\underline{N}$ = number in each sample

Example:

σ^2 (variance of population) = 12

$\underline{N}$ = 2: $\sigma_M^2 = \sigma^2/\underline{N} = 12/2 = 6$

$\underline{N}$ = 3: $\sigma_M^2 = \sigma^2/\underline{N} = 12/3 = 4$

$\underline{N}$ = 6: $\sigma_M^2 = \sigma^2/\underline{N} = 12/6 = 2$

$\underline{N}$ = 12: $\sigma_M^2 = \sigma^2/\underline{N} = 12/12 = 1$

$\underline{N}$ = 24: $\sigma_M^2 = \sigma^2/\underline{N} = 12/24 = 1/2$

$\underline{N}$ = 36: $\sigma_M^2 = \sigma^2/\underline{N} = 12/36 = 1/3$

σ^2 (variance of population) = 50

$\underline{N}$ = 1: $\sigma_M^2 = \sigma^2/\underline{N} = 50/1 = 50$

$\underline{N}$ = 2: $\sigma_M^2 = \sigma^2/\underline{N} = 50/2 = 25$

$\underline{N}$ = 5: $\sigma_M^2 = \sigma^2/\underline{N} = 50/5 = 10$

$\underline{N}$ = 100: $\sigma_M^2 = \sigma^2/\underline{N} = 50/100 = .5$

Aron/Aron
STATISTICS FOR PSYCHOLOGY

© 1994 by Prentice-Hall, Inc.
A Paramount Communications Company
Englewood Cliffs, New Jersey 07632

TRANSPARENCY 7.3

Formulas: $\sigma_M^2 = \sigma^2/N$ $\sigma_M = \sqrt{\sigma_M^2}$

In stress-reduction study, $\sigma = 10$. Therefore $\sigma^2 = 100$

If $N = 1$; $\sigma_M^2 = \sigma^2/N = 100/1 = 100$; $\sigma_M = 10$

If $N = 2$; $\sigma_M^2 = \sigma^2/N = 100/2 = 50$; $\sigma_M = 7.07$

If $N = 3$; $\sigma_M^2 = \sigma^2/N = 100/3 = 33.3$; $\sigma_M = 5.77$

If $N = 4$; $\sigma_M^2 = \sigma^2/N = 100/4 = 25$; $\sigma_M = 5$

If $N = 5$; $\sigma_M^2 = \sigma^2/N = 100/5 = 20$; $\sigma_M = 4.47$

If $N = 6$; $\sigma_M^2 = \sigma^2/N = 100/6 = 16.7$; $\sigma_M = 4.09$

If $N = 7$; $\sigma_M^2 = \sigma^2/N = 100/7 = 14.3$; $\sigma_M = 3.78$

If $N = 50$; $\sigma_M^2 = \sigma^2/N = 100/50 = 2$; $\sigma_M = 1.44$

If $N = 500$; $\sigma_M^2 = \sigma^2/N = 100/500 = .2$; $\sigma_M = .45$

Aron/Aron
STATISTICS FOR PSYCHOLOGY

© 1994 by Prentice-Hall, Inc.
A Paramount Communications Company
Englewood Cliffs, New Jersey 07632

TRANSPARENCY 7.4

Figure 7.4

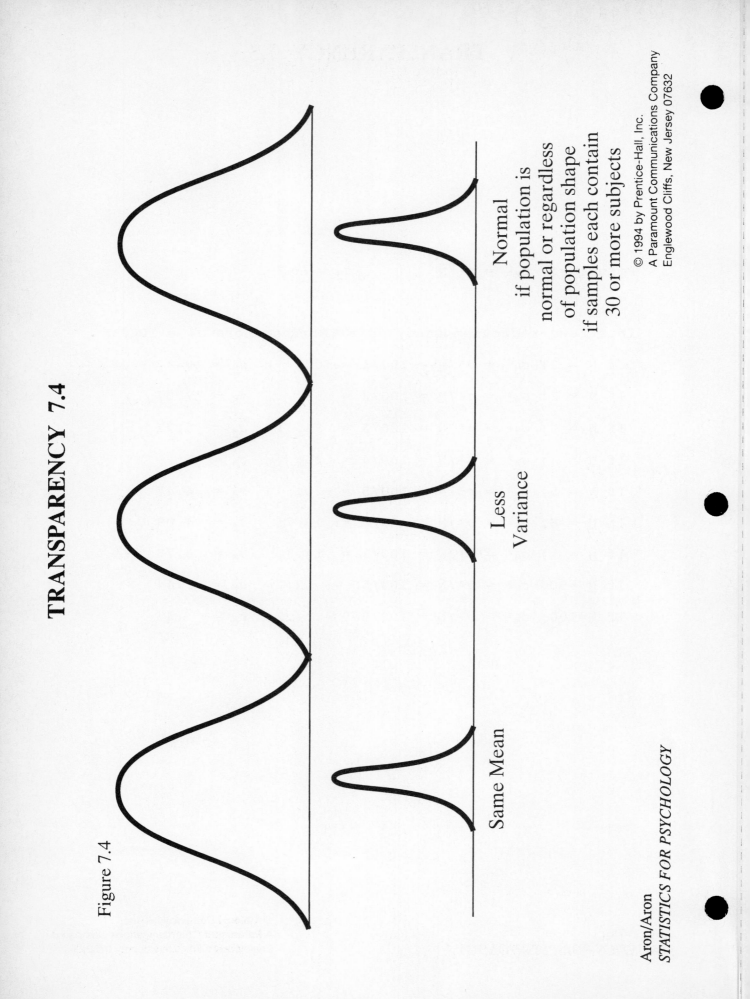

Same Mean

Less
Variance

Normal
if population is
normal or regardless
of population shape
if samples each contain
30 or more subjects

Aron/Aron
STATISTICS FOR PSYCHOLOGY

TRANSPARENCY 7.5

Population		Sample Size	Population's Shape	Distribution of Means			
μ	σ	N	Normal?	μ_M	σ_M^2	σ_M	Normal?
109.3	20	10	No	109.3	$400/10 = 40$	$\sqrt{40} = 6.32$	No
109.3	20	50	Yes	109.3	$400/50 = 8$	$\sqrt{8} = 2,83$	Yes
109.3	20	100	No	109.3	$400/100 = 4$	$\sqrt{4} = 2$	Yes
28.8	1.6	9	Yes	28.8	$25.56/9 = 2.84$	$\sqrt{2.84} = 1.69$	Yes
64	11	48	No	64	$121/48 = 2.52$	$\sqrt{2.52} = 1.59$	Yes

Aron/Aron
STATISTICS FOR PSYCHOLOGY

© 1994 by Prentice-Hall, Inc.
A Paramount Communications Company
Englewood Cliffs, New Jersey 07632

TABLE 7-1
Comparison of Three Types of Distributions

	Particular Sample's Distribution	Distribution of Means	Population's Distribution
Content	The subjects' scores in a single sample	Means of samples randomly drawn from the population	All individuals' scores in the population
Shape	Could be any shape	Normal if population is normal, approximately normal if samples contain ≥ 30 cases each	Could be any shape, often normal
Mean	$M = \Sigma X / N$, calculated from scores of those in the sample	$\mu_M = \mu$	μ
Variance	$SD^2 = \Sigma(X - M)^2 / N$, calculated from scores of those in the sample	$\sigma_M^2 = \sigma^2 / N$	σ^2
Standard deviation	$SD = \div SD^2$	$\sigma_M = \div \sigma_M^2 = \sigma / \div N$	σ

TRANSPARENCY 7.7

Stress-reduction-method example from previous lectures, but with 20 subjects. (Fictional data.)

Steps of hypothesis testing:

1. Reframe the question into a research hypothesis and a null hypothesis about populations.

   Population 1: People trained in the stress reduction method.
   Population 2: People in general.

   Research Hypothesis: Those trained in the stress reduction
                        method have lower stress scores.
   Null Hypothesis: Those trained in the method do not have lower
                    stress scores.

2. Determine characteristics of the comparison distribution.

   Population 2: shape = normal; μ = 40; σ = 10.

   Distribution of means: shape = normal;

   μ_M = 40; σ_M^2 = $\sigma^2/\underline{N}$ = 100/20 = 5; $\sigma_M = \sqrt{\sigma_M^2} = \sqrt{5}$ = 2.24.

3. Determine the cutoff sample score on comparison distribution at which the null hypothesis should be rejected.

   For 1% probability (bottom 1% of comparison distribution), $\underline{Z}$ needed is -2.33.

4. Determine the score of your sample on comparison distribution.

   Mean stress score of 20 people trained = 25.

   $Z = (\underline{M} - \mu_M)/\sigma_M$ = (25-40)/2.24 = -15/2.24 = -6.70.

5. Compare the scores obtained in Steps 3 and 4 to decide whether to reject the null hypothesis.

   Score on 4 (-6.70) is more extreme than score on 3 (-2.33).

   Therefore, reject the null hypothesis; the research hypothesis is supported.

Aron/Aron
STATISTICS FOR PSYCHOLOGY

© 1994 by Prentice-Hall, Inc.
A Paramount Communications Company
Englewood Cliffs, New Jersey 07632

TRANSPARENCY 7.8

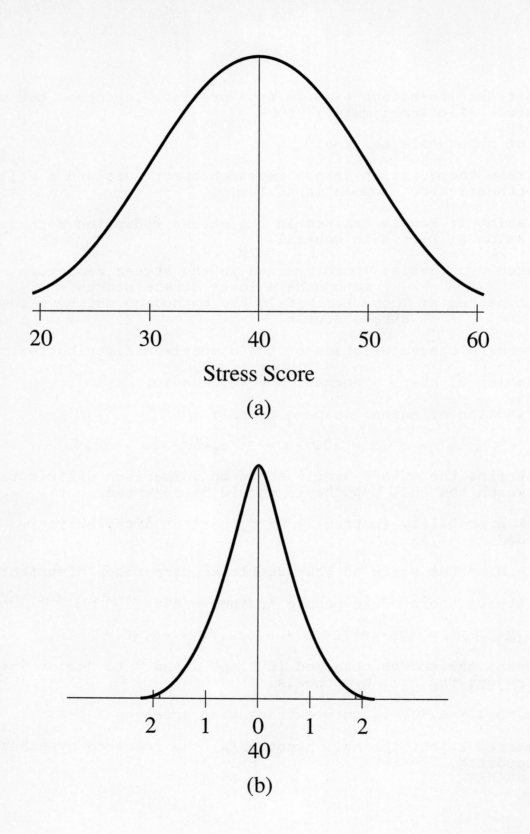

Stress Score

(a)

40

(b)

Aron/Aron
STATISTICS FOR PSYCHOLOGY

School Achievement Test Study with $\underline{N}$=64.
(Fictional example from the text.)

Steps of hypothesis testing:

1. Reframe the question into a research hypothesis and a null
   hypothesis about populations.

   Population 1: Fifth graders who get the special instructions.
   Population 2: Fifth graders in general.

   Research Hypothesis: Those who get the special instructions will
                        score higher.
   Null Hypothesis: Those who get the special instructions will not
                    score higher.

2. Determine characteristics of the comparison distribution.

   Population 2: shape = unknown; μ = 200; σ = 48.

   Distribution of means: shape = normal;
   $\mu_{\underline{M}}$ = 200; $\sigma_{\underline{M}}^2$ = $\sigma^2/\underline{N}$ = 2304/64 = 36; $\sigma_{\underline{M}}$ = $\sqrt{\sigma_{\underline{M}}^2}$ = $\sqrt{36}$ = 6.

3. Determine the cutoff sample score on comparison distribution
   at which the null hypothesis should be rejected.

   For 5% probability (top 5% of comparison distribution),
   $\underline{Z}$ needed is +1.64.

4. Determine the score of your sample on comparison distribution.

   Mean achievement test score of 64 fifth graders who received the
   special instructions = 220.

   $\underline{Z}$ = ($\underline{M}$ - μ) / $\sigma_{\underline{M}}$ = (220-200)/6 = 20/6 = 3.33.

5. Compare the scores obtained in Steps 3 and 4 to decide whether
   to reject the null hypothesis.

   Score on 4 (3.33) is more extreme than score on 3 (1.64).

   Therefore, reject the null hypothesis; the research hypothesis
   is supported.

Aron/Aron
STATISTICS FOR PSYCHOLOGY

© 1994 by Prentice-Hall, Inc.
A Paramount Communications Company
Englewood Cliffs, New Jersey 07632

TRANSPARENCY 7.10

Figure 7.8

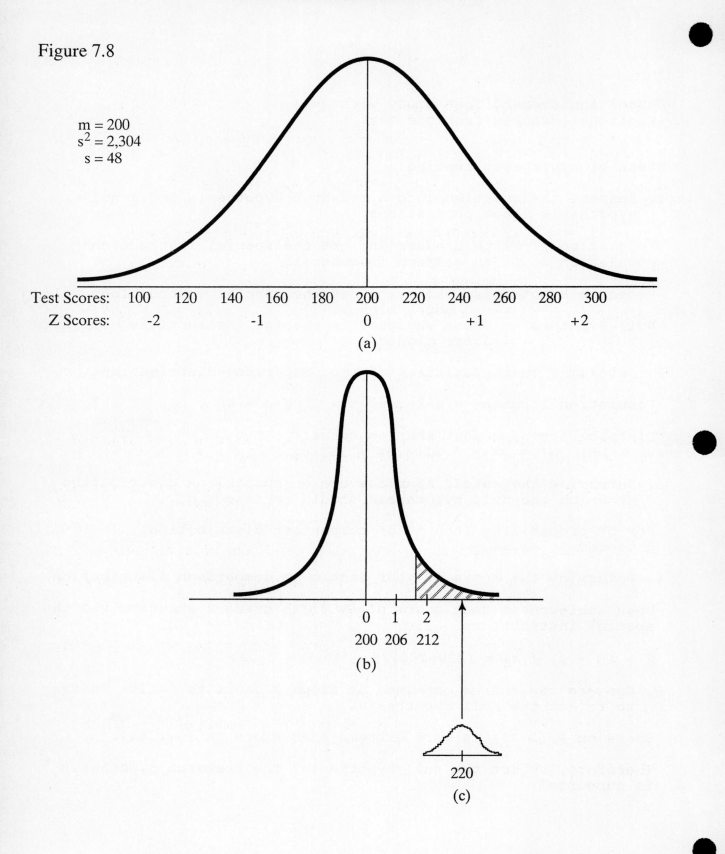

m = 200
s² = 2,304
s = 48

Test Scores:	100	120	140	160	180	200	220	240	260	280	300
Z Scores:	-2		-1			0		+1		+2	

(a)

0   1   2
200 206 212
(b)

220
(c)

TRANSPARENCY 7.11

Study of 50 people from a polluted region tested on amount of sleep. (Fictional data.)

Steps of hypothesis testing:

1. Reframe the question into a research hypothesis and a null hypothesis about populations.

   Population 1: People who live in polluted region.
   Population 2: People in general.

   Research Hypothesis: The two populations sleep different amounts.
   Null Hypothesis: The two populations are equal on amount of sleep.

2. Determine the characteristics of the comparison distribution.

   Population 2: shape = normal; $\mu = 8$; $\sigma = 1$.

   Distribution of means: shape = normal;

   $\mu_M = 8$; $\sigma_M^2 = 1/\underline{N} = 1/50 = .02$; $\sigma_M = \sqrt{\sigma_M^2} = \sqrt{.02} = .14$.

3. Determine cutoff sample score on the comparison distribution at which the null hypothesis should be rejected.

   For 5% probability 2-tailed (top and bottom 2.5% of comparison distribution), Z needed is above +1.96 or below -1.96.

4. Determine score of your sample on comparison distribution.

   Mean nightly sleep for 50 people in polluted region = 8.6 hours.

   $\underline{Z} = (\underline{M}-\mu_{\underline{M}})/\sigma_{\underline{M}} = (8.6-8) / .14 = .6/.14 = 4.29$

5. Compare the scores obtained in Steps 3 and 4 to decide whether to reject the null hypothesis.

   Score on 4 (4.29) is more extreme than score on 3 (1.96).

   Therefore, reject the null hypothesis; the research hypothesis is supported.

Aron/Aron
STATISTICS FOR PSYCHOLOGY

© 1994 by Prentice-Hall, Inc.
A Paramount Communications Company
Englewood Cliffs, New Jersey 07632

TRANSPARENCY 7.12

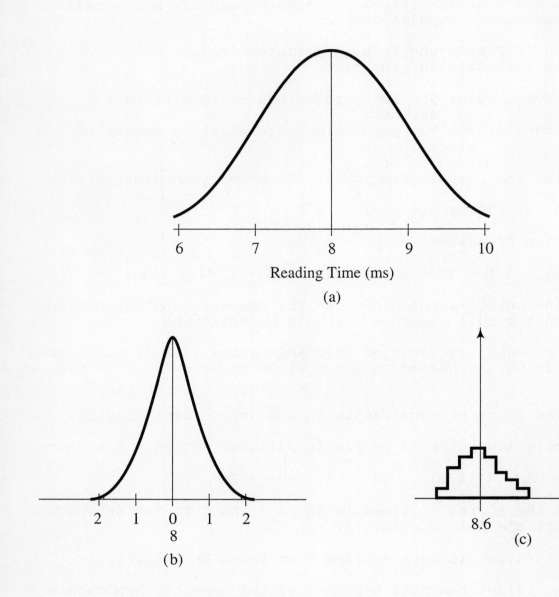

Reading Time (ms)

(a)

(b)

(c)

TABLE 7-2
Possible Correct and Erroneous Decisions in Hypothesis Testing

	Real Status of the Research Hypothesis (in practice, unknown)	
	True	*False*
Research hypothesis tenable (reject null hypothesis)	Correct decision	Error (Type I)
Study is inconclusive (fail to reject null hypothesis)	Error (Type II)	Correct decision

© 1994 by Prentice-Hall, Inc.
A Paramount Communications Company
Englewood Cliffs, New Jersey 07632

Aron/Aron
STATISTICS FOR PSYCHOLOGY

Chapter 8
Statistical Power and Effect Size

Instructor's Summary of Chapter

Power. The statistical power of a study is the probability that it will yield a significant result if the research hypothesis is true. The probability of not obtaining a significant result even though the research hypothesis is true (i.e., the probability of a Type II error) is called beta. Power is 1 - beta.

Calculating power. In the situation of a known population, a single sample (of $N \geq 1$), and a specific hypothesized mean, power can be computed as follows: First find the Z score on the Population 1 distribution corresponding to the cutoff score on the comparison distribution; the probability of exceeding this Z score (which is the power of the study) can then be found from the normal curve table.

Influences on power. Power is affected by the size of the predicted difference between means, the population standard deviation (σ), the sample size (N), the significance level (α), and whether a one- or a two-tailed test is used. Power is greatest when the predicted difference is large, σ is small, N is large, α is least stringent, and a one-tailed test is used.

Effect size combines two of the factors influencing power—the predicted difference between means (the greater the difference, the larger the effect size) and σ (the smaller this is, the larger the effect size). Effect size can be determined in advance of a study based on (a) theory or previous research findings, (b) determining the minimum difference between population means that would be meaningful, or (c) using rules of thumb of .2 for a small effect, .5 for a medium effect, and .8 for a large effect.

Practical ways of increasing the power of a planned experiment include increasing sample size or increasing effect size—by, for example, reducing variance with a less diverse population or more reliable measures, or increasing the intensity of the experimental treatment.

The importance of power in interpreting a completed experiment is that significant experimental findings with low power may not have practical importance, while nonsignificant experimental results with low·power make it possible that important, significant results might show up if power were increased. Further, while it is not possible to "prove" the null hypothesis, with sufficient power a nonsignificant finding may suggest that any true effect is very, very small.

Controversies. Psychologists disagree over the relative importance of significance versus effect size in interpreting experimental results, though more theoretically-oriented psychologists seem to emphasize significance while more applied researchers emphasize effect size. Meta-analysis is a recently developed procedure for systematically combining effects of independent studies, primarily on the basis of effect sizes.

How the procedures of this chapter are described in research articles. When power is directly mentioned in research articles, it is usually in the context of justifying the number of subjects used in a study or in the discussion of nonsignificant results. Effect size is more commonly mentioned, both when comparing results of different studies or different parts of the same study and in meta-analysis and other reviews of the literature.

Box 8.1. The Power of Typical Psychology Experiments. Reviews of psychology research articles over the years have shown that most studies are designed so that they would be likely to achieve significance only if the effect sizes of what they were studying were large.

Box 8.2. Effect Sizes for Relaxation and Meditation: A Restful Meta-Analysis. Summarizes an example of a meta-analysis study (Eppley, Abrams, & Shear, 1989).

Lecture 8.1: Power and Its Computation

Materials

Lecture outline
Transparencies 7.9, 7.10, 7.13, and 8.1 through 8.8

Outline for Blackboard

I.   **Review**
II. **What is Power?**
III. **Alpha and Beta**
IV. **Calculating Power**
V.   **Review this Class**

Instructor's Lecture Outline

I.   Review
  A. Idea of descriptive and inferential statistics.
  B. Hypothesis testing with $N > 1$. Show TRANSPARENCIES 7.9 and 7.10 (hypothesis testing for achievement test example, from text) and discuss step by step.
  C. Possible outcomes of hypothesis testing (Type I and Type II Error). Show TRANSPARENCY 7.13 (table of possible outcomes from text) and discuss.

II. What is Power?
  A. Ask class, "What is power?"
  B. Meanings of power in ordinary life.
    1. The ability to achieve your goals—presuming these goals are good for you.
    2. Thus, knowledge is power, and the more accurate the knowledge, the more power.
  C. In statistics, we speak of the power of an experiment.
  D. Show TRANSPARENCY 8.1 (definitions of statistical power) and discuss, relating each to the definitions of power in ordinary life.
  E. Why power is important (covered in depth in next class).
    1. It helps in planning studies—if power is too low, you can choose not to do the study at all, or to change it to increase power.
    2. It helps when evaluating results of studies, particularly of nonsignificant results.

III. Alpha and Beta
  A. Show TRANSPARENCY 8.2 (chart of four outcomes with alpha, beta, etc. from text) and discuss each cell, emphasizing new elements in this chart.
  B. Show TRANSPARENCY 8.3 (same chart as above but with outcomes from achievement-test-score experiment from text as example) and discuss.
  C. Show TRANSPARENCY 8.4 (same chart as above but with outcomes from stress-reduction study from Chapter 7 lectures as example) and discuss.

IV. Calculating Power

A. Show TRANSPARENCIES 8.5 and 8.6 (figure and steps of computation for achievement-test experiment from text) and discuss.

B. Show TRANSPARENCIES 8.7 and 8.8 (figure and steps of computation and for stress-reduction study) and discuss.

V. Review this Class: Use blackboard outline.

Lecture 8.2: Influences on Power

Materials

Lecture outline
Transparencies 7.9, 7.10, 7.13, 8.1 through 8.3, and 8.5 through 8.26

Outline for Blackboard

I.   **Review**
II. **Predicted Mean Difference**
III. **Population Standard Deviation**
IV. **Sample Size (N)**
V.   **Significance Level**
VI. **One- vs. Two-Tailed Tests**
VII. **Review this Class**

Instructor's Lecture Outline

I.   Review
   A. Idea of descriptive and inferential statistics.
   B. Hypothesis testing with $N > 1$. Show TRANSPARENCIES 7.9 and 7.10 (hypothesis testing for achievement test example, from text) and discuss step by step briefly.
   C. Possible outcomes of hypothesis testing (Type I and Type II error). Show TRANSPARENCY 7.13 (four outcomes chart from text) and discuss.
   D. Power. Show TRANSPARENCIES 8.1 through 8.3 (definition of power and charts of outcomes) and discuss.
   E. Computing power. Show TRANSPARENCIES 8.5 and 8.6 (figure and steps of computation for achievement-test experiment from text) and discuss step by step.

II. Predicted Mean Difference
   A. Principle.
      1. The greater the predicted mean difference, the more power.
      2. This is because the greater the predicted mean difference, the less overlap between the two distributions of means based on the known and hypothesized populations (they are farther apart).
      3. Show top of TRANSPARENCY 8.9 (overlaps of distributions for different mean differences from text) and discuss.
      4. Just how one determines the predicted mean difference is considered in the next lecture (Lecture 8.3).
   B. Examples of effect of different predicted means.
      1. Show TRANSPARENCIES 8.10 and 8.11 (achievement-test-score example with predicted mean 216 from text) and go through steps of computing power.
      2. Show TRANSPARENCIES 8.5 and 8.6 (same example with predicted mean = 208) and compare to above.
      3. Show TRANSPARENCIES 8.12 and 8.13 (stress-reduction example from Lecture 8.1 but with predicted mean = 216) and go through steps of computing power briefly.
      4. Show TRANSPARENCIES 8.7 and 8.8 (same example with predicted mean = 34) and compare to above.

III. Population Standard Deviation
A. Principle.
1. The smaller the population standard deviation, the more power.
2. This is because the smaller the population standard deviation, the smaller the standard deviation of the distribution of means, and thus the less overlap between the distributions of means based on the known and hypothesized populations (they are each narrower).
3. Show bottom of TRANSPARENCY 8.9 (illustration of effect of lower standard deviation from text) and discuss.
B. Examples of effect of different population standard deviations.
1. Show TRANSPARENCIES 8.14 and 8.15 (achievement-test-score example from text with $\sigma = 24$ and original predicted mean of 206) and go through steps of computing power.
2. Show TRANSPARENCIES 8.5 and 8.6 (same example with $\sigma = 48$) and compare to above.
3. Show TRANSPARENCIES 8.16 and 8.17 (stress-reduction example but with $\sigma = 5$) and go through steps of computing power briefly.
4. Show TRANSPARENCIES 8.7 and 8.8 (same example with $\sigma = 10$) and compare to above.
C. Note: The two influences on power considered so far, predicted mean difference and population standard deviation, are often combined in a figure called the effect size, which is considered in the next lecture (Lecture 8.3).

IV. Sample Size (N)
A. Principle.
1. The larger the N, the more power.
2. This is because the larger the N, the smaller the σ_M, and thus the less overlap between the distributions of means based on the known and hypothesized populations (they are each narrower).
3. Show bottom of TRANSPARENCY 8.9 (illustration of effect of lower standard deviation from text) again and discuss—this time pointing out how σ_M is affected by *both* σ^2 and N.
B. Examples of effect of different sample sizes.
1. Show TRANSPARENCIES 8.18 and 8.19 (achievement-test- score example from text as in original version of predicted mean of 206 and $\sigma = 48$, but with $N = 100$) and briefly go through steps of computing power.
2. Show TRANSPARENCIES 8.5 and 8.6 (same example with original sample size of 64) and compare to above.
3. Show TRANSPARENCIES 8.20 and 8.21 (stress-reduction example from Lecture 8.1 but with $N = 10$) and go through steps of computing power briefly.
4. Show TRANSPARENCIES 8.7 and 8.8 (same example with original N of 20) and compare to above.

V. Significance Level

A. Principle.

    1. The less stringent the significance level (for example .05 vs. .01), the more power.

    2. This is because the less stringent the significance level, the less extreme is the cutoff score on the lower distribution (the one for the known population), making the corresponding point on the upper distribution (the one for the hypothesized population) less extreme—so that more area is included in the power region.

    3. Show TRANSPARENCY 8.5 (original version of achievement-test experiment from text) and note how the line would move if a more stringent significance level were used.

B. Examples of effect of different significance levels:

    1. Show TRANSPARENCIES 8.22 and 8.23 (achievement-test-score example from text as in original version but with $p < .01$) and briefly go through power computation.

    2. Show TRANSPARENCIES 8.5 and 8.6 (same example but with original significance level of $p < .05$) and compare to above.

    3. Show TRANSPARENCIES 8.24 and 8.25 (stress-reduction example from Lecture 8.1 but with significance level of $p < .05$) and go through power computation briefly.

    4. Show TRANSPARENCIES 8.7 and 8.8 (same example with original significance level of $p < .01$) and compare to above.

VI. One- vs. Two-Tailed Tests

A. Principle.

    1. One-tailed tests have more power than two-tailed tests for a result in the predicted direction.

    2. This is because with a one-tailed test there is a less extreme cutoff score (in the predicted direction) on the lower distribution (for the known population), making the corresponding point on the upper distribution (for the hypothesized population) less extreme—so that more area is included in the power region.

    3. However, there is *zero* power for a result in the opposite-to-the-predicted direction.

B. Show TRANSPARENCY 8.26 (example of power with two-tailed test from text) and discuss.

VII. Review this Class

A. Use blackboard outline.

B. Emphasize the direction of influence on power of each of the influences considered.

Lecture 8.3: Power in Planning and Evaluating Studies

Materials

Lecture outline
Transparencies 8.1 through 8.3, 8.5, 8.6, and 8.27 through 8.31

Outline for Blackboard

I. **Review**
II. **Determining Predicted Mean**
III. **Effect Size**
IV. **Power in Planning Experiments**
V. **Power in Interpreting Experiments**
VI. **Review this Class**

Instructor's Lecture Outline

I. Review
  A. Idea of descriptive and inferential statistics.
  B. Principle of hypothesis testing.
  C. Principle of power. Show TRANSPARENCIES 8.1 through 8.3 (definition of power and charts of outcomes) and discuss.
  D. Computing power. Show TRANSPARENCIES 8.5 and 8.6 (figure and steps of computation for achievement-test experiment from text) and discuss step by step.
  E. Show TRANSPARENCY 8.27 (influences on power) and discuss.

II. Determining Predicted Mean
  A. Theory or previous experience.
  B. Minimum meaningful difference.
  C. Cohen's effect-size conventions (to be discussed shortly).
  D. Note: In most cases the predicted mean is very approximate and somewhat arbitrary.

III. Effect Size
  A. Effect size combines two of the influences on power—predicted mean difference and population standard deviation.
  B. Show TRANSPARENCY 8.28 (formula and symbols for effect size with a known population and a single sample).
     1. Discuss formula and symbols.
     2. Notice it is NOT influenced by sample size.
     3. It is like a Z score in that it gives amount of difference between populations without regard to unit of measure.
     4. In this way it provides a standard metric that can be used to compare results of different studies even if they have different sample sizes, and even if they use different measures.

TRANSPARENCY 8.1

Statistical power is probability that

1. if the research hypothesis is true, the results of your experiment will support the research hypothesis.

OR

2. you will not make a Type II Error.

Aron/Aron
STATISTICS FOR PSYCHOLOGY

© 1994 by Prentice-Hall, Inc.
A Paramount Communications Company
Englewood Cliffs, New Jersey 07632

TABLE 8-1
Possible Outcomes of Hypothesis Testing

	Real Status of the Research Hypothesis (in Practice, Unknown)	
	True	*False*
Research hypothesis tenable (reject null hypothesis)		
Study is inconclusive (fail to reject null hypothesis)		

Aron/Aron
STATISTICS FOR PSYCHOLOGY

TRANSPARENCY 8.3

Possible outcomes of experiment on increasing achievement test scores.

Real Status of the
Research Hypothesis
(In practice, unknown)

		TRUE	FALSE
Decision Using Hypothesis Testing Procedure	Research Hypothesis Tenable (Reject Null Hypothesis)	Method increases scores on test and study result is significant (p = power = 1 - beta)	Method does not increase scores but study result is significant (Type I Error p = alpha)
	Study Is Inconclusive (Do Not Reject Null Hypothesis)	Method increases scores on test but study result is inconclusive (Type II Error, p = beta)	Method does not increase scores and result is inconclusive

Aron/Aron
STATISTICS FOR PSYCHOLOGY

TRANSPARENCY 8.4

Possible outcomes of experiment on stress-reduction.

Real Status of the
Research Hypothesis
(In practice, unknown)

		TRUE	FALSE
Decision Using Hypothesis Testing Procedure	Research Hypothesis Tenable (Reject Null Hypothesis)	Method reduces stress and result of study is significant (p = power = 1 - beta)	Method does not reduce stress but result of study is signif. (Type I Error p = alpha)
	Study Is Inconclusive (Do Not Reject Null Hypothesis)	Method reduces stress but result is inconclusive (Type II Error, p = beta)	Method does not reduce stress and result is inconclusive

Aron/Aron
STATISTICS FOR PSYCHOLOGY

TRANSPARENCY 8.5

Figure 8.2

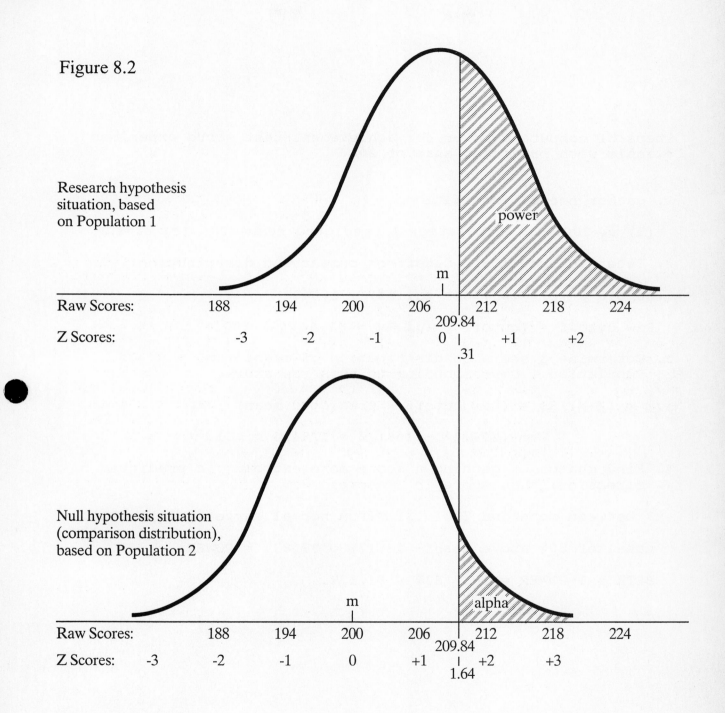

Research hypothesis
situation, based
on Population 1

power

Raw Scores: 188   194   200   206   m   212   218   224

209.84

Z Scores: -3   -2   -1   0 | +1   +2

.31

Null hypothesis situation
(comparison distribution),
based on Population 2

m   alpha

Raw Scores: 188   194   200   206   209.84   212   218   224

Z Scores: -3   -2   -1   0   +1 | +2   +3

1.64

TRANSPARENCY 8.6

Steps of computing power for achievement-test score experiment example with predicted mean of 208.

1. Gather needed information:

   (a) μ_M=200; (b) Population 1 predicted mean=208; (c) σ_M=6.

2. Determine raw-score cutoff on comparison distribution:

   Cutoff $\underline{Z}$ ($\underline{p}$ < .05, one-tailed) = 1.64.

   Raw cutoff = (cutoff $\underline{Z}$)(σ_M)+μ_M = (1.64)(6)+200 = 209.84.

3. Determine $\underline{Z}$ score on distribution of means for Population 1 corresponding to this raw score.

   $\underline{Z}$ = ($\underline{X}$-$\underline{M}$)/$\underline{SD}$ = (Raw cutoff - predicted mean) / σ_M

   $$= (209.84 - 208)/6 = 1.84/6 = .31.$$

4. Find chance of getting a score more extreme (in predicted direction) than score in 3 above.

   % between mean and $\underline{Z}$ of .31 (from normal curve table) = 12.17

   total of 50% above mean - 12.17% = 37.83% = POWER.

   BETA = 1-POWER = 1-37.83% = 62.17%.

Aron/Aron
STATISTICS FOR PSYCHOLOGY

© 1994 by Prentice-Hall, Inc.
A Paramount Communications Company
Englewood Cliffs, New Jersey 07632

TRANSPARENCY 8.7

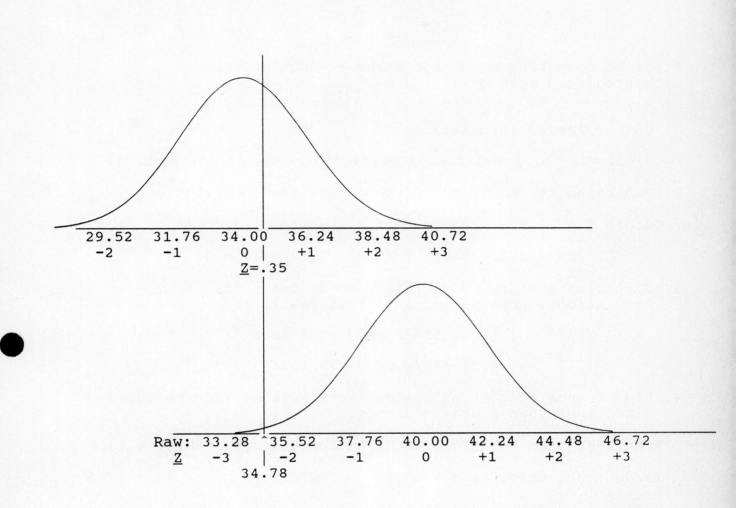

```
        29.52     31.76     34.00     36.24     38.48     40.72
         -2        -1         0    |    +1        +2        +3
                                  Z=.35
```

```
   Raw: 33.28  ^35.52   37.76    40.00    42.24    44.48    46.72
    Z     -3    |  -2      -1       0       +1       +2       +3
              34.78
```

TRANSPARENCY 8.8

Steps of computing power for stress-reduction example
with predicted mean of 34.

1. Gather needed information:

   (a) μ_M=40; (b) Population 1 predicted mean=34; (c) σ_M=2.24.

2. Determine raw-score cutoff on comparison distribution:

   Cutoff $\underline{Z}$ = -2.33.   Raw cutoff = (cutoff $\underline{Z}$)(σ_M) + μ_M

   $\qquad\qquad$ = (-2.33)(2.24) + 40 = - 5.22 + 40 = 34.78.

3. Determine $\underline{Z}$ score on distribution of means for
   Population 1 corresponding to this raw score.

   $\underline{Z}$ = ($\underline{X}$-$\underline{M}$)/$\underline{SD}$ = (Raw cutoff - predicted mean) / σ_M

   $\qquad\qquad$ = (34.78-34)/2.24 = .78/2.24 = .35.

4. Find chance of getting a score more extreme (in predicted
   direction) than score in 3 above.

   % between mean and $\underline{Z}$ of .35 (from normal curve table) = 13.68.

   13.68 + 50% below mean = 63.68% = POWER.

   BETA = 1-POWER = 1-63.68% = 36.32%.

Aron/Aron
STATISTICS FOR PSYCHOLOGY

© 1994 by Prentice-Hall, Inc.
A Paramount Communications Company
Englewood Cliffs, New Jersey 07632

TRANSPARENCY 8.9

Figure 8.4

Research hypothesis
situation, based
on Population 1

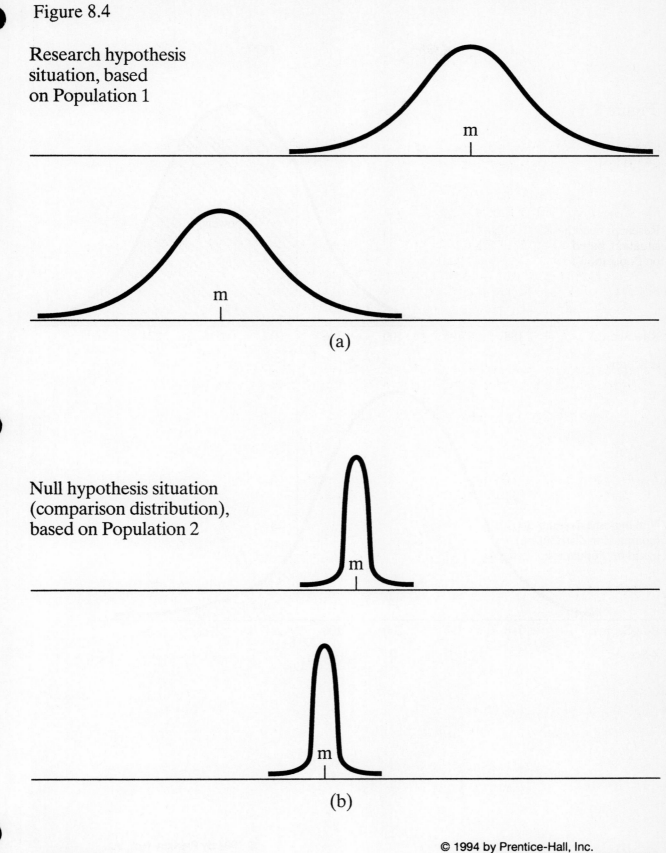

(a)

Null hypothesis situation
(comparison distribution),
based on Population 2

(b)

TRANSPARENCY 8.10

Figure 8.5

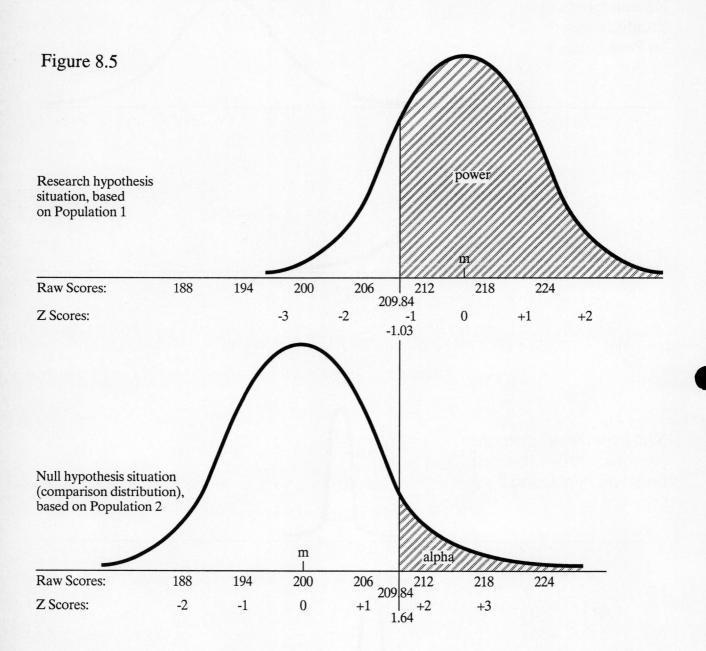

Research hypothesis
situation, based
on Population 1

power

m

Raw Scores: 188   194   200   206   212   218   224

209.84

Z Scores: -3   -2   -1   0   +1   +2

-1.03

Null hypothesis situation
(comparison distribution),
based on Population 2

m

alpha

Raw Scores: 188   194   200   206   212   218   224

209.84

Z Scores: -2   -1   0   +1   +2   +3

1.64

Aron/Aron
STATISTICS FOR PSYCHOLOGY

© 1994 by Prentice-Hall, Inc.
A Paramount Communications Company
Englewood Cliffs, New Jersey 07632

TRANSPARENCY 8.11

Steps of computing power for achievement-test-score example with predicted mean of 216.

1. Gather needed information:

    (a) μ_M=200 (b) Population 1 predicted mean=216; (c) σ_M=6.

2. Determine raw-score cutoff on comparison distribution:

    Cutoff $\underline{Z}$ ($\underline{p}$ < .05, one-tailed) = 1.64.
    Raw cutoff = (cutoff $\underline{Z}$)(σ_M)+μ_M = (1.64)(6)+200 = 209.84.

3. Determine $\underline{Z}$ score on distribution of means for Population 1 corresponding to this raw score.

    $\underline{Z}$ = ($\underline{X}$-$\underline{M}$)/$\underline{SD}$ = (Raw cutoff - predicted mean) / σ_M

                  = (209.84 - 216)/6 = -6.16/6 = -1.03.

4. Find chance of getting a score more extreme (in predicted direction) than score in 3 above.

    % between mean and $\underline{Z}$ of -1.03 (from normal curve table) = 34.85.

    total of 50% above mean + 34.85% = 84.85% = POWER.

    BETA = 1-POWER = 1-84.85% = 15.15%.

Predicted Mean = 32

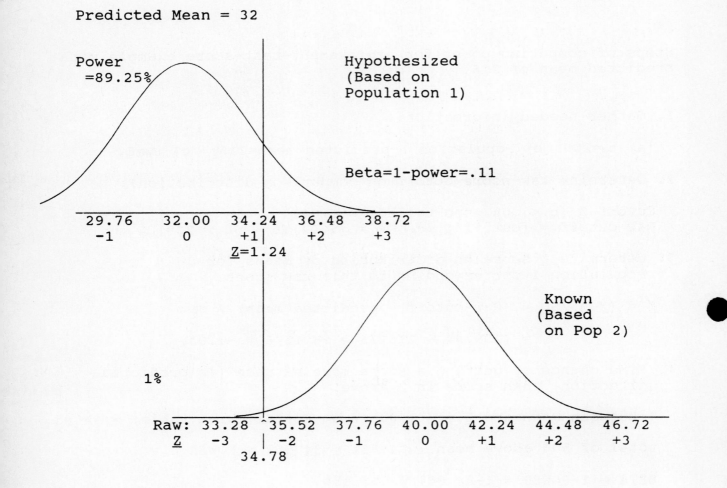

Power
=89.25%

Hypothesized
(Based on
Population 1)

Beta=1-power=.11

29.76	32.00	34.24	36.48	38.72
-1	0	+1	+2	+3

$\underline{Z}$=1.24

Known
(Based
on Pop 2)

1%

Raw:	33.28	35.52	37.76	40.00	42.24	44.48	46.72
$\underline{Z}$	-3	-2	-1	0	+1	+2	+3

34.78

Aron/Aron
STATISTICS FOR PSYCHOLOGY

© 1994 by Prentice-Hall, Inc.
A Paramount Communications Company
Englewood Cliffs, New Jersey 07632

TRANSPARENCY 8.13

Steps of computing power for stress-reduction example
with predicted mean of 32.

1. Gather needed information:

   (a) μ_M=40; (b) Population 1 predicted mean=32; (c) σ_M=2.24.

2. Determine raw-score cutoff on comparison distribution:

   Cutoff $\underline{Z}$ = -2.33.   Raw cutoff = (cutoff $\underline{Z}$)(σ_M) + μ_M

   $$= (-2.33)(2.24) + 40 = - 5.22 + 40 = 34.78.$$

3. Determine $\underline{Z}$ score on distribution of means for
   Population 1 corresponding to this raw score.

   $\underline{Z}$ = ($\underline{X}$-$\underline{M}$)/$\underline{SD}$ = (Raw cutoff - predicted mean) / σ_M

   $$= (34.78-32)/2.24 = 2.78/2.24 = 1.24.$$

4. Find chance of getting a score more extreme (in predicted
   direction) than score in 3 above.

   % between mean and $\underline{Z}$ of 1.24 (from normal curve table) = 39.25

   39.25 + 50% below mean = 89.25% = POWER.

   BETA = 1-POWER = 1-89.25% = 10.75%.

Aron/Aron
STATISTICS FOR PSYCHOLOGY

© 1994 by Prentice-Hall, Inc.
A Paramount Communications Company
Englewood Cliffs, New Jersey 07632

TRANSPARENCY 8.14

Figure 8.7

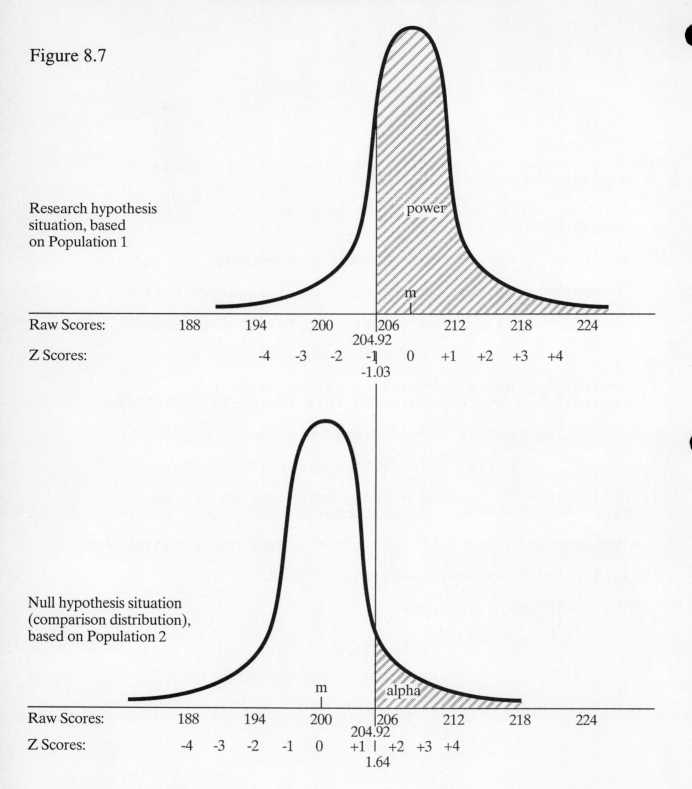

Research hypothesis
situation, based
on Population 1

power

m

Raw Scores: 188   194   200   206   212   218   224
              204.92

Z Scores: -4   -3   -2   -1   0   +1   +2   +3   +4
                                    -1.03

Null hypothesis situation
(comparison distribution),
based on Population 2

m   alpha

Raw Scores: 188   194   200   206   212   218   224
                            204.92

Z Scores: -4   -3   -2   -1   0   +1 | +2   +3   +4
                                    1.64

Aron/Aron
STATISTICS FOR PSYCHOLOGY

TRANSPARENCY 8.15

Steps of computing power for achievement-test-score example but with population standard deviation of 24.

1. Gather needed information:

   (a) μ_M=200; (b) Population 1 predicted mean=208;
   (c) $\sigma_{\underline{M}}$= 3 (σ=24; σ^2=576; $\sigma_M{}^2$=576/64=9; σ_M=$\sqrt{9}$=3).

2. Determine raw-score cutoff on comparison distribution:

   Cutoff $\underline{Z}$ ($\underline{p}$ < .05, one-tailed) = 1.64.

   Raw cutoff = (cutoff $\underline{Z}$)(σ_M)+μ_M = (1.64)(3)+200 = 204.92.

3. Determine $\underline{Z}$ score on distribution of means for Population 1 corresponding to this raw score.

   $\underline{Z}$ = ($\underline{X}$-$\underline{M}$)/$\underline{SD}$ = (Raw cutoff - predicted mean) / σ_M

   $\qquad\qquad$ = (204.92 - 208)/6 = -3.08/3 = -1.03.

4. Find chance of getting a score more extreme (in predicted direction) than score in 3 above.

   % between mean and $\underline{Z}$ of -1.03 (from normal curve table) = 34.85.

   total of 50% above mean + 34.85% = 84.85% = POWER.

   BETA = 1-POWER = 1-84.85% = 15.15%.

Aron/Aron
STATISTICS FOR PSYCHOLOGY

© 1994 by Prentice-Hall, Inc.
A Paramount Communications Company
Englewood Cliffs, New Jersey 07632

TRANSPARENCY 8.16

Population standard deviation (σ) = 5

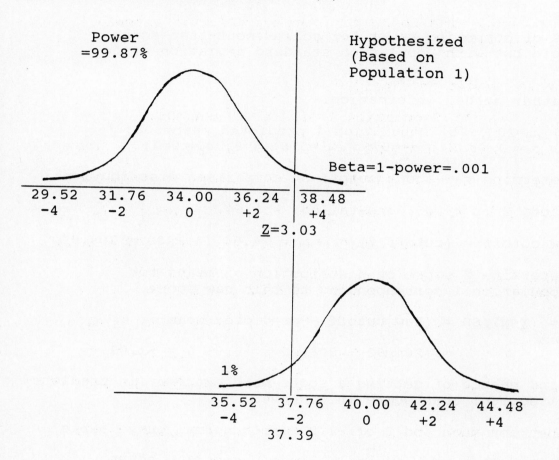

Power
=99.87%

Hypothesized
(Based on
Population 1)

Beta=1-power=.001

| 29.52 | 31.76 | 34.00 | 36.24 | 38.48 |
| -4 | -2 | 0 | +2 | +4 |

$\underline{Z}$=3.03

1%

| 35.52 | 37.76 | 40.00 | 42.24 | 44.48 |
| -4 | -2 | 0 | +2 | +4 |

37.39

Aron/Aron
STATISTICS FOR PSYCHOLOGY

© 1994 by Prentice-Hall, Inc.
A Paramount Communications Company
Englewood Cliffs, New Jersey 07632

TRANSPARENCY 8.17

Steps of computing power for stress-reduction example with population standard deviation of 5.

1. Gather needed information:

   (a) μ_M=40; (b) Population 1 predicted mean=34;

   (c) σ_M=1.12 (σ=5 σ^2=25 σ_M^2=25/20=1.25).

2. Determine raw-score cutoff on comparison distribution:

   Cutoff $\underline{Z}$ = -2.33.   Raw cutoff = (cutoff $\underline{Z}$)(σ_M) + μ_M

   $= (-2.33)(1.12) + 40 = - 2.61 + 40 = 37.39.$

3. Determine $\underline{Z}$ score on distribution of means for Population 1 corresponding to this raw score.

   $\underline{Z}$ = ($\underline{X}$ - $\underline{M}$)/ $\underline{SD}$ = (Raw cutoff - predicted mean) / σ_M

   $= (37.39-34)/1.12 = 3.39/1.12 = 3.03.$

4. Find chance of getting a score more extreme (in predicted direction) than score in 3 above.

   % between mean and $\underline{Z}$ of 1.24 (from normal curve table) = 39.25.

   39.25 + 50% below mean = 89.25% = POWER.

   BETA = 1-POWER = 1-99.87% = .13%.

Aron/Aron
STATISTICS FOR PSYCHOLOGY

© 1994 by Prentice-Hall, Inc.
A Paramount Communications Company
Englewood Cliffs, New Jersey 07632

TRANSPARENCY 8.18

Figure 8.10

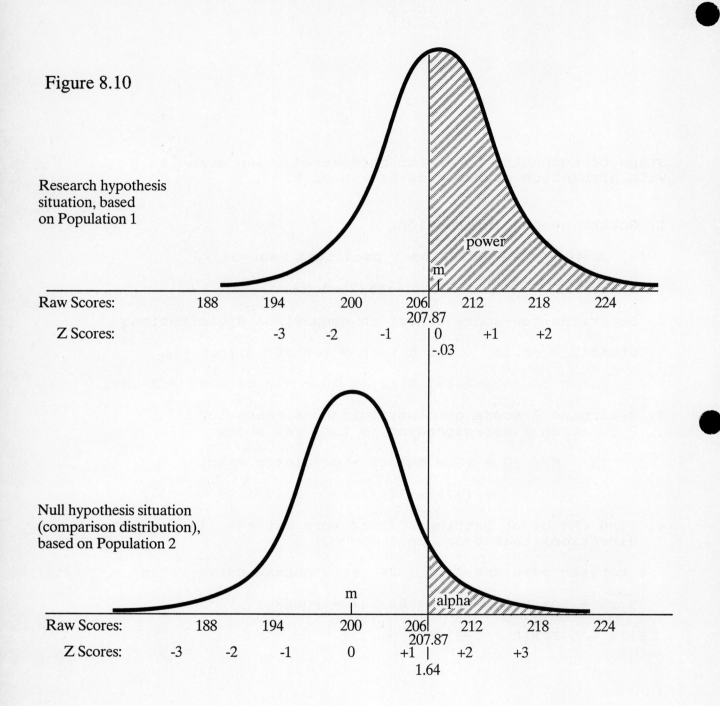

Research hypothesis
situation, based
on Population 1

power

m

Raw Scores: 188   194   200   206| 212   218   224
                                          207.87

Z Scores: -3   -2   -1   0   +1   +2
                                                       -.03

Null hypothesis situation
(comparison distribution),
based on Population 2

m

alpha

Raw Scores: 188   194   200   206| 212   218   224
                                                207.87

Z Scores: -3   -2   -1   0   +1 | +2   +3
                                                   1.64

Aron/Aron
STATISTICS FOR PSYCHOLOGY

TRANSPARENCY 8.19

Steps of computing power for achievement-test score example but with sample size = 100.

1. Gather needed information:

    (a) μ_M=200; (b) Population 1 predicted mean=208;
    (c) σ_M= 4.8 (σ=48, σ^2=2304 σ_M^2=2304/100=23.04 σ_M=$\sqrt{23.04}$=4.8).

2. Determine raw-score cutoff on comparison distribution:

    Cutoff $\underline{Z}$ ($\underline{p}$ < .05, one-tailed) = 1.64.

    Raw cutoff = (cutoff $\underline{Z}$)(σ_M)+μ_M = (1.64)(4.8)+200 = 207.87.

3. Determine $\underline{Z}$ score on distribution of means for Population 1 corresponding to this raw score.

    $\underline{Z}$ = ($\underline{X-M}$)/$\underline{SD}$ = (Raw cutoff - predicted mean) / σ_M

    $= (207.87 - 208)/4.8 = -.13/4.8 = -.03.$

4. Find chance of getting a score more extreme (in predicted direction) than score in 3 above.

    % between mean and $\underline{Z}$ of -.03 (from normal curve table) = 1.20.

    total of 50% above mean + 1.20% = 51.20% = POWER.

    BETA = 1-POWER = 1-51.20% = 48.8%.

Aron/Aron
STATISTICS FOR PSYCHOLOGY

© 1994 by Prentice-Hall, Inc.
A Paramount Communications Company
Englewood Cliffs, New Jersey 07632

TRANSPARENCY 8.20

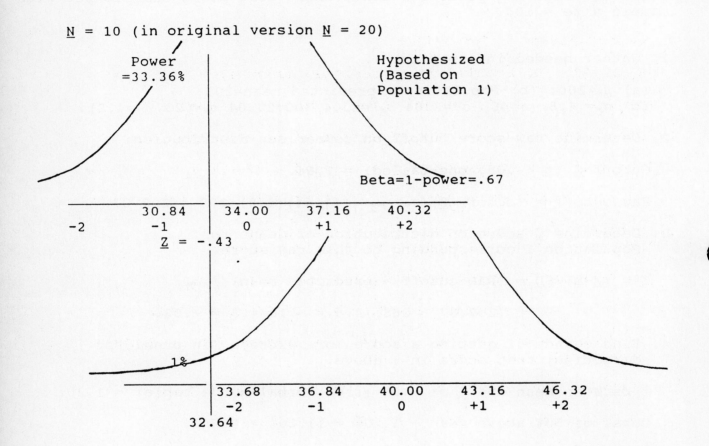

$\underline{N}$ = 10 (in original version $\underline{N}$ = 20)

Power
=33.36%

Hypothesized
(Based on
Population 1)

Beta=1-power=.67

	30.84	34.00	37.16	40.32
-2	-1	0	+1	+2

$\underline{Z}$ = -.43

1%

	33.68	36.84	40.00	43.16	46.32
-2	-1	0	+1	+2	

32.64

Aron/Aron
STATISTICS FOR PSYCHOLOGY

© 1994 by Prentice-Hall, Inc.
A Paramount Communications Company
Englewood Cliffs, New Jersey 07632

TRANSPARENCY 8.21

Steps of computing power for stress-reduction example but with $\underline{N}$ = 10.

1. Gather needed information:

   (a) $\mu_{\underline{M}}$=40; (b) Population 1 predicted mean=34;

   (c) $\sigma_{\underline{M}}$=3.16 (σ=10; σ^2=100; $\sigma_{\underline{M}}^2$=100/10=10; $\sigma_{\underline{M}}$=$\sqrt{10}$=3.16).

2. Determine raw-score cutoff on comparison distribution:

   Cutoff $\underline{Z}$ = -2.33; Raw cutoff = (cutoff $\underline{Z}$)($\sigma_{\underline{M}}$) + $\mu_{\underline{M}}$

   $\quad\quad\quad$ = (-2.33)(3.16) + 40 = - 7.36 + 40 = 32.64.

3. Determine $\underline{Z}$ score on distribution of means for Population 1 corresponding to this raw score.

   $\underline{Z}$ = ($\underline{X}$-$\underline{M}$)/$\underline{SD}$ = (Raw cutoff - predicted mean) / $\sigma_{\underline{M}}$

   $\quad\quad\quad$ = (32.64 - 34)/3.16 = -1.36/3.16 = -.43.

4. Find chance of getting a score more extreme (in predicted direction) than score in 3 above.

   % between mean and $\underline{Z}$ of -.43 (from normal curve table) = 16.64.

   50% below mean - 16.64% = 33.36% = POWER.

   BETA = 1-POWER = 1-33.36% = 66.64%.

Aron/Aron
STATISTICS FOR PSYCHOLOGY

TRANSPARENCY 8.22

Figure 8.12

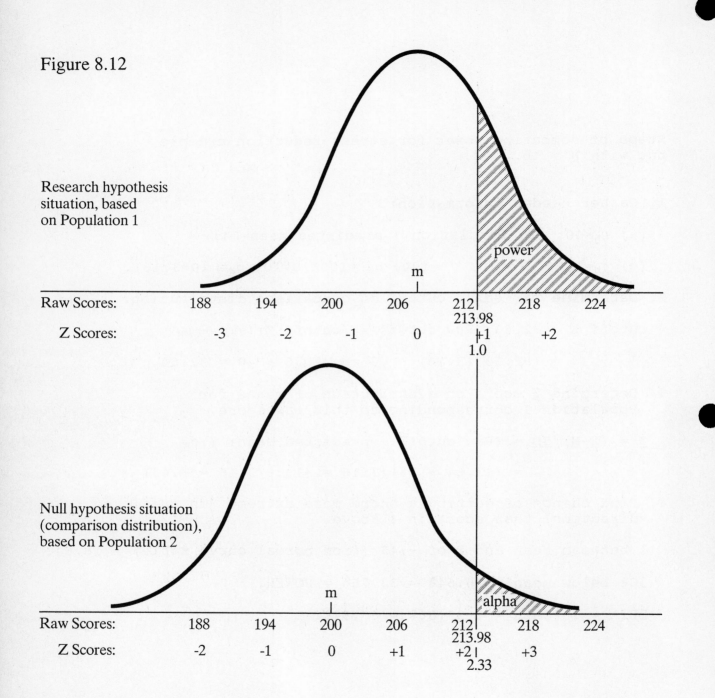

Research hypothesis
situation, based
on Population 1

Null hypothesis situation
(comparison distribution),
based on Population 2

TRANSPARENCY 8.23

Steps of computing power for achievement-test-score example but using $p < .01$ (and with predicted mean of 208)

1. Gather needed information:

   (a) $\mu_M = 200$; (b) Population 1 predicted mean = 208; (c) $\sigma_M = 6$.

2. Determine raw-score cutoff on comparison distribution:

   Cutoff Z ($p < .01$, one-tailed) = 2.33.

   Raw cutoff = (cutoff Z)(σ_M) + μ_M = (2.33)(6) + 200 = 213.98.

3. Determine Z score on distribution of means for Population 1 corresponding to this raw score.

   $Z = (X - M)/SD$ = (Raw cutoff - predicted mean) / σ_M

   $$= (213.98 - 208)/6 = 5.98/6 = 1.00.$$

4. Find chance of getting a score more extreme (in predicted direction) than score in 3 above.

   % between mean and Z of 1.00 (from normal curve table) = 34.13.

   total of 50% above mean - 34.13% = 15.87% = POWER.

   BETA = 1 - POWER = 1 - 15.87% = 84.13%.

Aron/Aron
STATISTICS FOR PSYCHOLOGY

TRANSPARENCY 8.24

alpha = .05 (vs. .01 before)

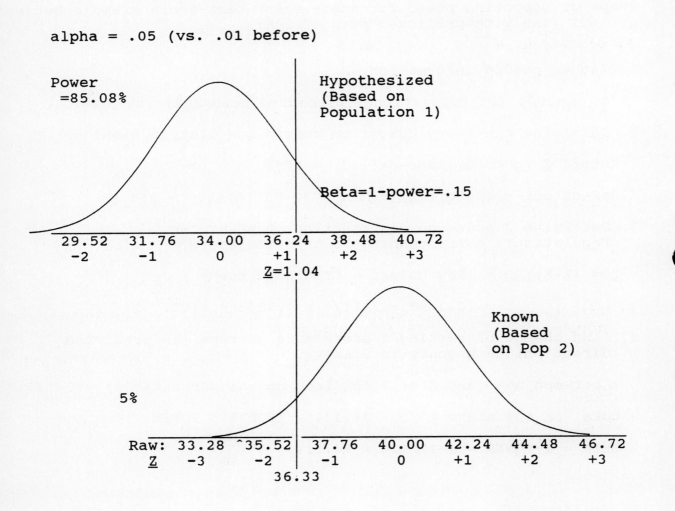

Power
=85.08%

Hypothesized
(Based on
Population 1)

Beta=1-power=.15

29.52	31.76	34.00	36.24	38.48	40.72
-2	-1	0	+1	+2	+3

$\underline{Z}$=1.04

Known
(Based
on Pop 2)

5%

Raw:	33.28	^35.52	37.76	40.00	42.24	44.48	46.72
$\underline{Z}$	-3	-2	-1	0	+1	+2	+3

36.33

Aron/Aron
STATISTICS FOR PSYCHOLOGY

© 1994 by Prentice-Hall, Inc.
A Paramount Communications Company
Englewood Cliffs, New Jersey 07632

TRANSPARENCY 8.25

Steps of computing power for stress-reduction example but using $p < .05$

1. Gather needed information:

   (a) $\mu_M = 40$; (b) Population 1 predicted mean $= 34$; (c) $\sigma_M = 2.24$.

2. Determine raw-score cutoff on comparison distribution:

   Cutoff $Z = -1.64$. Raw cutoff $= $ (cutoff Z)(σ_M) $+ \mu_M$

   $$= (-1.64)(2.24) + 40 = -3.67 + 40 = 36.33.$$

3. Determine Z score on distribution of sample means for Population 1 which corresponding to this raw score.

   $Z = (X-M)/SD = $ (Raw cutoff - predicted mean) $/ \sigma_M$

   $$= (36.33-34)/2.24 = 2.33/2.24 = 1.04.$$

4. Find chance of getting a score more extreme (in predicted direction) than score in 3 above.

   % between mean and Z of 1.04 (from normal curve table) $= 35.08$.

   $35.08 + 50\%$ below mean $= 85.08\% = $ POWER.

   BETA $= 1-$POWER $= 1-85.08\% = 14.92\%$.

Aron/Aron
STATISTICS FOR PSYCHOLOGY

© 1994 by Prentice-Hall, Inc.
A Paramount Communications Company
Englewood Cliffs, New Jersey 07632

TRANSPARENCY 8.26

Figure 8.13

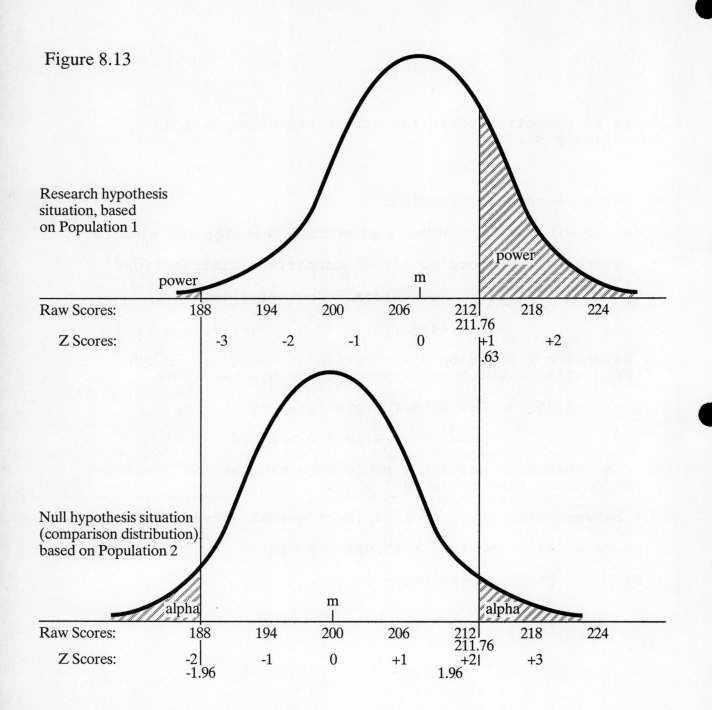

Research hypothesis
situation, based
on Population 1

power

power

Raw Scores:	188	194	200	206	212	218	224
					211.76		
Z Scores:	-3	-2	-1	0	+1	+2	
					.63		

m

Null hypothesis situation
(comparison distribution)
based on Population 2

alpha

alpha

m

Raw Scores:	188	194	200	206	212	218	224
					211.76		
Z Scores:	-2	-1	0	+1	+2	+3	
	-1.96				1.96		

Aron/Aron
STATISTICS FOR PSYCHOLOGY

Influences on Power

Feature of the Study	Increases Power	Decreases Power
Hypothesized Difference Between Population Means ($\mu_1 - \mu_2$)	Large Differences	Small Differences
Population Standard Deviation (σ)	Small σ	Large σ
Sample Size (<u>N</u>)	Large <u>N</u>	Small <u>N</u>
Significance Level (α)	Lenient, High Alpha (5% or 10%)	Stringent, Low Alpha (1% or .1%)
One vs. Two-Tailed Test	One-Tailed	Two-Tailed

Aron/Aron
STATISTICS FOR PSYCHOLOGY

© 1994 by Prentice-Hall, Inc.
A Paramount Communications Company
Englewood Cliffs, New Jersey 07632

TRANSPARENCY 8.28

Effect size for known population and single sample:

Effect size = difference between predicted mean and
mean of known population, divided by
population standard deviation.

Effect Size

Formula: $\underline{d} = (\mu_1 - \mu_2) / \sigma$

   $\underline{d}$ (Cohen's $\underline{d}$) = effect size

   μ_1 = Mean of Population 1 (hypothesized mean
for the population that is subjected to
the experimental manipulation)

   μ_2 = Mean of Population 2 (which is also the
mean of the comparison distribution).

   σ = Standard deviation of Population 2
(assumed to be the standard deviation of
both populations)

Aron/Aron
STATISTICS FOR PSYCHOLOGY

© 1994 by Prentice-Hall, Inc.
A Paramount Communications Company
Englewood Cliffs, New Jersey 07632

TABLE 8-2
Summary of Cohen's Effect Size Conventions for Mean Differences

Verbal Description	Effect Size (d)
Small	.2
Medium	.5
Large	.8

Aron/Aron
STATISTICS FOR PSYCHOLOGY

© 1994 by Prentice-Hall, Inc.
A Paramount Communications Company
Englewood Cliffs, New Jersey 07632

TABLE 8-4
Summary of Practical Ways of Increasing the Power of a Planned Experiment

Feature of the Study	Practical Way of Raising Power	Disadvantages
Hypothesized difference between population means $(\mu_1 - \mu_2)$	Increase the intensity of experimental manipulation.	May not be practical or may distort study's meaning.
Standard deviation (σ)	Use a less diverse population.	May not be available; decreases generalizability.
	Use standardized, controlled circumstances of testing or more precise measurement.	Not always practical.
Sample size (N)	Use a larger sample size.	Not always practical; can be costly.
Significance level (α)	Use a more lenient level of significance.	Raises alpha, the probability of Type I error.
One-tailed versus two-tailed test	Use a one-tailed test.	May not be appropriate to the logic of the study.
Type of hypothesis-testing procedure	Use a more sensitive procedure.	None may be available or appropriate.

Aron/Aron
STATISTICS FOR PSYCHOLOGY

© 1994 by Prentice-Hall, Inc.
A Paramount Communications Company
Englewood Cliffs, New Jersey 07632

TABLE 8-5
Role of Significance and Sample Size in Interpreting Experimental Results

Outcome Statistically Significant	Sample Size	Conclusion
Yes	Small	Important result
Yes	Large	Might or might not have practical importance
No	Small	Inconclusive
No	Large	Research hypothesis probably false

© 1994 by Prentice-Hall, Inc.
A Paramount Communications Company
Englewood Cliffs, New Jersey 07632

Aron/Aron
STATISTICS FOR PSYCHOLOGY

Chapter 9
The *t* Test for Dependent Means

Instructor's Summary of Chapter

One-sample t test. The same five steps of hypothesis testing described in Chapter 7 apply when the variance of the population is not known, except for the following: (a) The population variance is estimated from the information in the sample using a formula that divides the sum of squared deviations (*SS*) by the degrees of freedom (*df* = *N* - 1), $S^2 = SS/df$. (b) The comparison distribution will have the shape of a *t* distribution (for which cutoffs are found in a *t* table). It has more area at the extremes than a normal curve (just how much more depends on how few degrees of freedom were used in estimating the population variance). (c) The number of standard deviations from the mean that a sample's mean falls on the *t* distribution is called a *t* score, $t = (M - \mu)/S_M$.

t test for dependent means. A study in which there is a group of subjects each having two scores, such as a before score and an after score, are often analyzed using a *t* test for dependent means. In this *t* test you first compute a difference score for each subject, then proceed with the five-step hypothesis testing procedure in the same way as if you had only a single sample with an unknown variance, assuming that the null hypothesis is about a population of difference scores with a mean of zero (no difference).

Assumptions. An assumption of the *t* test is that the population distribution is normal. However, even when the population distribution departs substantially from normal, the *t* test is usually sufficiently accurate. The major exception in the case of a *t* test for dependent means is when the population of difference scores appears to be highly skewed and a one-tailed test is being used.

Effect size and power. The effect size of a study using a *t* test for dependent means can be computed as the mean of the difference scores divided by the estimated population standard deviation of the difference scores. Power and needed sample size for a given effect size can be determined from special tables. The power of studies using difference scores is typically much higher than studies using other designs with the same number of subjects.

Controversies and limitations. Research involving a single group tested before and after some intervening event, without a control group, is a weak approach that permits many alternative explanations of any observed changes.

How the procedures of this chapter are described in research articles. Research articles typically describe *t* tests either (a) in the text, using a standard format, such as "*t*(24) = 2.8, *p* < .05," or (b) in a table listing a series of results in which the *t* score itself may be omitted and only an asterisk used to indicate which comparisons are significant.

Box 9.1. William S. Gosset: Alias "Student," Not a Mathematician, but a "Practical Man." Briefly describes the life and character of Gosset, and how he developed the *t* test.

Box 9.2. The Power of Studies Using Difference Scores: How The Lankeshire Milk Experiment Could Have Been Milked for More. Describes a historical controversy in which Gosset argued that a major experiment involving 20,000 school children, comparing those who did and did not drink milk regularly, could have been done using only 50 identical twins, analyzing data using a *t* test for dependent means.

Lecture 9.1: Introduction to the *t* Test

Materials

Lecture outline

Transparencies 7.1, 7.9, 7.10, and 9.1 through 9.10

    (If using transparencies based on your class's questionnaires, replace 9.5 and 9.6 with 9.5R and 9.6R.)

Outline for Blackboard

  I.   Review
  II. Estimated Population Variance (S^2)
  III. *t* Distribution
  IV. *t* Table
  V. *t* Test
  VI. Review this Class

Instructor's Lecture Outline

I.   Review

  A. Idea of descriptive and inferential statistics.

  B. Basic logic of hypothesis testing with a single sample and a known population.

     1. Show TRANSPARENCIES 7.9 and 7.10 (hypothesis testing example for a single sample and known population) and discuss.

     2. Show TRANSPARENCY 7.1 (three kinds of distributions) and discuss.

     3. Emphasize that in determining the variance of the distribution of means, we have always begun with a known population variance.

II.   Estimated Population Variance (S^2)

  A. Show TRANSPARENCY 9.1 (hours/day studied from text).

     1. Note that this is still a somewhat unrealistic research situation, but we are getting closer to very common ones.

     2. Logic of hypothesis testing is generally same as before—discuss Step 1.

     3. Go to top of Step 2 and note that while it may be reasonable to assume that the population shape is normal, we have no information on its variance.

  B. If the population variance is not known, a solution is to estimate it from the sample's variance.

     1. The only information we have available in this example (and usually in practice) about such a population is the sample.

     2. We assume here (as we have throughout the course so far) that Populations 1 and 2 have the same variance.

     3. If the sample is representative of its population, then its variance should tell us something about the variance of the population it represents:

       a. If the sample has little variance, probably the population it came from has little variance.

       b. If the sample has a lot of variance, probably the population it came from has a lot of variance.

  C. However, the sample's variance can not be used directly as an estimate of the population variance. It can be shown mathematically that a sample's variance will on the average be a bit smaller than its population's.

  D. Thus, we make an adjustment to compute the estimated population variance.

     1. Ordinarily, variance is calculated as the sum of squared deviations from the mean divided by the number of cases: $\sigma^2 = SS/N$.

     2. But the estimated population variance is computed as the sum of squared deviations from the mean divided by the number of cases *minus one*: $S^2 = SS/(N - 1)$.

3. N-1 is called the *degrees of freedom* (*df*). It represents the number of scores in the sample that are free to vary when calculating the variance. There are $N - 1$ because when figuring the deviations each score is subtracted from the mean. Thus, if you knew all the deviation scores but one, you could calculate the last one given the mean.

   STUDENTS: The idea just presented of *why* degrees of freedom is $N - 1$ is an advanced idea. It is not necessary to understand fully in order to grasp the rest of the material.

4. Show top section of TRANSPARENCY 9.2 (computations of S^2—*not* from ongoing example, in order to present a simple illustration of the process with few numbers) and discuss symbols and the computation of estimated population variance.

5. Show TRANSPARENCY 9.2 bottom (hours studied computations) and discuss.

E. Having estimated the population variance, we can proceed to find the characteristics of the distribution of means.

    1. Show rest of Step 2 (up to shape of distribution of means) in TRANSPARENCY 9.1 and discuss computations and symbols.

    2. NOTE: Remember that when estimating population variance, you divide the sum of squared deviations by the degrees of freedom ($N - 1$), but when computing the variance of the distribution of means you divide the estimated population variance by the full sample size (N).

III. *t* Distribution

A. Estimating the population variance loses some accuracy.

B. We make up for this by setting the cutoff score for significance a little more extreme.

C. In fact, an exact distribution takes this into account. Show TRANSPARENCY 9.3 (*t* distribution) and discuss.

    1. Similar to a normal curve, but with fatter tails, making more cases in the extremes so that to include any particular percent of cases (say 5%), the cutoff has to be farther out.

    2. This distribution is called a *t* distribution.

D. There is a different *t* distribution for each number of degrees of freedom.

    1. The more degrees of freedom, the closer the *t* distribution is to the normal curve. This is because you are estimating with increasing amounts of information.

    2. When there are infinite degrees of freedom, the *t* distribution is the same as the normal curve. This is because your sample is infinitely large and is thus the same as the population, so the estimate is perfect.

IV. *t* Table

A. To determine the cutoffs on the *t* distribution, you use a *t* table.

B. Show TRANSPARENCY 9.4 (*t* table) and discuss.

    1. For the hours-studied example there are 15 *df* with a one-tailed test at the .05 level.

    2. Thus, cutoff *t* needed is 1.753.

    3. If using a normal curve for the distribution of means, the cutoff Z would be 1.64—a more extreme cut-off is required when using a *t* distribution.

C. Show TRANSPARENCY 9.1, bottom of Step 2 and Step 3, and discuss.

V. *t* Test

A. Show rest of TRANSPARENCY 9.1 and discuss.

B. Additional examples.

    1. Show TRANSPARENCIES 9.5 and 9.6 or 9.5R and 9.6R (hypothesis testing and computations for whether distribution of morning-night person ratings in population represented by class, from class questionnaire, is different from population with average of 4 on this question) and discuss.

    2. Show TRANSPARENCIES 9.7 and 9.8 (hypothesis testing and computations for fictional study of whether distribution of time to fill out questionnaire about childhood experiences is longer for adults with troubled childhoods) and discuss.

VI. Review this Class: Use blackboard outline and TRANSPARENCIES 9.9 (comparison of *t* test to situation with known σ) and 9.10 (steps of conducting a *t* test for a single sample and known μ).

Lecture 9.2: The *t* Test for Dependent Means

Materials

Lecture outline

Transparencies 9.1 through 9.4 and 9.9 through 9.19

    (If using transparencies based on your class's questionnaires, replace 9.13 and 9.14 with 9.13R and 9.14R.)

Outline for Blackboard

   I.   **Review**
   II. **Repeated-Measures Designs**
   III. **Change (or Difference) Scores**
   IV. **t Test for Dependent Means**
   V. **Additional Examples**
   VI. **Assumptions**
   VII.**Effect Size and Power**
   VIII.**Review This Class**

Instructor's Lecture Outline

I.   Review

   A. Idea of descriptive and inferential statistics.

   B. Hypothesis testing with a single sample and a known population requires knowing population variance.

   C. When population variance is not known, it can be estimated from the scores in the sample. Show TRANSPARENCY 9.2 (computations of S^2) and discuss.

   D. When using an estimated population variance, the shape of the comparison distribution is a *t* distribution and not a normal curve.

      1. Show TRANSPARENCY 9.3 (normal and *t* curves) and discuss.

      2. Show TRANSPARENCY 9.4 (*t* table) and discuss.

   E. Otherwise the hypothesis testing proceeds as usual, but is called a *t* test and the cutoff and sample Z scores are called *t* scores.

      1. Show TRANSPARENCY 9.1 (hours studied hypothesis testing example) and discuss each step.

      2. Show TRANSPARENCY 9.9 (comparison of *t* test to situation with known σ) and discuss each step.

      3. Show TRANSPARENCY 9.10 (steps of *t* test with known μ) and discuss each step.

II.   Repeated-Measures Designs

   A. Also called "within-subjects" designs.

   B. When same people are tested twice.

      1. Before and after some procedure (sometimes called a "pretest-posttest" design). For example:

        a. Before and after psychotherapy.

        b. Before and after taking a medication.

        c. Before and after exercise.

        d. Before and after leaving home.

      2. When there are two conditions to an experiment and the same people are tested under both conditions. For example:

        a. Reading ability under bright versus dim light.

        b. Alertness when alone versus when with a friend.

c. Answering questions about how anxious you are at home versus about how anxious you are at school.

C. Also considered "repeated-measures" when different people can be matched into pairs, as with identical twins, one being tested in each of two conditions. The situation is treated as if the two of them were the same person tested at two different times.

D. Refer students to Appendix A of the text for a discussion of some of the validity problems of repeated-measures designs (for pretest-posttest, problems of no control group; for same people under two conditions, problems of order effects).

III. Change (or Difference) Scores

A. In these situations, the whole process is much simplified by using change (or difference) scores.

B. For example, for each subject subtract the before from after score.
   1. Examples:
      a. If before = 6 and after = 8, change = 2.
      b. If before = 39 and after = 10, change = -29.
      c. If before = 60 and after = 80, change = 20.
   2. Ask class:
      a. If before = 30 and after = 40, what is change? [10]
      b. If before = 30 and after = 50, what is change? [20]
      c. If before = 30 and after = 25, what is change? [-5]
      d. If before = 30 and after = 30, what is change? [0]
      e. If before = 100 and after = 100.5, what is change? [.5]

C. In this way you end up with one score per subject (a change score) rather than two.

D. In statistical analysis of repeated-measures designs we think entirely in terms of these changes scores:
   1. Population distribution of change scores.
   2. Sample of change scores.
   3. Comparison distribution of change scores.

IV. *t* Test for Dependent Means

A. A *t* test for dependent means tests hypotheses about repeated-measures designs of this kind (involving change or difference scores).

B. Usually the situation is that we are interested in whether there is a significant change (or difference).

C. The null hypothesis is that there is no change and the research hypothesis is there is some change (or change in a particular direction).

D. Computation is identical to *t* test for a single sample, except:
   1. You first convert everything to change scores and do everything from then on working with change scores.
   2. The assumed mean of the Population 2 change scores is ordinarily zero—that is, a population in which there is on the average no change.

E. An example.
   1. Show TRANSPARENCY 9.11 (communication quality before and after marriage from text), Step 1, and note that emphasis is right away on change.
   2. Show TRANSPARENCY 9.11, Step 2, and note:
      a. Population 2 is assumed to have a zero mean (that is, we are comparing to a population in which there is no change).
      b. The population variance must be estimated since we do not know it. This requires some computations, and first we need change scores.
   4. Show TRANSPARENCY 9.12 (computations for communication study example) and discuss.
   5. Show TRANSPARENCY 9.11 again and finish discussing each step.

V. Additional Examples

A. Show TRANSPARENCIES 9.13 and 9.14 or 9.13R and 9.14R (closeness to mother versus father, from class questionnaire) and discuss.

B. Show TRANSPARENCIES 9.15 and 9.16 (fictional study of change in water use for households exposed to advertising urging them to reduce their use during a drought) and discuss.

VI. Assumptions

A. Assumptions are circumstances that must be true for the logic and mathematics of a statistical procedure to work properly.

B. Major assumption of the *t* test for dependent means is that the population distribution (of change scores) must be normal.

C. However, usually you need not worry about this unless:
   1. There is reason to expect a very large discrepancy from normal in the population of change scores.
   2. You have reason to expect that the population of change scores is quite skewed and you are using a one-tailed test.

D. Later in the course (Chapter 15) we consider what to do when normality can not be reasonably assumed.

VII. Effect Size and Power

A. Effect size for the *t* test for dependent means.
   1. It is computed in the same way as we did before: $d = (\mu_1 - \mu_2) / \sigma$.
   2. However, remember we are dealing with change scores and the mean of Population 2 is almost always going to be 0. Thus, in effect, the formula is the mean change divided by the standard deviation of the population of change scores.
   NOTE: This standard deviation is not the same as S_M — rather, it is S. This is because effect size is intended to be a measure that is not influenced by N, and S_M is much influenced by N.
   3. For example, if the average change is 6 and estimated population standard deviation is 12, then $d = 1/2$.
   4. As before: A small effect size is .20, a medium effect size is .50, and a large effect size is .80.
   5. Ask class:
      a. If average change is 5 and estimated population standard deviation is 10, what is *d*? [.5] Is this a small, medium, or large effect? [medium]
      b. If average change is 2 and estimated population standard deviation is 10, what is *d*? [.2] Is this a small, medium, or large effect? [small]
      c. If average change is 200 and estimated population standard deviation is 400, what is *d*? [.5] Is this a small, medium, or large effect? [medium]
      d. If average change is 12 and estimated population standard deviation is 10, what is *d*? [1.2] Is this a small, medium, or large effect? [large]

B. Power.
   1. Power can be computed directly from tables. Show TRANSPARENCY 9.17 (power table) and discuss.
   2. There is also a table that shows how many subjects are needed to achieve 80% power (a standard desired level). Show TRANSPARENCY 9.18 (number of subjects needed table) and discuss.

VIII. Review this Class

A. Use blackboard outline.

B. When summarizing first sections, show TRANSPARENCY 9.19 (steps of conducting a *t* test for dependent means) and discuss.

TRANSPARENCY 9.1

Hours per day studied by students in your dorm versus students in general at your college. (Example from text)

Steps of hypothesis testing:

1. Reframe the question into a research hypothesis and a null hypothesis about populations.

   Population 1: The kind of students who live in your dorm.
   Population 2: The kind of students at your college generally.

   Research Hypothesis: Population 1 students study more than Population 2 students.
   Null Hypothesis: Population 1 students do not study more than Population 2 students.

2. Determine the characteristics of the comparison distribution.

   Population 2: shape=assumed normal; μ=2.5; σ^2=unknown; $\underline{S}^2$=.64.

   Distribution of means: shape = $\underline{t}$ ($\underline{df}$=15); μ_M=2.5; $\underline{S}_M$=.2.

   ($\underline{S}_M{}^2$ = $\underline{S}^2$/N =.64/16 = .04; $\underline{S}_M$ = $\sqrt{\underline{S}_M{}^2}$ = $\sqrt{.04}$ = .2.)

3. Determine cutoff sample score on the comparison distribution at which the null hypothesis should be rejected.

   5% level, 1-tailed, t distribution of $\underline{df}$=15: t needed=1.753.

4. Determine score of your sample on comparison distribution.

   Mean hours/studied in your dorm = 3.2.

   t = ($\underline{M}-\mu_M$)/$\underline{S}_M$ = (3.2-2.5)/.2 = .7/.2 = 3.5.

5. Compare the scores obtained in Steps 3 and 4 to decide whether to reject the null hypothesis.

   t on 4 (3.5) is more extreme than cutoff t on 3 (1.753).

   Therefore, reject the null hypothesis; the research hypothesis is supported (students in your dorm study more than at your college in general).

Aron/Aron
STATISTICS FOR PSYCHOLOGY

© 1994 by Prentice-Hall, Inc.
A Paramount Communications Company
Englewood Cliffs, New Jersey 07632

TRANSPARENCY 9.2

Computation of estimated population variance
based on sample data:

$$\underline{S}^2 = \Sigma(\underline{X}-\underline{M})^2/\underline{N}-1 = \underline{SS} / \underline{N}-1 = \underline{SS}/\underline{df}$$

EXAMPLE:

Subject	Score X	Deviation (X-M)	Deviation Squared (X-M)²
1	8	2	4
2	6	0	0
3	4	-2	4
4	9	3	9
5	3	-3	9
	Σ = 30		SS = 26
	M = 6		

$\underline{df}=\underline{N}-1=5-1=4$ $\underline{S}^2=\underline{SS}/\underline{df}=26/4=6.5$

+++

Example: Hours per day studied by students in
your dorm versus students in general
at your college. (Example from text)

Computation of estimated population variance
based on sample data:

Subject	Score X	Deviation (X-M)	Deviation Squared (X-M)²
1	4	.8	.64
2	5	1.8	3.24
.	.	.	.
.	.	.	.
16	3	.2	.04
	Σ = 51		SS= 9.60
	M = 3.2		

$\underline{df}=\underline{N}-1=16-1=15$ $\underline{S}^2=\underline{SS}/\underline{df}=9.6/15=.64$

Aron/Aron
STATISTICS FOR PSYCHOLOGY

TRANSPARENCY 9.3

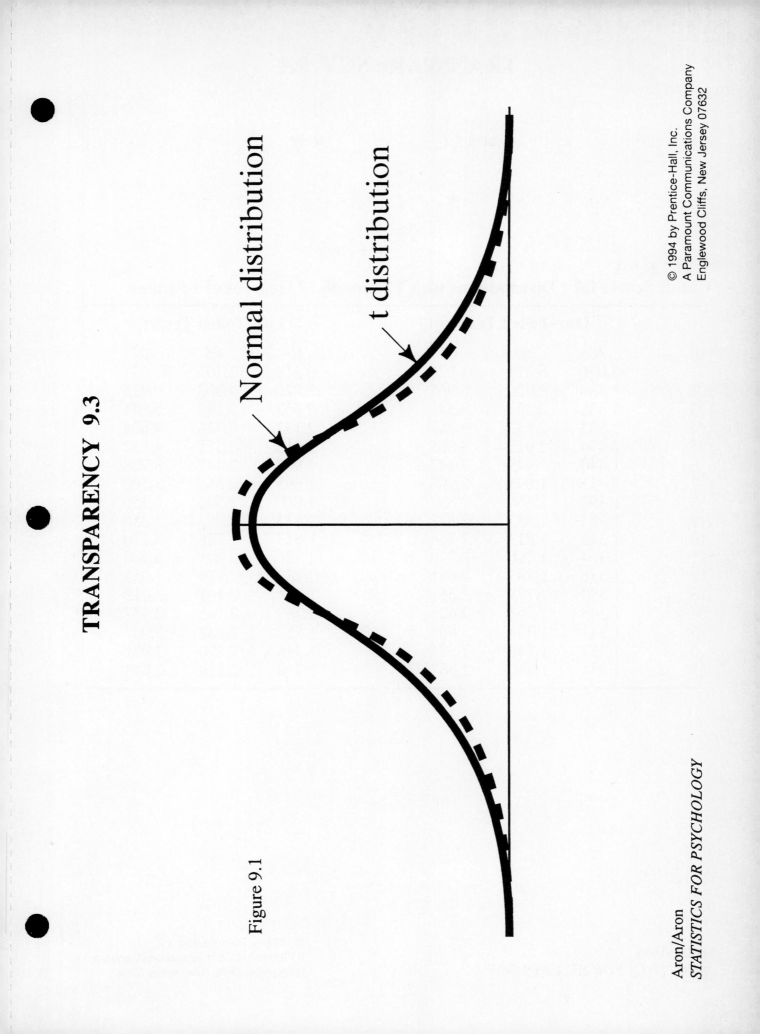

Normal distribution

t distribution

Figure 9.1

Aron/Aron
STATISTICS FOR PSYCHOLOGY

TABLE 9-1
Cutoff Scores for *t* Distributions with 1 Through 17 Degrees of Freedom

	One-Tailed Tests				Two-Tailed Tests		
df	*.10*	*.05*	*.01*		*.10*	*.05*	*.01*
1	3.078	6.314	31.821		6.314	12.706	63.657
2	1.886	2.920	6.965		2.920	4.303	9.925
3	1.638	2.353	4.541		2.353	3.182	5.841
4	1.533	2.132	3.747		2.132	2.776	4.604
5	1.476	2.015	3.365		2.015	2.571	4.032
6	1.440	1.943	3.143		1.943	2.447	3.708
7	1.415	1.895	2.998		1.895	2.365	3.500
8	1.397	1.860	2.897		1.860	2.306	3.356
9	1.383	1.833	2.822		1.833	2.262	3.250
10	1.372	1.813	2.764		1.813	2.228	3.170
11	1.364	1.796	2.718		1.796	2.201	3.106
12	1.356	1.783	2.681		1.783	2.179	3.055
13	1.350	1.771	2.651		1.771	2.161	3.013
14	1.345	1.762	2.625		1.762	2.145	2.977
15	1.341	1.753	2.603		1.753	2.132	2.947
16	1.337	1.746	2.584		1.746	2.120	2.921
17	1.334	1.740	2.567		1.740	2.110	2.898

Aron/Aron
STATISTICS FOR PSYCHOLOGY

TRANSPARENCY 9.5

Do students taking psychological statistics differ from 4 on how much they are morning people? (From class questionnaire.)

Steps of hypothesis testing:

1. Reframe the question into a research hypothesis and a null hypothesis about populations.

   Population 1: People like those in the statistics class taking the questionnaire.
   Population 2: People averaging 4 on the morning-person rating.

   Null Hypothesis: The two populations are the same.
   Research Hypothesis: The two populations are different.

2. Determine the characteristics of the comparison distribution.

   Population 2: shape=assumed normal; μ=4; σ^2=unknown; $\underline{S}^2$=3.10.

   Distribution of means: shape=t ($\underline{df}$=104); μ_M=4; $\underline{S}_M$=.17.

   $(\underline{S}_M^2 = \underline{S}^2/\underline{N} = 3.10/105 = .03; \underline{S}_M = \sqrt{\underline{S}_M^2} = \sqrt{.03} = .17.)$

3. Determine the cutoff sample score on the comparison distribution at which the null hypothesis should be rejected.

   5% level, 2-tailed, t distribution of $\underline{df}$=104: t needed = ±1.984

4. Determine score of your sample on comparison distribution.

   Mean rating for this class was 3.46.

   $t = (\underline{M}-\mu_M)/\underline{S}_M = (3.46-4)/.17 = -.56/.17 = -3.29.$

5. Compare the scores obtained in Steps 3 and 4 to decide whether to reject the null hypothesis.

   Score on 4 (-3.29) is more extreme than score on 3 (±1.984).

   Therefore, reject the null hypothesis; the research hypothesis is supported (those in this class are less likely to be morning people).

Aron/Aron
STATISTICS FOR PSYCHOLOGY

© 1994 by Prentice-Hall, Inc.
A Paramount Communications Company
Englewood Cliffs, New Jersey 07632

TRANSPARENCY 9.6

Do students taking psychological statistics differ from 4 on
how much they are morning people? (From class questionnaire.)

Computation of estimated population variance based on sample:

Subject	Score X	Deviation (X-M)	Deviation Squared $(X-M)^2$
1	2	-1.46	2.132
2	6	2.54	6.452
.	.	.	.
.	.	.	.
105	4	0.54	0.292

Σ = 363   SS=322.058

$\underline{M}$ =   3.46

$\underline{df} = \underline{N} - 1 = 105 - 1 = 104$ $\underline{S}^2 = \underline{SS}/\underline{df} = 322.058/104 = 3.10$

Aron/Aron
STATISTICS FOR PSYCHOLOGY

© 1994 by Prentice-Hall, Inc.
A Paramount Communications Company
Englewood Cliffs, New Jersey 07632

TRANSPARENCY 9.7

It is known that it takes ordinary adults an average of 26 minutes to fill out a particular questionnaire about their childhood experiences and that the distribution of times is normal. Twelve adults with troubled childhoods are tested. (Fictional study.)

Steps of hypothesis testing:

1. Reframe the question into a research hypothesis and a null hypothesis about populations.

   Population 1: People who had a particularly troubled childhood.
   Population 2: People in general (who are known to take 26 minutes to fill out the questionnaire)

   Research Hypothesis: People who had a particularly troubled childhood take longer to fill out the questionnaire.
   Null Hypothesis: People who had a particularly troubled childhood do not take longer to fill out the questionnaire.

2. Determine the characteristics of the comparison distribution.

   Population 2: shape=normal; μ=28 minutes; σ^2=unknown; $\underline{S}^2$=41.18.

   Distribution of means: shape=t ($\underline{df}$=11); μ_M=28; $\underline{S}_M$=1.85.

   ($\underline{S}_M{}^2$ = $\underline{S}^2$/N = 41.18/12 = 3.43; $\underline{S}_M$ =$\sqrt{\underline{S}_M{}^2}$ = $\sqrt{3.43}$ = 1.85.)

3. Determine the cutoff sample score on the comparison distribution at which the null hypothesis should be rejected.

   5% level, 1-tailed, t distribution of $\underline{df}$=11: t needed = 1.796.

4. Determine score of your sample on comparison distribution.

   Mean number of minutes for the 12 with troubled childhoods = 31.

   t = ($\underline{M}-\mu_M$)/$\underline{S}_M$ = (31-28)/1.85 = 3/1.85 = 1.62.

5. Compare the scores obtained in Steps 3 and 4 to decide whether to reject the null hypothesis.

   t on 4 (1.62) is NOT more extreme than cutoff t on 3 (1.796).

   Therefore, do not reject the null hypothesis; the result is inconclusive (it is not clear whether it takes adults with troubled childhoods longer to fill out the questionnaire).

Aron/Aron
STATISTICS FOR PSYCHOLOGY

© 1994 by Prentice-Hall, Inc.
A Paramount Communications Company
Englewood Cliffs, New Jersey 07632

TRANSPARENCY 9.8

Time to fill out questionnaire. (Fictional data.)

Computation of estimated population variance
based on sample:

Subject	Score X	Deviation (X−M)	Deviation Squared (X−M)²
1	32	1	1
2	25	−6	36
.	.	.	.
.	.	.	.
12	36	5	25
	Σ = 372		SS= 453
	M = 31		

df=N−1=12−1=11   S²=SS/df=453/11=41.18

Aron/Aron
STATISTICS FOR PSYCHOLOGY

© 1994 by Prentice-Hall, Inc.
A Paramount Communications Company
Englewood Cliffs, New Jersey 07632

TABLE 9-2
Hypothesis Testing Involving a Single Sample Mean When Population Variance Is Unknown (*t* Test) Compared to When Population Variance Is Known

Step in Hypothesis Testing	Difference From When Population Variance Is Known
1. Reframe the question into a research hypothesis and a null hypothesis about the populations.	No difference in method.
2. Determine the characteristics of the comparison distribution:	
Population mean	No difference in method.
Population variance	Estimate from the sample.
Standard deviation of the distribution of sample means	No difference in method (but based on estimated population variance).
Shape of the comparison distribution	Use the *t* distribution with $df = N - 1$.
3. Determine the significance cutoff.	Use the *t* table.
4. Determine score of your sample on the comparison distribution.	No difference in method (but called a *t* score).
5. Compare Steps 3 and 4 to determine whether to reject the null hypothesis.	No difference in method.

Aron/Aron
STATISTICS FOR PSYCHOLOGY

TRANSPARENCY 9.10

TABLE 9-4
Steps for Conducting a *t*-Test for a Single Sample

1. Reframe the question into a research hypothesis and a null hypothesis about the populations.

2. Determine the characteristics of the comparison distribution.

   a. The mean is the same as the known population mean.

   b. The standard deviation is computed as follows:

      i. Compute the estimated population variance: $S^2 = SS/df$.

      ii. Compute the variance of the distribution of means: $S_M^2 = S^2/N$.

      iii. Compute the standard deviation: $S_M = \sqrt{S_M^2}$.

   c. The shape will be a *t* distribution with $N - 1$ degrees of freedom.

3. Determine the cutoff sample score on the comparison distribution at which the null hypothesis should be rejected.

   a. Determine the degrees of freedom, desired significance level, and number of tails in the test (one or two).

   b. Look up the appropriate cutoff in a *t* table.

4. Determine the score of your sample on the comparison distribution: $t = (M - \mu)/S_M$.

5. Compare the scores obtained in Steps 3 and 4 to decide whether or not to reject the null hypothesis.

Aron/Aron
STATISTICS FOR PSYCHOLOGY

© 1994 by Prentice-Hall, Inc.
A Paramount Communications Company
Englewood Cliffs, New Jersey 07632

TRANSPARENCY 9.11

Example: Olthoff (1989) data of husbands' reported communication
         quality before and after marriage. (Example from text)

Steps of hypothesis testing:

1. Reframe the question into a research hypothesis and a null
   hypothesis about populations.

   Population 1: Husbands of the kind included in this study.
   Population 2: Husbands who do not change on their communication
                 quality from before to after marriage.

   Research Hypothesis: Populations are different.
   Null Hypothesis: Populations are the same.

2. Determine the characteristics of the comparison distribution.

   Population 2: shape=assumed normal; μ=0; σ^2=unknown; $\underline{S}^2$=154.05.

   Distribution of means: shape=t ($\underline{df}$=18); μ_M=0; $\underline{S}_M$=2.85.

3. Determine the cutoff sample score on the comparison
   distribution at which the null hypothesis should be rejected.

   5% level, 2-tailed, t distribution of $\underline{df}$=18: t needed = ±2.101.

4. Determine score of your sample on comparison distribution.

   Mean change from before to after marriage of -12.05.

   $t = (\underline{M}-\mu_M)/\underline{S}_M = (-12.05-0)/2.85 = -12.05/2.85 = -4.23.$

5. Compare the scores obtained in Steps 3 and 4 to decide whether
   to reject the null hypothesis.

   t on 4 (-4.23) is more extreme than cutoff t on 3 (±2.101).

   Therefore, reject the null hypothesis; the research hypothesis
   is supported (husbands' reported communication quality
   decreases from before to after marriage).

Aron/Aron
STATISTICS FOR PSYCHOLOGY

TRANSPARENCY 9.12

TABLE 9-5
Data and *t*-Test Analysis for Communication Quality Scores Before and After Marriage for 19 Husbands Who Received No Special Communication Training

Husband	Communication Quality		Difference (After – Before)	Deviation of Differences From the Mean of Differences	Squared Deviation
	Before	*After*			
A	126	115	– 11	1.05	1.1
B	133	125	– 8	4.05	16.4
C	126	96	– 30	–17.95	322.2
D	115	115	0	12.05	145.2
E	108	119	11	23.05	531.3
F	109	82	– 27	–14.95	233.5
G	124	93	– 31	–18.95	359.1
H	98	109	11	23.05	531.3
I	95	72	– 23	–10.95	119.9
J	120	104	– 16	– 3.95	15.6
K	118	107	– 11	1.05	1.1
L	126	118	– 8	4.05	16.4
M	121	102	– 19	– 6.95	48.3
N	116	115	– 1	11.05	122.1
O	94	83	– 11	1.05	1.1
P	105	87	– 18	– 5.95	35.4
Q	123	121	– 2	10.05	101.0
R	125	100	– 25	–12.95	167.7
S	128	118	– 10	2.05	4.2
Σ:	2,210	1,981	–229	– .05	2,772.9

For difference scores:

$M = -229/19 = -12.05$.

$\mu = 0$ (assumed as a no-change baseline of comparison).

$S^2 = SS/df = 2{,}772.9/(19 - 1) = 154.05$.

$S_M^2 = S^2/N = 154.05/19 = 8.11$.

$S_M = \div S_M^2 = \div 8.11 = 2.85$.

t with $df = 18$ needed for 5% level, two-tailed $= \pm 2.101$.

$t = (M - \mu)/S_M = (-12.05 - 0)/2.85 = -4.23$.

Decision: Reject the null hypothesis.

Note. Data from Olthoff & Aron (1993).

Aron/Aron
STATISTICS FOR PSYCHOLOGY

© 1994 by Prentice-Hall, Inc.
A Paramount Communications Company
Englewood Cliffs, New Jersey 07632

TRANSPARENCY 9.13

How close to mother versus to father.   (From class questionnaire.)

Steps of hypothesis testing:

1. Reframe the question into a research hypothesis and a null hypothesis about populations.

 Population 1: Students like those in class studied.
 Population 2: Students who were equally close to both parents.

 Research Hypothesis: Populations are different.
 Null Hypothesis: Populations are the same.

2. Determine the characteristics of the comparison distribution.

 Population 2: shape=assumed normal; $\mu=0$; σ^2=unknown; $\underline{S}^2$=5.71.

 Distribution of means: shape=t ($\underline{df}$=108); μ_M=0; $\underline{S}_M$=.23.

   ($\underline{S}_M{}^2$ = $\underline{S}^2$/N = 5.71/109 = .052; $\underline{S}_M$=$\sqrt{\underline{S}_M{}^2}$ = $\sqrt{.052}$ = .23.)

3. Determine the cutoff sample score on the comparison distribution at which the null hypothesis should be rejected.

 1% level, 2-tailed, t distribution of $\underline{df}$=108: t needed = ±2.626.

4. Determine score of your sample on comparison distribution.

 Mean difference (mother minus father) in this sample is 1.44.

 t = ($\underline{M}-\mu_M$)/$\underline{S}_M$ = (1.44-0) / .23 = 1.44/.23 = 6.26.

5. Compare the scores obtained in Steps 3 and 4 to decide whether to reject the null hypothesis.

 t on 4 (6.26) is more extreme than cutoff t on 3 (±2.626).

 Therefore, reject the null hypothesis; the research hypothesis is supported (students like those in this class report having been closer to their mothers than to their fathers).

TRANSPARENCY 9.14

How close to mother versus to father.
(From class questionnaire.)

Computation of difference scores and of estimated
population variance (S^2) of difference scores:

Stu-dent	Closeness to Mother	Father	Difference (D=Aft-Bef)	Dev (D-M)	Dev² (D-M)²
1	7	7	0	-1.44	2.074
2	4	3	1	-0.44	0.194
.	.	.	.	.	.
.	.	.	.	.	.
108	6	1	5	3.56	12.674
109	7	5	2	0.56	0.314
Σ	606	449	157		SS=616.862
M	5.56	4.12	1.44		

$df=N-1=109-1=108$; $S^2 = SS/df = 616.862/108 = 5.71$

Aron/Aron
STATISTICS FOR PSYCHOLOGY

© 1994 by Prentice-Hall, Inc.
A Paramount Communications Company
Englewood Cliffs, New Jersey 07632

TRANSPARENCY 9.15

Water use of six households exposed to experimental advertizing campaign about drought. (Fictional data.)

Steps of hypothesis testing:

1. Reframe the question into a research hypothesis and a null hypothesis about populations.

   Population 1: Households exposed to advertising.
   Population 2: Households in general (no change).

   Research Hypothesis: Population 2 decreases more.
   Null Hypothesis: Population 1 decreases no more than
                    Population 2.

2. Determine the characteristics of the comparison distribution.

   Population 2: shape=assumed normal; μ=0; σ^2=unknown; $\underline{S}^2$=18.

   Distribution of means: shape=t ($\underline{df}$=5); $\mu_{\underline{M}}$=0; $\underline{S}_{\underline{M}}$=2.03.

   ($\underline{S}_{\underline{M}}^2$ = $\underline{S}^2$/N = 24.8/6 = 4.13; $\underline{S}_{\underline{M}}$=$\sqrt{\underline{S}_{\underline{M}}^2}$ = $\sqrt{4.13}$ = 2.03.)

3. Determine the cutoff sample score on the comparison distribution at which the null hypothesis should be rejected.

   5% level, 1-tailed, t distribution of $\underline{df}$=5: needed t = -2.015.

4. Determine score of your sample on comparison distribution.

   Mean change (after-before) in water usage = -7

   t = ($\underline{M}-\mu_{\underline{M}}$)/$\underline{S}_{\underline{M}}$ = (-7-0) / 2.03 = -7/2.03 = -3.45

5. Compare the scores obtained in Steps 3 and 4 to decide whether to reject the null hypothesis.

   t on 4 (-3.45) is more extreme than cutoff t on 3 (-2.015).

   Therefore, reject the null hypothesis; the research hypothesis is supported (households exposed to advertizing reduce their water usage.)

Aron/Aron
STATISTICS FOR PSYCHOLOGY

© 1994 by Prentice-Hall, Inc.
A Paramount Communications Company
Englewood Cliffs, New Jersey 07632

TRANSPARENCY 9.16

Water use of six households exposed to experimental
advertising campaign about drought. (Fictional data.)

Computation of change scores and of estimated
population variance ($\underline{S}^2$) of change scores:

House hold	Water Use Before	After	Difference (D=Aft-Bef)	Dev (D-M)	Dev² (D-M)²
A	485	473	-12	-5	25
B	561	556	- 5	2	4
C	430	432	2	9	81
D	675	665	-10	-3	9
E	522	514	- 8	-1	1
F	489	480	- 9	-2	4
Σ	3162	3120	-42	SS =	124
M	527	520	- 7		

$$\underline{S}^2 = \underline{SS}/\underline{df} = 124/(6-1) = 24.8$$

Aron/Aron
STATISTICS FOR PSYCHOLOGY

© 1994 by Prentice-Hall, Inc.
A Paramount Communications Company
Englewood Cliffs, New Jersey 07632

TABLE 9-9
Approximate Power for Studies Using the *t* Test for Dependent Means in Testing Hypotheses at the .05 Significance Level

Scores in Sample (*N*)	Effect Size		
	Small (*d* = .20)	Medium (*d* = .50)	Large (*d* = .80)
Two-tailed test			
10	.09	.32	.66
20	.14	.59	.93
30	.19	.77	.99
40	.24	.88	*
50	.29	.94	*
100	.25	*	*
One-tailed test			
10	.15	.46	.78
20	.22	.71	.96
30	.29	.86	*
40	.35	.93	*
50	.40	.97	*
100	.63	*	*

*Power is nearly 1.

© 1994 by Prentice-Hall, Inc.
A Paramount Communications Company
Englewood Cliffs, New Jersey 07632

Aron/Aron
STATISTICS FOR PSYCHOLOGY

TABLE 9-10
Approximate Number of Subjects Needed to Achieve 80% Power for the t Test for Dependent Means in Testing Hypotheses at the .05 Significance Level

	Effect Size		
	Small ($d = .20$)	Medium ($d = .50$)	Large ($d = .80$)
Two-tailed	196	33	14
One-tailed	156	26	12

TABLE 9-8
Steps for Conducting a *t* Test for Dependent Means

1. Reframe the question into a research hypothesis and a null hypothesis about the populations.

2. Determine the characteristics of the comparison distribution.

   a. Convert each subject's two scores into a difference score. Carry out the remaining steps using these difference scores.

   b. Compute the mean of the difference scores.

   c. Assume a population mean of 0: $\mu = 0$.

   d. Compute the estimated population variance of difference scores: $S^2 = SS/df$.

   e. Compute the variance of the distribution of means of difference scores: $S_M^2 = S^2/N$.

   f. Compute the standard deviation of the distribution of means of difference scores: $S_M = \div S_M^2$.

   g. Note that it will be a t distribution with $df = N - 1$.

3. Determine the cutoff sample score on the comparison distribution at which the null hypothesis should be rejected.

   a. Determine the desired significance level and whether to use a one-tailed or a two-tailed test.

   b. Look up the appropriate cutoff in a t table.

4. Determine the score of your sample on the comparison distribution: $t = (M - \mu)/S_M$.

5. Compare the scores obtained in Steps 3 and 4 to decide whether to reject the null hypothesis.

Aron/Aron
STATISTICS FOR PSYCHOLOGY

© 1994 by Prentice-Hall, Inc.
A Paramount Communications Company
Englewood Cliffs, New Jersey 07632

Chapter 10
The *t* Test for Independent Means

Instructor's Summary of Chapter

Distribution of differences between means. The main difference in procedure between a *t* test for independent means and a *t* test for a single sample is that the comparison distribution is now a distribution of differences between means of samples. This distribution can be thought of as arising in two steps: (a) each population of individual cases produces a distribution of means, and then (b) a new distribution is created consisting of differences between pairs of means selected from these two distributions of means.

Mean and shape of the distribution of means. The distribution of differences between means has a mean of 0 and will be a *t* distribution with degrees of freedom equal to the total degrees of freedom contributed by each sample.

Standard deviation of the distribution of means and the pooled estimate of the population variance. Its standard deviation is determined in several steps: (a) Each sample is used to estimate the population variance. (b) Since the populations are assumed to have the same variance, a pooled estimate is computed by simple averaging of the two estimates if the numbers in each sample are equal, or by a weighted average if they are not (multiplying each estimate by the proportion of the total degrees of freedom its sample contributes and adding up the products). (c) The pooled estimate is divided by each sample's number of cases to determine the variances of their associated distribution of means. (d) These two variances are added to produce the variance of the distribution of differences between means. (e) The square root is taken.

Assumptions. The *t* test for independent means assumes the populations are normal and have equal variances, although it is robust to moderate violations of these assumptions.

Effect size and power. Effect size (*d*) for a *t* test for independent means is the difference between the means divided by the population standard deviation. For a given number of subjects, power is greatest when sample sizes of the two groups are equal.

Controversies. When many significance tests are conducted in the same study, such as a series of *t* tests comparing two groups on various measures, the possibility that any one of the comparisons may turn out significant by chance is greater than .05 (or whatever level is being used to test each comparison). There is a lot of controversy about just how to adjust for this problem, though all agree that results should be interpreted cautiously in a situation of this kind.

How the procedures of this chapter are reported in research articles. *t* tests for independent means are reported in a standard format—e.g., $t(29)=3.41$, $p < .01$, where the number in parentheses is the degrees of freedom. Results may also appear in a table, each significant difference indicated merely by an asterisk.

Box 10.1. Monte Carlo Methods, or When Mathematics Becomes Just an Experiment and Statistics Depend on a Game of Chance. Monte Carlo studies are procedures in which a large number of random numbers are generated to test some mathematical or statistical question—such as the robustness of a statistical test under various conditions of violating its assumptions. This box describes the history and some applications of these procedures.

Lecture 10.1: Introduction to the t Test for Independent Means

Materials

Lecture outline

Transparencies 9.10, 9.19, and 10.1 through 10.9
(If using transparencies based on your class's questionnaires, replace 10.5 and 10.6 with 10.5R and 10.6R.)

Outline for Blackboard

    I. Review
    II. Between-Subject Designs
    III. Estimating Population Variance ($S_P{}^2$)
    IV. Distribution of Differences Between Means
    V. Conducting the Hypothesis Test
    VI. Additional Examples
    VII. Review this Class

Instructor's Lecture Outline

I. Review

    A. Idea of descriptive and inferential statistics.
    B. Show TRANSPARENCY 9.10 (steps of conducting a t test for a single sample) and discuss.
    C. Show TRANSPARENCY 9.19 (steps of conducting a t test for dependent means) and discuss.
    D. Show top of TRANSPARENCY 10.1 (basic calculations for t test for dependent means) and discuss.

II. Between-Subject Designs

    A. Principle: A study with two different groups of subjects, such as an experimental and a control group.
    B. Show TRANSPARENCY 10.2 (performance under quiet vs. noisy conditions) and discuss (without considering computations now).
        1. This is a new situation. You ca not just take difference scores as they are not the same subjects.
        2. In fact, in the example there are not even the same number of subjects in the two conditions.
    C. The hypothesis testing question is whether the means of these two groups are different enough to permit us to conclude that the two populations they represent have different means.
    D. The procedure for testing such a hypothesis is the t test for independent means.
    E. Populations and hypotheses: Show TRANSPARENCY 10.3 (hypothesis testing, quiet-noisy study) and discuss Step 1.

III. Estimating Population Variance ($S_P{}^2$)

    A. This is part of Step 2 of the hypothesis testing process.
    B. We assume that both populations have the same variance when doing a t test for independent means.
    C. Thus we can estimate the variance from each sample.
    D. We then average the two estimates.
    E. However, if the sample sizes are different, the larger sample provides a more accurate estimate.

246

F. Thus, we use a weighted average, giving emphasis in proportion to the degrees of freedom each sample contributes.

G. This estimate based on the weighted average of the two estimates is called the pooled estimate.

H. Show TRANSPARENCY 10.2 (computations for quiet-noisy example) and discuss computations of S_P^2.

IV. Distribution of Differences Between Means

A. The crucial result of a study comparing two groups is a difference between the means of the two samples.

B. Thus, the comparison distribution has to be a distribution of differences between means of all possible samples from the two populations.

C. This is called a distribution of differences between means.

D. Show TRANSPARENCY 10.4 (illustration for logic of constructing a distribution of differences between means) and discuss each step.
   1. Create two distributions of means, one for each population.
   2. Take a mean from each distribution of means, compute the difference.
   3. Do this many times and make a distribution of these differences.

E. Its mean will be 0—since under the null hypothesis, the means being subtracted from each other come from populations with the same mean.

F. Its variance is computed as follows:
   1. Find variance of each distribution of means in usual way.
   2. Sum these variances to get S_{DIF}^2.
   3. Show TRANSPARENCY 10.2 (computations for quiet-noisy example) and discuss computations of S_{DIF}^2.

F. It will be a t distribution because it is based on estimated population variance. Its df is the total df on which that estimate is based.

G. Show TRANSPARENCY 10.3 (hypothesis testing for quiet-noisy example) and discuss Step 2.

V. Conducting the Hypothesis Test: Show TRANSPARENCY 10.3 (hypothesis testing for quiet-noisy example) and review each step, noting slight changes from previous t test procedures, and emphasizing how much is the same.

VI. Additional Examples

A. Show TRANSPARENCIES 10.5 and 10.6 or 10.5R and 10.6R (sensitivity scale scores for women vs. men, from class questionnaire) and discuss.

B. Show TRANSPARENCIES 10.7 and 10.8 (fictional study of effectiveness of a new job-skills program, from text) and discuss.

VII. Review this Class: Use blackboard outline and show TRANSPARENCIES 10.3 (hypothesis testing, quiet-noisy example), 10.4 (illustration of logic of distribution of differences between means), and 10.9 (computations for a t test for independent means).

Lecture 10.2: Applying the *t* Test for Independent Means

Materials

Lecture outline

Transparencies 10.2 through 10.4 and 10.9 through 10.18

(If using transparencies based on your class's questionnaires, replace 10.13 and 10.14 with 10.13R and 10.14R.)

Outline for Blackboard

I.   **Review**

II. **Additional Examples**

III. **Assumptions**

IV. **Effect Size and Power**

V. **Review this Class**

Instructor's Lecture Outline

I. Review

A. Idea of descriptive and inferential statistics.

B. Basic logic of the *t* test for independent means.

    1. Show TRANSPARENCIES 10.4 (logic of construction of the distribution of differences between means) and discuss.

    2. Show TRANSPARENCY 10.10 (steps of conducting a *t* test for independent means) and discuss.

C. Show TRANSPARENCIES 10.2 and 10.3 (quiet-noisy example) and discuss.

D. Show TRANSPARENCY 10.9 (computations for *t* test for independent means) and discuss.

II. Additional Examples

A. Show TRANSPARENCIES 10.11 and 10.12 (men and women swimmers' optimism, from text) and discuss.

B. Show TRANSPARENCIES 10.13 and 10.14 or 10.13R and 10.14R (first vs. later borns on ratings of being prone to fears, from class questionnaire) and discuss.

III. Assumptions

A. As with *t* test for dependent means, populations must be normal.

    1. However, Monte Carlo studies indicate test is robust to moderate violations.

    2. Especially robust if N is large (about > 30 each group).

    3. A problem if the two populations are skewed in opposite directions.

B. In addition, populations must have same variance.

    1. This is assumed in producing a pooled estimate.

    2. However, Monte Carlo studies indicate test is robust to moderate violations.

    3. A problem if variance of samples are 3 or 4 to 1.

C. A problem if both assumptions are violated.

D. Show TRANSPARENCY 10.15 (situations in which ordinary *t* test vs. alternatives should be used) and discuss.

IV. **Effect Size and Power**

    A. Effect size.

        1. $d = (\mu_1 - \mu_2) / \sigma$.

        2. For a completed study, estimated as $d = (M_1 - M_2) / S_p$.

        3. As before, a small effect size is .20, a medium effect size is .50, and a large effect size is .80.

    B. Power.

        1. Power can be determined directly from table in the text. Show TRANSPARENCY 10.16 (power table) and discuss.

        2. Needed N for 80% power for a given effect size can also be determined from a table in the text. Show TRANSPARENCY 10.17 (needed number of subjects table) and discuss.

        3. For the same overall number of subjects, power is greatest with equal Ns.

          a. For unequal Ns, effective N for computing power is closer to the lower N (it is the harmonic mean).

          b. Show TRANSPARENCY 10.18 (harmonic mean formula and example computations) and discuss.

V.   **Review this Class:** Use blackboard outline and TRANSPARENCY 10.10 (steps of conducting a t test for independent means).

TRANSPARENCY 10.1

Necessary calculations for a $\underline{t}$-test for dependent means.

1. Compute difference scores:

   Each difference score =
   each after score - its before score

2. Compute $\underline{M}$ (mean of sample of difference scores):

   $\underline{M} = \Sigma\underline{X}/\underline{N}$

3. Using difference scores, compute S^2 (estimated variance of population of difference scores):

   $S^2 = \Sigma(\underline{X}-\underline{M})^2 / (\underline{N}-1) = \underline{SS}/\underline{df}$

4. Compute $\underline{S}_M$ (standard deviation of comparison distribution--the distribution of means of samples of difference scores):

   $\underline{S}_M{}^2 = S^2/\underline{N}$ $\qquad\qquad$ $\underline{S}_M = \sqrt{\underline{S}_M{}^2}$

5. Calculate $\underline{t}$ score for sample's mean of difference scores:

   $\underline{t} = (\underline{M}-0)/\underline{S}_M$

Aron/Aron
STATISTICS FOR PSYCHOLOGY

© 1994 by Prentice-Hall, Inc.
A Paramount Communications Company
Englewood Cliffs, New Jersey 07632

TRANSPARENCY 10.2

Experiment comparing task performance of subjects
tested under quiet conditions versus of subjects
tested under noisy conditions. (Fictional data.)

Data and computations:

Quiet Condition		Noisy Condition	
Subject	Score	Subject	Score
RX	22	KA	17
BL	20	BI	15
JC	23	OF	16
DM	21	BK	20
FM	19		

$$M_1 = 21 \qquad\qquad M_2 = 17$$
$$S_1^2 = 2.5 \qquad\qquad S_2^2 = 4.67$$
$$df_1 = 4 \qquad\qquad df_2 = 3$$

$$d_{TOT} = df_1 + df_2 = 4 + 3 = 7$$

$$S_P^2 = \frac{df_1}{df_{TOT}}(S_1^2) + \frac{df_2}{df_{TOT}}(S_2^2)$$

$$= \frac{4}{7}(2.5) + \frac{3}{7}(4.67)$$

$$= (.57)(2.5) + (.43)(4.67)$$

$$= \quad 1.43 \quad + \quad 2.01 \quad = 3.44$$

$$S_{M1}^2 = S_P^2/N_1 = 3.44/5 = .69$$

$$S_{M2}^2 = S_P^2/N_2 = 3.44/4 = .86$$

$$S_{DIF}^2 = S_{M1}^2 + S_{M2}^2 = .69 + .86 = 1.55$$

$$S_{DIF} = \sqrt{(S_{DIF}^2)} = \sqrt{1.55} = 1.24$$

Aron/Aron
STATISTICS FOR PSYCHOLOGY

© 1994 by Prentice-Hall, Inc.
A Paramount Communications Company
Englewood Cliffs, New Jersey 07632

TRANSPARENCY 10.3

Experiment comparing task performance of subjects tested under quiet conditions versus of subjects tested under noisy conditions. (Fictional data.)

Steps of hypothesis testing:

1. Reframe into a research hypothesis and a null hypothesis about populations.

   Population 1: People tested under quiet conditions
   Population 2: People tested under noisy conditions

   Research Hypothesis: Those tested under quiet conditions score better (that is, Pop. 1 has higher mean.)
   Null Hypothesis: Those tested under quiet conditions do not score better.

2. Determine the characteristics of the comparison distribution.

Estimated population variance = Weighted Average $\underline{S}^2$ = $\underline{S}_p^2$ = 3.44.

Comparison distribution (distribution of differences between means):

   Mean = 0; $\underline{S}_{DIF}$ = 1.24; Shape = $\underline{t}$ ($\underline{df}$ = 7).

3. Determine the cutoff sample score on the comparison distribution at which the null hypothesis should be rejected.

 5% level, 1-tailed, t distribution of $\underline{df}$=7: t needed=1.895.

4. Determine score of your sample on comparison distribution.

 $\underline{t}$ = ($\underline{M}_1$-$\underline{M}_2$)/$\underline{S}_{DIF}$ = (21-17)/1.24 = 4/1.24 = 3.23.

5. Compare scores obtained in Steps 3 and 4 to decide whether to reject the null hypothesis.

 $\underline{t}$ on 4 (3.23) is more extreme than cutoff $\underline{t}$ on 3 (1.895).

 Therefore, reject the null hypothesis; the research hypothesis is supported (people perform better under quiet conditions).

Aron/Aron
STATISTICS FOR PSYCHOLOGY

© 1994 by Prentice-Hall, Inc.
A Paramount Communications Company
Englewood Cliffs, New Jersey 07632

TRANSPARENCY 10.4

Figure 10.1

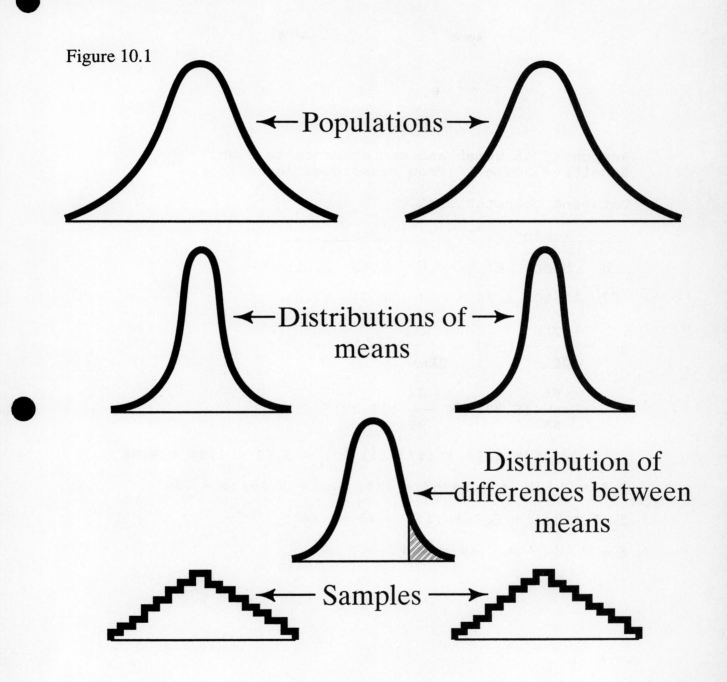

Populations

Distributions of means

Distribution of differences between means

Samples

Aron/Aron
STATISTICS FOR PSYCHOLOGY

TRANSPARENCY 10.5

Responses of women and men students to Highly
Sensitive Scale. (From class questionnaire.)

Data and computations:

Women			Men		
N	Mean	S^2	N	Mean	S^2
67	11.90	8.88	34	10.74	11.83

$$S_P{}^2 = \frac{df_1}{df_{TOT}} (S_1{}^2) + \frac{df_2}{df_{TOT}} (S_2{}^2)$$

$$= \frac{66}{99} (8.88) + \frac{33}{99} (11.83)$$

$$= (2/3)(8.88) + (1/3)(11.83) = 5.92 + 3.94 = 9.86$$

$$S_{M1}{}^2 = S_P{}^2/N_1 = 9.86/67 = .15; \; S_{M2}{}^2 = 9.86/34 = .29$$

$$S_{DIF}{}^2 = S_{M1}{}^2 + S_{M2}{}^2 = .15 + .29 = .44$$

$$S_{DIF} = \sqrt{S_{DIF}{}^2} = \sqrt{.44} = .66$$

Aron/Aron
STATISTICS FOR PSYCHOLOGY

© 1994 by Prentice-Hall, Inc.
A Paramount Communications Company
Englewood Cliffs, New Jersey 07632

TRANSPARENCY 10.6

Responses of women and men students to Highly Sensitive Scale. (From class questionnaire.)

Steps of hypothesis testing:

1. Reframe into a research hypothesis and a null hypothesis about populations.

   Population 1: Women psychology majors.
   Population 2: Men psychology majors.

   Research Hypothesis: Two populations have different means.
   Null Hypothesis: Two populations have the same mean.

2. Determine the characteristics of the comparison distribution.

   Estimated population variance = Weighted Average = $\underline{S}_P{}^2$ = 9.86.

   Comparison distribution (distribution of differences between means):

      Mean=0; $\underline{S}_{DIF}$=.66; Shape = $\underline{t}$(99).

3. Determine the cutoff sample score on the comparison distribution at which the null hypothesis should be rejected.

   5% level, 2-tailed, $\underline{t}$(99) needed= ±1.986.

4. Determine score of your sample on comparison distribution.

   $\underline{t}$ = $(\underline{M}_1-\underline{M}_2)/\underline{S}_{DIF}$ = (11.90-10.74)/.66 = 1.16/.66 = 1.76.

5. Compare scores obtained in Steps 3 and 4 to decide whether to reject the null hypothesis.

   $\underline{t}$ on 4 (1.76) is not more extreme than cutoff $\underline{t}$ on 3 (±1.986).

   Therefore, do not reject the null hypothesis; the study is inconclusive.

Aron/Aron
STATISTICS FOR PSYCHOLOGY

© 1994 by Prentice-Hall, Inc.
A Paramount Communications Company
Englewood Cliffs, New Jersey 07632

TABLE 10-3
Computations for a t Test for Independent Means for an Experiment Examining the Effectiveness (Using Employers' Ratings) of a New Job Skills Program for People Who Have Previously Not Been Able to Hold Jobs (Fictional Data)

Experimental Group (Receiving Special Program)			Control Group (Receiving Standard Program)		
Score	Deviation from mean	Squared deviation from mean	Score	Deviation from mean	Squared deviation from mean
6	0	0	6	3	9
4	–2	4	1	–2	4
9	3	9	5	2	4
7	1	1	3	0	0
7	1	1	1	–2	4
3	–3	9	1	–2	4
6	0	0	4	1	1
Σ: 42	0	24	21	0	26

$M_1 = 6; S_1^2 = 24/6 = 4; M_2 = 3; S_2^2 = 26/6 = 4.33$

$N_1 = 7; df_1 = N_1 - 1 = 6; N_2 = 7; df_2 = N_2 - 1 = 6$

$df_T = df_1 + df_2 = 6 + 6 = 12$

$S_P^2 = \dfrac{df_1}{df_T}(S_1^2) + \dfrac{df_2}{df_T}(S_2^2) = \dfrac{6}{12}(4) + \dfrac{6}{12}(4.33) = .5(4) + .5(4.33) = 2.00 + 2.17 = 4.17$

$S_{M1}^2 = S_P^2/N_1 = 4.17/7 = .60$

$S_{M2}^2 = S_P^2/N_2 = 4.17/7 = .60$

$S_{DIF}^2 = S_{M1}^2 + S_{M2}^2 = .60 + .60 = 1.20$

$S_{DIF} = \div S_{DIF}^2 = \div 1.20 = 1.10$

Needed t with $df = 12$, 5% level, two-tailed $= \pm 2.179$

$t = (M_1 - M_2)/S_{DIF} = (6.00 - 3.00)/1.10 = 3.00/1.10 = 2.73$

Conclusion: Reject the null hypothesis; the research hypothesis is supported.

Aron/Aron
STATISTICS FOR PSYCHOLOGY

© 1994 by Prentice-Hall, Inc.
A Paramount Communications Company
Englewood Cliffs, New Jersey 07632

TRANSPARENCY 10.8

Effectiveness of a new job-skills program. (Example from text.)

Steps of hypothesis testing:

1. Reframe into a research hypothesis and a null hypothesis about populations.

   Population 1: Participants in a <u>special</u> job-skills program.
   Population 2: Participants in an <u>ordinary</u> job-skills program.

   Research Hypothesis: Two populations have different means.
   Null Hypothesis: Two populations have the same mean.

2. Determine the characteristics of the comparison distribution.

   Estimated population variance = Weighted Average = $\underline{S}_p{}^2$ = 4.17.

   Comparison distribution (distribution of differences between means):

      Mean=0; $\underline{S}_{DIF}$=1.10; Shape = $\underline{t}$(12).

3. Determine the cutoff sample score on the comparison distribution at which the null hypothesis should be rejected.

   5% level, 2-tailed, $\underline{t}$(12) needed= ±2.179.

4. Determine score of your sample on comparison distribution.

   $\underline{t}$ = $(\underline{M}_1-\underline{M}_2)/\underline{S}_{DIF}$ = (6.00-3.00)/1.10 = 3.00/1.10 = 2.70.

5. Compare scores obtained in Steps 3 and 4 to decide whether to reject the null hypothesis.

   $\underline{t}$ on 4 (2.70) is more extreme than cutoff $\underline{t}$ on 3 (±2.179).

   Therefore, reject the null hypothesis; the research hypothesis is supported (the new job-skills program is effective).

Aron/Aron
STATISTICS FOR PSYCHOLOGY

© 1994 by Prentice-Hall, Inc.
A Paramount Communications Company
Englewood Cliffs, New Jersey 07632

TRANSPARENCY 10.9

Necessary calculations for a $\underline{t}$-test for independent means.

1. Compute $\underline{M}_1$ and $\underline{M}_2$ (means of each sample):

$$\underline{M}_1 = \Sigma \underline{X}_1 / \underline{N}_1 \qquad\qquad \underline{M}_2 = \Sigma \underline{X}_2 / \underline{N}_2$$

2. Compute $\underline{S}_1{}^2$ and $\underline{S}_2{}^2$ (estimated variance of population) using scores in each sample:

$$\underline{S}_1{}^2 = \Sigma(\underline{X}_1 - \underline{M}_1)^2 / (\underline{N}_1 - 1) = \underline{SS}_1 / \underline{df}_1$$
$$\underline{S}_2{}^2 = \Sigma(\underline{X}_2 - \underline{M}_2)^2 / (\underline{N}_2 - 1) = \underline{SS}_2 / \underline{df}_2$$

3. Compute $\underline{S}_P{}^2$ (pooled estimate of population variance):

$$\underline{S}_P{}^2 = \frac{\underline{df}_1}{\underline{df}_{TOT}}(\underline{S}_1{}^2) \;+\; \frac{\underline{df}_2}{\underline{df}_{TOT}}(\underline{S}_2{}^2)$$

4. Compute $\underline{S}_{M1}{}^2$ and $\underline{S}_{M2}{}^2$ (variances of distribution of means corresponding to each population):

$$\underline{S}_{M1}{}^2 = \underline{S}_P{}^2 / \underline{N}_1 \qquad \underline{S}_{M2}{}^2 = \underline{S}_P{}^2 / \underline{N}_2$$

5. Compute $\underline{S}_{DIF}$ (standard deviation of distribution of differences between means):

$$\underline{S}_{DIF}{}^2 = \underline{S}_{M1}{}^2 + \underline{S}_{M2}{}^2 \qquad \underline{S}_{DIF} = \sqrt{(\underline{S}_{DIF}{}^2)}$$

6. Calculate $\underline{t}$ for the difference between two sample means:

$$\underline{t} = (\underline{M}_1 - \underline{M}_2) / \underline{S}_{DIF}$$

Aron/Aron
STATISTICS FOR PSYCHOLOGY

© 1994 by Prentice-Hall, Inc.
A Paramount Communications Company
Englewood Cliffs, New Jersey 07632

TABLE 10-4
Steps for Conducting a t Test for Independent Means

1. Reframe the question into a research hypothesis and a null hypothesis about the populations.

2. Determine the characteristics of the comparison distribution.

   a. Its mean will be 0.

   b. Compute its standard deviation.

      i. Compute estimated population variances based on each sample (that is, compute two estimates).

      ii. Compute a pooled estimate of population variance:

$$S_P^2 = \frac{df_1}{df_T}(S_1^2) + \frac{df_2}{df_T}(S_2^2)$$

   $(df_1 = N_1 - 1$ and $df_2 = N_2 - 1$; $df_T = df_1 + df_2)$

      iii. Compute the variance of each distribution of means: $S_{M1}^2 = S_P^2/N_1$ and $S_{M2}^2 = S_P^2/N_2$.

      iv. Compute the variance of the distribution of differences between means:
$S_{DIF}^2 = S_{M1}^2 + S_{M2}^2$.

      v. Compute the standard deviation of the distribution of differences between means: $S_{DIF} = \div S_{DIF}^2$.

   c. Determine its shape: It will be a t distribution with df_T degrees of freedom.

3. Determine the cutoff sample score on the comparison distribution at which the null hypothesis should be rejected.

   a. Determine the degrees of freedom (df_T), desired significance level, and tails in the test (one or two).

   b. Look up the appropriate cutoff in a t table. If the exact df is not given, use the df below.

4. Determine the score of the sample on the comparison distribution: $t = (M_1 - M_2)/S_{DIF}$.

Aron/Aron
STATISTICS FOR PSYCHOLOGY

© 1994 by Prentice-Hall, Inc.
A Paramount Communications Company
Englewood Cliffs, New Jersey 07632

TABLE 10-1
Computations for a t Test for Independent Means for a Study of Optimism Comparing Women and Men Varsity Swimmers

Women: $N_1 = 26; df_1 = N_1 - 1 = 25; M_1 = 15.96; S_1^2 = 2.07$

Men: $N_2 = 21; df_2 = N_2 - 1 = 20; M_2 = 18.03; S_2^2 = 3.75$

$df_T = df_1 + df_2 = 25 + 20 = 45$

$S_P^2 = \dfrac{df_1}{df_T}(S_1^2) + \dfrac{df_2}{df_T}(S_2^2) = \dfrac{25}{45}(2.07) + \dfrac{20}{45}(3.75) = .56(2.07) + .44(3.75) = 1.16 + 1.65 = 2.81$

$S_{M1}^2 = S_P^2/N_1 = 2.81/26 = .11$

$S_{M2}^2 = S_P^2/N_2 = 2.81/21 = .13$

$S_{DIF}^2 = S_{M1}^2 + S_{M2}^2 = .11 + .13 = .24$

$S_{DIF} = \div S_{DIF}^2 = \div .24 = .49$

Needed t with $df = 45$, 1% level, two-tailed $= \pm 2.690$

$t = (M_1 - M_2)/S_{DIF} = (15.96 - 18.03)/.49 = -2.07/.49 = -4.22$

Conclusion: Reject the null hypothesis; the results suggest that women swimmers are less optimistic than men swimmers.

Note. Data from Seligman, Nolen-Hoeksema, Thornton, & Thornton (1990).

Aron/Aron
STATISTICS FOR PSYCHOLOGY

TRANSPARENCY 10.12

Optimism of women versus men varsity swimmers. (Example from text.)

Steps of hypothesis testing:

1. Reframe into a research hypothesis and a null hypothesis about populations.

   Population 1: Women varsity swimmers.
   Population 2: Men varsity swimmers.

   Research Hypothesis: Two populations have different means.
   Null Hypothesis: Two populations have the same mean.

2. Determine the characteristics of the comparison distribution.

   Estimated population variance = Weighted Average = S_P^2 = 2.81.

   Comparison distribution (distribution of differences between means):

   Mean=0; S_{DIF}=.49; Shape = $t(45)$.

3. Determine the cutoff sample score on the comparison distribution at which the null hypothesis should be rejected.

   1% level, 2-tailed, $t(45)$ needed = ±2.014.

4. Determine score of your sample on comparison distribution.

   $t = (M_1-M_2)/S_{DIF} = (15.96-18.03)/.49 = -2.07/.49 = -4.22$.

5. Compare scores obtained in Steps 3 and 4 to decide whether to reject the null hypothesis.

   t on 4 (-4.22) is more extreme than cutoff t on 3 (±2.014).

   Therefore, reject the null hypothesis; the research hypothesis is supported (women swimmers are less optimistic than men swimmers).

Aron/Aron
STATISTICS FOR PSYCHOLOGY

© 1994 by Prentice-Hall, Inc.
A Paramount Communications Company
Englewood Cliffs, New Jersey 07632

TRANSPARENCY 10.13

Responses to "Do you tend to fall in love hard?"
of first born and later born students.
(From class questionnaire.)

Data and computations:

First Borns			Later Borns		
N	Mean	S^2	N	Mean	S^2
42	4.21	4.27	57	4.93	3.25

$$S_P{}^2 = \frac{df_1}{df_{TOT}} (S_1{}^2) + \frac{df_2}{df_{TOT}} (S_2{}^2)$$

$$= \frac{41}{97} (4.27) + \frac{56}{97} (3.25)$$

$$= (.42)(4.27) + (.58)(3.25)$$

$$= \qquad 1.79 \qquad + 1.89 \qquad = 3.68$$

$$\underline{S}_{M1}{}^2 = \underline{S}_P{}^2 / \underline{N}_1 = 3.68/42 = .088$$

$$\underline{S}_{M2}{}^2 = 3.68/57 = .065$$

$$\underline{S}_{DIF}{}^2 = \underline{S}_{M1}{}^2 + \underline{S}_{M2}{}^2 = .088 + .065 = .153$$

$$\underline{S}_{DIF} = \sqrt{\underline{S}_{DIF}{}^2} = \sqrt{.153} = .39$$

Aron/Aron
STATISTICS FOR PSYCHOLOGY

© 1994 by Prentice-Hall, Inc.
A Paramount Communications Company
Englewood Cliffs, New Jersey 07632

TRANSPARENCY 10.14

Responses to "Do you tend to fall in love hard?" of first born and later born students. (From class questionnaire.)

Steps of hypothesis testing:

1. Reframe into a research hypothesis and a null hypothesis about populations.

   Population 1: First born psychology majors.
   Population 2: Later born psychology majors.

   Research Hypothesis: Two populations have different means.
   Null Hypothesis: Two populations have the same mean.

2. Determine the characteristics of the comparison distribution.

   Estimated population variance = Weighted Average = S_P^2 = 3.68.

   Comparison distribution (distribution of differences between means):

   Mean=0; S_{DIF}=.39; Shape = t(97).

3. Determine the cutoff sample score on the comparison distribution at which the null hypothesis should be rejected.

   5% level, 2-tailed, t(97) needed = ±1.986.

4. Determine score of your sample on comparison distribution.

   t = $(M_1-M_2)/S_{DIF}$ = (4.21-4.93)/.39 = .72/.39 = 1.85.

5. Compare scores obtained in Steps 3 and 4 to decide whether to reject the null hypothesis.

   t on 4 (1.85) is not more extreme than cutoff t on 3 (±1.986).

   Therefore, do not reject the null hypothesis; the result is inconclusive.

Aron/Aron
STATISTICS FOR PSYCHOLOGY

© 1994 by Prentice-Hall, Inc.
A Paramount Communications Company
Englewood Cliffs, New Jersey 07632

TABLE 10-5
Situations in Which the Ordinary *t* Test for Independent Means or Alternatives Should Be Used

	Populations Follow a Normal Curve	Populations Do Not Follow a Normal Curve
Population variances are equal ($\sigma_1^2 = \sigma_2^2$)	*t* test	*t* test if nonnormality is not extreme. If extreme, use Chapter 15 procedures.
Population variances are unequal ($\sigma_1^2 \neq \sigma_2^2$):		
Sample sizes are equal ($N_1 = N_2$)	*t* test if unequal σ^2 not extreme	Chapter 15 procedures
Sample sizes are unequal ($N_1 \neq N_2$)	Special separate variance estimates *t* test	Chapter 15 procedures

Aron/Aron
STATISTICS FOR PSYCHOLOGY

TABLE 10-6
Approximate Power for Studies Using the *t* Test for Independent Means
Testing Hypotheses at the .05 Significance Level

Number of Subjects (*N*)	Effect Size		
	Small (d = .20)	*Medium* (d = .50)	*Large* (d = .80)
One-tailed test			
10	.11	.29	.53
20	.15	.46	.80
30	.19	.61	.92
40	.22	.72	.97
50	.26	.80	.99
100	.41	.97	*
Two-tailed test			
10	.07	.18	.39
20	.09	.33	.69
30	.12	.47	.86
40	.14	.60	.94
50	.17	.70	.98
100	.29	.94	*

*Nearly 1.
Note. Based on Cohen (1988), pp. 28–39.

Aron/Aron
STATISTICS FOR PSYCHOLOGY

© 1994 by Prentice-Hall, Inc.
A Paramount Communications Company
Englewood Cliffs, New Jersey 07632

TABLE 10-7
Approximate Number of Subjects Needed in Each Group (Assuming Equal Sample Sizes) to Achieve 80% Power for the *t* Test for Independent Means, Testing Hypotheses at the .05 Significance Level

	Effect Size		
	Small (d = .20)	Medium (d = .50)	Large (d = .80)
One-tailed	310	50	20
Two-tailed	393	64	26

Harmonic mean.

$$\underline{N}' = \frac{(2)(\underline{N}_1)(\underline{N}_2)}{\underline{N}_1 + \underline{N}_2}$$

Example with 6 subjects in one group and 34 in the other.

$$\underline{N}' = \frac{(2)(6)(34)}{6 + 34} = \frac{408}{40} = 10.2$$

Example with 3 subjects in one group and 37 in the other.

$$\underline{N}' = \frac{(2)(3)(37)}{3 + 37} = \frac{222}{40} = 5.55$$

Aron/Aron
STATISTICS FOR PSYCHOLOGY

Chapter 11
Introduction to the Analysis of Variance

Instructor's Summary of Chapter

Analysis of variance (ANOVA) is used to test hypotheses involving differences among means of several samples.

The two variance estimates. ANOVA compares two estimates of population variance: (a) a "within-group estimate," determined by pooling the variance estimates from each of the samples, and (b) a "between-group estimate," based on the variation among the means of the samples.

The F ratio is computed by dividing the between-group by the within-group estimate. If the null hypothesis is true, so that all the samples come from populations with the same mean, the *F* ratio should be about 1, since the two population-variance estimates are based on the same variation, the variation within each of the samples. But if the research hypothesis is true, so that samples come from populations with different means, then the *F* ratio should be larger than 1, since the between-group estimate is now influenced by both the variation within the samples and between them, while the within-group estimate is still only affected by the variation within each of the samples.

Computations of the two variance estimates. When there are equal sample sizes, the within-group population-variance estimate is the average of the estimates of the population variance computed from each sample; the between-group population variance estimate is computed by first finding the estimate of the variance of the distribution of means (computed by using the ordinary formula for S^2 but computing it with the means), then multiplying this estimate by the sample size to make it comparable to the variance of a distribution of individual cases (which is what the within-group estimate is about).

The F distribution and the F table. The distribution of *F* ratios under the null hypothesis is known, and significance cutoff values are available in tables which are used by providing the degrees of freedom for each population variance estimate—the between (or numerator) estimate being based on the number of groups minus one, and the within-group (or denominator) estimate being based on the sum of the degrees of freedom in each sample.

Assumptions. The assumptions for ANOVA are the same as for the *t* test—the populations must be normally distributed, with equal variances—and like the *t* test, ANOVA is considered robust to moderate violations of these assumptions.

Effect size and power. Effect size in ANOVA can be estimated from the elements involved in computing ANOVA or from the *F* and number of subjects in each group in a completed study. Power depends on effect size, number of subjects, significance level, and number of groups.

Controversy: Systematic matching by groups. Systematically assigning subjects to experimental groups in order to assure similar averages on background variables generally reduces power since the procedure reduces the contribution of random variance to the between-group estimate but not to the within-group estimate. However, under certain conditions this procedure can increase power.

How the procedures of this chapter are reported in research articles. ANOVA results are reported in research articles using a standard format—e.g., $F(3,38)=3.41, p < .05$, where the first number in parentheses is the numerator *df* (the *df* for the between-group's estimate) and the second number in parentheses is the denominator *df* (the *df* for the within-group's estimate).

Box 11.1. Sir Ronald Fisher, Caustic Genius of Statistics. Summarizes the life and work of Fisher, inventor of ANOVA.

Lecture 11.1: Introduction to the Analysis of Variance

Materials

Lecture outline

Transparencies 10.10 and 11.1 through 11.11

(If using transparencies based on your class's questionnaires, replace 11.7 and 11.8 with 11.7R and 11.8R.)

Outline for Blackboard

I.   **Review**
II. **Basic Logic**
III. **Computations**
IV. **The *F* Distribution**
V.   **Examples**
VI. **Assumptions**
VII. **Effect Size and Power**
VIII.**Review this Class**

Instructor's Lecture Outline

I.   Review

A. Idea of descriptive and inferential statistics.

B. Basic logic of the *t* test for independent means. Show TRANSPARENCY 10.10 (steps of computing a *t* test for independent means) and discuss, emphasizing:

    1. Overall logic of hypothesis testing.

    2. *t* distribution and *t* table.

    3. Pooled population variance estimate.

C. Show TRANSPARENCY 11.1 (central logic of hypothesis testing process) and discuss.

II.   Basic Logic

A. Procedure capable of analyzing studies with more than two groups.

B. Example (from text): Subjects rate guilt of defendant either knowing defendant has a criminal record, has no criminal record, or not having any information about the defendant's criminal record.

C. Null and research hypothesis. Show top of TRANSPARENCY 11.2 (logic of ANOVA) and discuss.

D. Population variance can be estimated based on scores in each sample—the *within-group population variance estimate*.

    1. Estimate from each sample uses usual procedure: $S^2 = SS/df$.

    2. We assume all populations have equal variances.

    3. Thus we can average estimates. (In this class we are dealing only with situations with equal *N*s, so no weighting is necessary.)

E. Population variance can also be estimated based on variation among means of groups—the *between-group population variance estimate*.

    1. If null hypothesis is true, all populations have same mean. Thus, variation among means of samples can only reflect variation within populations.

    2. If null hypothesis is not true, populations have different means. Thus, variation among means of samples reflects both variation within populations and variation between them.

F. If null hypothesis is true, both estimates are reflecting same variance. Their ratio would be 1 to 1.

G. If null hypothesis is false, the between-group estimate is reflecting more than the within-group, and thus the ratio of the between to the within will be more than 1 to 1.

H. Show rest of TRANSPARENCY 11.2 (logic of ANOVA) and discuss each step.

III. **Computations:** Show top of TRANSPARENCY 11.3 (main ANOVA computations with equal Ns) and discuss.

IV. **The F Distribution**

A. Can be thought of as constructed this way.
   1. Assume the null hypothesis is true and all populations have same mean.
   2. Draw a sample from each population and compute the F for the set of samples.
   3. Repeat a large number of times and make a distribution of the resulting Fs.

B. Can also be derived mathematically.

C. Shape. Show top of TRANSPARENCY 11.4 (F distribution) and discuss:
   1. Exact shape depends on numerator and denominator degrees of freedom.
   2. Numerator degrees of freedom is degrees of freedom in between-group variance estimate: $df_B = N_G - 1$.
   3. Denominator degrees of freedom is degrees of freedom in within-group variance estimate:
      $df_W = df_1 + df_2 + \ldots + df_{Last}$

D. Table of F values requires knowing significance level and two degrees-of-freedom values. (Note that there are no one-tailed tests possible.) Show bottom of TRANSPARENCY 11.4 (F table) and discuss.

V. **Examples**

A. Show TRANSPARENCIES 11.5 and 11.6 (criminal-record example from text) and discuss.

B. Show TRANSPARENCIES 11.7 and 11.8 or 11.7R and 11.8R (ratings on prone to fears as a function of self-rating of attachment style, from class questionnaire) and discuss. (NOTE: For purposes of this example, subsets of each group were randomly selected so that there are equal numbers in each group.)

C. Show TRANSPARENCIES 11.9 and 11.10 (fictional experiment in which people's willingness to volunteer to help is measured under three conditions—when alone, with best friend present, in classroom setting) and discuss.

VI. **Assumptions**

A. All populations normally distributed.
   1. Robust to moderate violations, especially if N is not small.
   2. A problem if populations skewed in different directions.

B. All populations have same variance.
   1. Robust to moderate violations.
   2. A problem when the largest variance estimate of any group is 4 or 5 times that of the smallest.

VII. Effect Size and Power

A. Effect size.

1. $f = \sigma_M / \sigma$.
2. For a completed study, estimated as $f = S_M / S_W$.
3. For a study in an article in which only F and N are available: $f = (\sqrt{F})/(\sqrt{n})$.
4. Cohen's conventions: small $f = .10$; medium $f = .25$; large $f = .40$.

B. Power.

1. Power depends on the usual considerations (effect size, sample size, significance level) *plus* number of groups.
2. Power can be determined directly from table in the text.
3. Needed N for 80% power for a given effect size can also be determined from a table in the text.

VIII. Review this Class: Use blackboard outline and TRANSPARENCY 11.11 (summary of steps of conducting an analysis of variance).

TRANSPARENCY 11.1

Central logic of the hypothesis testing process.

1. "Construct" a comparison distribution representing the null hypothesis situation in the form appropriate to your sample situation.

2. Compare your actual sample data to that comparison distribution.

Types of sample situations and corresponding comparison distributions:

Sample Situation to be Tested	Population	Comparison Distribution			
		Mean	Var	Shape	Name
1 case	known	μ	σ^2	normal	Population Distribution
1 mean	known	μ	σ_M^2	normal	Distribution of Means
1 mean	σ unknown	μ	S_M^2	$\underline{t}$	Distribution of Means
1 Mean of difference scores	σ unknown	0	S_M^2	$\underline{t}$	Distribution of Means (of Difference Scores)
Difference between 2 means	σ unknown	0	S_{DIF}^2	$\underline{t}$	Distribution of Differences Between Means

Aron/Aron
STATISTICS FOR PSYCHOLOGY

© 1994 by Prentice-Hall, Inc.
A Paramount Communications Company
Englewood Cliffs, New Jersey 07632

TRANSPARENCY 11.2

Logic of the analysis of variance.

Null Hypothesis: All groups are randomly drawn
                   from identical populations

Research Hypothesis: Groups are drawn from population
                      with different means

IF null hypothesis is true, two equally good ways to
   estimate population variance:

1. WITHIN-GROUPS ESTIMATE: Based on the variation of
                             the scores within each group

2. BETWEEN-GROUPS ESTIMATE: Based on the variation of
                             the means of the groups

	VARIATION WITHIN POPULATIONS	VARIATION BETWEEN POPULATIONS

Null Hypothesis True

Within-group estimate reflects	X	
Between-group estimate reflects	X	

Research Hypothesis True

Within-group estimate reflects	X	
Between-group estimate reflects	X	X

Therefore:

If null hypothesis true: Two estimates are equal
                          Ratio of Between/Within = About 1

If research hyopth true: Between group estimate is larger
                          Ratio of Between/Within > 1

Ratio of Between/Within called F ratio

Aron/Aron
STATISTICS FOR PSYCHOLOGY

© 1994 by Prentice-Hall, Inc.
A Paramount Communications Company
Englewood Cliffs, New Jersey 07632

TRANSPARENCY 11.3

Main analysis of variance computations with equal $\underline{N}$s.

Within-Group Population Variance Estimate:

1. Estimate variance of each population: $\underline{S}^2 = \underline{SS}/\underline{df}$

2. S_W^2 or $\underline{MS}_W = \dfrac{\underline{S}_1^2 + \underline{S}_2^2 + \ldots + \underline{S}^2{}_{Last}}{\underline{N}_G}$

Between-Group Population Variance Estimate:

1. $\underline{S}_M^2 = (\underline{M} - \underline{GM})^2 / (\underline{N}_G - 1)$ (same as $\underline{SS}/\underline{df}$)

2. $\underline{S}_B^2$ or $\underline{MSB} = \underline{S}_M^2 \times \underline{n}$

F Ratio

$\underline{F} = \underline{S}_B^2 / \underline{S}_W^2$ or $\underline{MS}_B / \underline{MS}_W$

Aron/Aron
STATISTICS FOR PSYCHOLOGY

TRANSPARENCY 11.4

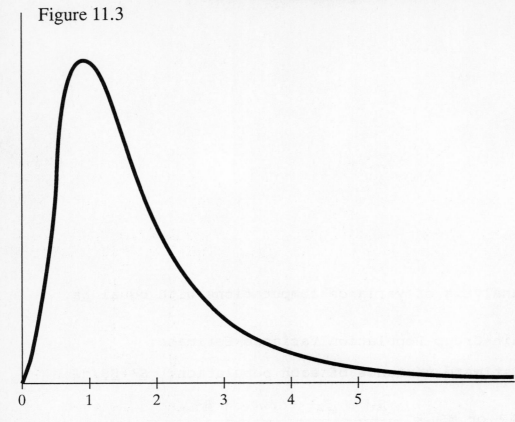

Figure 11.3

TABLE 11-4
Cutoff Scores for the *F* Distribution (Portion)

Denominator Degrees of Freedom	Significance Level	Numerator Degrees of Freedom					
		1	*2*	*3*	*4*	*5*	*6*
10	.01	10.05	7.56	6.55	6.00	5.64	5.39
	.05	4.97	4.10	3.71	3.48	3.33	3.22
	.10	3.29	2.93	2.73	2.61	2.52	2.46
11	.01	9.65	7.21	6.22	5.67	5.32	5.07
	.05	4.85	3.98	3.59	3.36	3.20	3.10
	.10	3.23	2.86	2.66	2.54	2.45	2.39
12	.01	9.33	6.93	5.95	5.41	5.07	4.82
	.05	4.75	3.89	3.49	3.26	3.11	3.00
	.10	3.18	2.81	2.61	2.48	2.40	2.33
13	.01	9.07	6.70	5.74	5.21	4.86	4.62
	.05	4.67	3.81	3.41	3.18	3.03	2.92
	.10	3.14	2.76	2.56	2.43	2.35	2.28

Aron/Aron
STATISTICS FOR PSYCHOLOGY

© 1994 by Prentice-Hall, Inc.
A Paramount Communications Company
Englewood Cliffs, New Jersey 07632

TRANSPARENCY 11.5

Criminal-record study. (From text.)

Computations:

$\underline{df}_B = \underline{N}_G - 1 = 3-1 = 2$

$\underline{df}_W = \underline{df}_1 + \underline{df}_2 + \ldots + \underline{df}_{Last} = 4 + 4 + 4 = 12$

$\underline{F}$ needed for significance at .05 level with $\underline{df}=2,12 = 3.89$.

Criminal-Record Group			Clean-Record Group			No Information Group		
Rating	Dev	Dev²	Rating	Dev	Dev²	Rating	Dev	Dev²
10	2	4	5	1	1	4	-1	1
7	-1	1	1	-3	9	6	1	1
5	-3	9	3	-1	1	9	4	16
10	2	4	7	3	9	3	-2	4
8	0	0	4	0	0	3	-2	4
40		18	20		20	25		26

$\underline{M} = 40/5 = 8$ $\underline{M} = 20/5 = 4$ $\underline{M} = 25/5 = 5$

$\underline{S}^2 = 18/4 = 4.5$ $\underline{S}^2 = 20/4 = 5.0$ $\underline{S}^2 = 26/4 = 6.5$

$$\underline{S}_W^2 \text{ or } \underline{MS}_W = \frac{\underline{S}_1^2 + \underline{S}_2^2 + \ldots + \underline{S}^2_{Last}}{\underline{N}_G} = \frac{4.5+5.0+6.5}{3} = \frac{16.3}{3} = 5.33$$

Sample means	Deviations from	Squared Deviations
M	Grand Mean (M-GM)	from Grand Mean (M-GM)²
4	-1.67	2.79
8	2.33	5.43
5	- .67	.45
17 ($\underline{GM} = \Sigma\underline{M}/\underline{N}_G = 17/3 = 5.67$)		8.67

$\underline{S}_M^2 = \Sigma(\underline{M}-\underline{GM})/\underline{df}_B = 8.67/2 = 4.34$.

$\underline{S}_B^2$ or $\underline{MS}_B = (\underline{S}_M^2)(\underline{n}) = (4.34)(5) = 21.7$

$\underline{F} = \underline{S}_B^2/\underline{S}_W^2$ (or $\underline{MS}_B/\underline{MS}_W$) $= 21.7/5.33 = 4.07$

Aron/Aron
STATISTICS FOR PSYCHOLOGY

TRANSPARENCY 11.6

Criminal-record study. (From text.)

Steps of hypothesis testing:

1. Reframe into a research hypothesis and a null hypothesis about populations.

   Population 1: Jurors told defendant has a criminal record.
   Population 2: Jurors told defendant has a clean record.
   Population 3: Jurors given no information on defendants' record.

   Research hypothesis: The three populations have different means.
   Null hypothesis: The three populations have the same mean.

2. Determine the characteristics of the comparison distribution.

   F distribution with 2 and 12 degrees of freedom.

3. Determine the cutoff sample score on the comparison distribution at which the null hypothesis should be rejected.

   5% level, $F(2,12)$ needed = 3.89.

4. Determine score of your sample on comparison distribution.

   Within-group population variance estimate =   5.33.
   Between-group population variance estimate = 21.70.
   F ratio = 21.7/5.33 = 4.07

5. Compare scores obtained in Steps 3 and 4 to decide whether to reject the null hypothesis.

   F on 4 (4.08) is more extreme than cutoff F on 3 (3.89).

   Therefore, reject the null hypothesis; the research hypothesis is supported (information available to subjects about criminal record makes a difference in ratings of guilt).

Aron/Aron
STATISTICS FOR PSYCHOLOGY

© 1994 by Prentice-Hall, Inc.
A Paramount Communications Company
Englewood Cliffs, New Jersey 07632

TRANSPARENCY 11.7

Responses of subjects of three attachment styles to
"Are you prone to fears?" (From class questionnaire.)
(Note: 18 randomly selected from each, to create equal Ns.)

Data and computations:

	Attachment Style		
	Secure	Avoidant	Anxious/Ambivalent
M	2.78	3.94	4.39
S^2	2.42	1.47	2.72
n	18	18	18

$df_B = N_G - 1 = 3-1 = 2$
$df_W = df_1 + df_2 + . . . + df_{Last} = 17 + 17 + 17 = 51$
F needed for significance at .05 level with $df=2,51 = 5.06$.

	M	Deviation	Deviation^2
Secure	2.78	- .92	.85
Avoidant	3.94	.24	.06
Anxious/Ambiv	4.39	.69	.48
Σ	11.11		$\Sigma(M-GM)^2 = 1.39$
GM	3.70	$S_M^2 = \Sigma(M-GM)/df_B = .70$	

$$S_B^2 = (S_M^2)(n) = (.70)(18) = 12.6$$

$$S_W^2 = \frac{S_1^2 + S_2^2 + . . . + S^2_{Last}}{N_G} = \frac{2.42+1.47+2.72}{3} = \frac{6.61}{3} = 2.20$$

$$F = S_B^2/S_W^2 \text{ (or } MS_B/MS_W) = 12.6/2.2 = 5.73.$$

Aron/Aron
STATISTICS FOR PSYCHOLOGY

© 1994 by Prentice-Hall, Inc.
A Paramount Communications Company
Englewood Cliffs, New Jersey 07632

TRANSPARENCY 11.8

Responses of subjects of three attachment styles to
"Are you prone to fears?" (From class questionnaire.)
(Note: 18 randomly selected from each, to create equal Ns.)

Steps of hypothesis testing:

1. Reframe into a research hypothesis and a null hypothesis about
   populations.

   Population 1:   Secure psychology students.
   Population 2:   Avoidant psychology students.
   Population 3:   Anxious/Ambivalent psychology students.

   Research hypothesis: The three population means differ.
   Null hypothesis: The three populations have the same mean.

2. Determine the characteristics of the comparison distribution.

   F distribution with 2 and 51 degrees of freedom.

3. Determine the cutoff sample score on the comparison
   distribution at which the null hypothesis should be rejected.

   1% level, $F(2,51)$ needed = 5.06.

4. Determine score of your sample on comparison distribution.

   Within-group population variance estimate =   2.2.
   Between-group population variance estimate = 12.6.
   F ratio = 12.6/2.2 = 5.73.

5. Compare scores obtained in Steps 3 and 4 to decide whether to
   reject the null hypothesis.

   F on 4 (5.73) is more extreme than cutoff F on 3 (5.06).

   Therefore, reject the null hypothesis; the research hypothesis is
   supported--those of different attachment styles appear to differ on
   reported proneness to fears.

Aron/Aron
STATISTICS FOR PSYCHOLOGY

© 1994 by Prentice-Hall, Inc.
A Paramount Communications Company
Englewood Cliffs, New Jersey 07632

TRANSPARENCY 11.9

Willingness to volunteer under different social conditions.
(Fictional data.)

Data and computations

Circumstances When Request Made

	Alone	Friend Present	Asked in Class
	4	8	8
	5	7	9
	3	4	6
	3	6	6
	6	5	9
	3	6	4
M	4	6	7
S^2	1.6	2.0	4.0
n	6	6	6

$\underline{df}_B = \underline{N}_G - 1 = 3-1 = 2$

$\underline{df}_W = \underline{df}_1 + \underline{df}_2 + . . . + \underline{df}_{Last} = 5 + 5 + 5 = 15$

$\underline{F}$ needed for significance at .05 level with $\underline{df}$=2,15 = 3.68.

	m	Deviation	Deviation^2
Alone	4	-1.67	2.79
Friend	6	.33	.11
Classroom	7	1.33	1.77
Σ	17		$\Sigma(\underline{M}-\underline{GM})^2 = 4.67$
$\underline{GM}$	5.67	$\underline{S}_M^2=\Sigma(\underline{M}-\underline{GM})/\underline{df}_B=4.67/2=$	2.33

$$\underline{S}_B^2 = (\underline{S}_M^2)(\underline{n}) = (2.33)(6) = 14$$

$$\underline{S}_W^2 = \frac{\underline{S}_1^2 + \underline{S}_2^2 + . . . + \underline{S}^2_{Last}}{\underline{N}_G} = \frac{1.6 + 2 + 4}{3} = \frac{7.6}{3} = 2.53$$

$$\underline{F} = \underline{S}_B^2/\underline{S}_W^2 \text{ (or } \underline{MS}_B/\underline{MS}_W) = 14/2.53 = 5.53.$$

Aron/Aron
STATISTICS FOR PSYCHOLOGY

© 1994 by Prentice-Hall, Inc.
A Paramount Communications Company
Englewood Cliffs, New Jersey 07632

TRANSPARENCY 11.10

Willingness to volunteer under different social conditions.
(Fictional data.)

Steps of hypothesis testing:

1. Reframe into a research hypothesis and a null hypothesis about
   populations.

   Population 1: People asked to volunteer when alone.
   Population 2: People asked to volunteer when with their best friend.
   Population 3: People asked to volunteer in a classroom.

   Research hypothesis: The three population means differ.
   Null hypothesis: The three populations have the same mean.

2. Determine the characteristics of the comparison distribution.

   F distribution with 2 and 15 degrees of freedom.

3. Determine the cutoff sample score on the comparison
   distribution at which the null hypothesis should be rejected.

   5% level, $F(2,15)$ needed = 3.68

4. Determine score of your sample on comparison distribution.

   Within-group population variance estimate =   2.53
   Between-group population variance estimate = 14.00.
   F ratio = 14/2.53 = 5.53.

5. Compare scores obtained in Steps 3 and 4 to decide whether to
   reject the null hypothesis.

   F on 4 (5.53) is more extreme than cutoff F on 3 (3.68).

   Therefore, reject the null hypothesis; the research hypothesis is
   supported (willingness to volunteer appears to be different under
   the different social conditions tested).

Aron/Aron
STATISTICS FOR PSYCHOLOGY

TRANSPARENCY 11.11

TABLE 11-6
Steps for Conducting an Analysis of Variance (When Sample Sizes Are Equal)

1. Reframe the question into a research hypothesis and a null hypothesis about the populations.

2. Determine the characteristics of the comparison distribution.

    a. The comparison distribution will be an F distribution.
    b. The numerator degrees of freedom is the number of groups minus 1: $df_B = N_G - 1$.
    c. The denominator degrees of freedom is the sum of the degrees of freedom in each group (the number of cases in the group minus 1): $df_W = df_1 + df_2 + \ldots + df_{Last}$.

3. Determine the cutoff sample score on the comparison distribution at which the null hypothesis should be rejected.

    a. Determine the desired significance level.
    b. Look up the appropriate cutoff in an F table, using the degrees of freedom calculated in Step 2.

4. Determine the score of the sample on the comparison distribution. (This will be an F ratio.)

    a. Compute the between-group population variance estimate (S_B^2 or MS_b).

        i. Compute the means of each group.
        ii. Compute a variance estimate based on the means of the groups: $S_M^2 = \Sigma(M - GM)^2/df_B$.
        iii. Convert this estimate of the variance of a distribution of means to an estimate of the variance of a population of individual scores by multiplying by the number of cases in each group: S_B^2 or $MS_B = (S_M^2)(n)$.

    b. Compute the within-group population variance estimate S_W^2 or MS_W.

        i. Compute population variance estimates based on each group's scores: For each group, $S^2 = SS/df$.
        ii. Average these variance estimates: $S_W^2 = (S_1^2 + S_2^2 + \ldots + S_{Last}^2)/N_G$.

    c. Compute the F ratio: $F = S_B^2/S_W^2$ (or $F = MS_B/MS_W$).

5. Compare the scores obtained in Steps 3 and 4 to decide whether to reject the null hypothesis.

Aron/Aron
STATISTICS FOR PSYCHOLOGY

© 1994 by Prentice-Hall, Inc.
A Paramount Communications Company
Englewood Cliffs, New Jersey 07632

Chapter 12
The Structural Model in the Analysis of Variance

Instructor's Summary of Chapter

Structural model. In the structural model approach to ANOVA, the deviation of each score from the grand mean is divided into two parts: (a) the score's difference from its group's mean, and (b) its group's mean's difference from the grand mean. These deviations, when squared, summed, and divided by the appropriate degrees of freedom, yield the same within- and between-group estimates as were obtained using the method described in Chapter 11. However, the structural model is more flexible and can be applied to situations such as unequal numbers of subjects.

Analysis of variance table. Computations using the structural model are usually summarized in an ANOVA table, with columns for source of variation (between and within), sums of squared deviations (SS), degrees of freedom (df), population-variance estimates (MS—which equals SS/df), and F (which equals MS_B/MS_W).

Assumptions are unchanged. But ANOVA is less robust with unequal sample sizes.

Multiple comparisons. An ANOVA is usually followed up by multiple comparisons, either planned or post-hoc, that examine differences between specific pairs or subgroups of means. Such comparisons have to protect against the possibility of getting some significant results just by chance since a great many comparisons can be made.

Proportion of variance accounted for (R^2) is a readily calculable measure of ANOVA effect size: $R^2 = SS_B / SS_T$.

Controversy: Diffuse overall F versus targeted, planned comparisons. Rosenthal and Rubin recommend that instead of using ANOVA to make "diffuse," overall comparisons among several means, researchers should plan in advance to conduct specific planned comparisons, targeted directly to their theoretical questions. Linear contrasts are often made for this purpose.

How multiple comparisons are reported in research articles. Planned comparisons are reported as such. Post-hoc comparisons are often reported in tables of means, such that means with different subscripted letters are not significantly different.

Box 11.1. Sir Ronald Fisher, Caustic Genius of Statistics. Summarizes the life and work of Fisher, inventor of ANOVA.

Box 12.1. Analysis of Variance as a Way of Thinking about the World. Summarizes Kelley's view that cognitions of characteristics of social situations are identified by distinguishing information into that which is figural (analogous to the between-group variance in ANOVA) versus that which is background (analogous to the within-group variance in ANOVA).

Lecture 12.1:   The Structural Model Approach to the Analysis of Variance

Materials

Lecture outline

Transparencies 11.1, 11.2, 11.5, 11.6, 11.11, and 12.1 through 12.10
   (If using transparencies based on your class's questionnaires, replace 12.6 and 12.7 with 12.6R and 12.7R.)

Outline for Blackboard

I.   Review
II.   Limits of Method Just Reviewed
III.   Logic of Structural Model Approach
IV.   Computation and the ANOVA Table
V.   Additional Computational Examples
VI.   Assumptions for ANOVA with Unequal Sample Sizes
VII. Review this Class

Instructor's Lecture Outline

I.   Review
   A. Idea of descriptive and inferential statistics.
   B. Show TRANSPARENCY 11.1 (central logic of hypothesis testing process) and discuss.
   C. Show TRANSPARENCY 11.2 (logic of ANOVA) and discuss.
   D. Show TRANSPARENCY 11.11 (ANOVA procedure, Chapter 11 method) and discuss.

II.   Limits of Method Just Reviewed
   A. Does not handle unequal sample sizes easily.
   B. Is not traditional way ANOVA has been done using hand calculations.

III.   Logic of Structural Model Approach
   A. Show TRANSPARENCY 12.1 (text figure of division of deviation from grand mean into two parts) and discuss how you can divide each score's deviation from grand mean into two parts:
      1. Score's deviation from its group's mean.
      2. Score's group's mean's deviation from grand mean.
   B. Show TRANSPARENCY 12.2 (formulas) and discuss:
      1. $\Sigma(X - GM)^2 = \Sigma(X - M)^2 + \Sigma(M - GM)^2$ or $SS_T = SS_W + SS_B$.
      2. Dividing each SS by its appropriate df gives population variance estimates: $MS_B = SS_B/df_B$; $MS_W = SS_W/df_W$.
   C. Show TRANSPARENCY 12.3 (division of deviation showing corresponding variance estimates, from text) and discuss.
      1. Population variance estimates correspond to the two types of deviation scores.
      2. If the null hypothesis is true, the division of the overall deviation into two parts should be random, making population estimates producing an F ratio of about 1.
      3. If the research hypothesis is true, the deviations of the group means from the grand mean should be greater than the deviations of the scores from their group's mean, making population estimates producing an F ratio greater than 1.

IV. Computation and the ANOVA Table

A. Show TRANSPARENCIES 12.4 (table of computations for criminal-record study from text) and discuss each step up to ANOVA table.

B. Show TRANSPARENCY 11.5 (criminal-record ANOVA computations using Chapter 11 method) and note how results come out the same.

C. Show TRANSPARENCY 12.4 again, bottom, and discuss ANOVA table.

D. Show TRANSPARENCY 12.5 (general ANOVA table from text) and discuss each step.

E. Show TRANSPARENCY 11.6 (criminal-record study, hypothesis testing steps—which are identical under the structural model approach to what was found in Chapter 11)—and review each step, noting how Step 4 is the same, even though computed through a different method.

V. Additional Computational Examples

A. Show TRANSPARENCIES 12.6 and 12.7 or 12.6R and 12.7R (prone-to-fears and attachment-style example using structural model approach, from class questionnaire) and discuss.

   1. Note that in the last class we did same analysis but selecting only a subset of cases in order to have equal numbers of subjects.

   2. Now we can use all subjects because structural model approach can handle unequal sample sizes.

   3. Discuss each step.

B. Show TRANSPARENCIES 12.8 and 12.9 (identification of emotion with different modes of information available) and discuss.

VI. Assumptions for ANOVA with Unequal Sample Sizes

A. Same as with equal sample sizes.

B. Less robust to violations of equal population variances.

VII. Review this Class

A. Use blackboard outline.

B. Show TRANSPARENCY 12.3 (division of deviation) and discuss.

C. Show TRANSPARENCY 12.10 (summary of structural model approach to ANOVA) and discuss.

D. Show TRANSPARENCY 12.5 (general ANOVA table) and discuss.

Lecture 12.2: Proportion of Variance Accounted for and Multiple Comparisons

Materials

Lecture outline
Transparencies 12.10 through 12.13

Outline for Blackboard

I. **Review**
II. **Proportion of Variance Accounted for (R^2) as ANOVA Effect Size**
III. **Multiple Comparisons**
IV. **Review this Class**

Instructor's Lecture Outline

I. **Review**
   A. Idea of descriptive and inferential statistics.
   B. Show TRANSPARENCY 12.10 (structural model approach to ANOVA) and discuss.

II. **Proportion of Variance Accounted for (R^2) as ANOVA Effect Size**
   A. Alternate effect-size measure to f (more familiar because same as R^2 in multiple regression).
   B. Meaning.
      1. Proportion of overall variation from grand mean accounted for by variation of groups' means from the grand mean.
      2. Show top half of TRANSPARENCY 12.11 (formula $R^2 = SS_B/SS_T$ and worked out example) and discuss.
   C. Minimum 0, maximum 1. Square root (R) is a correlation.
   D. Can be determined directly from F and degrees of freedom given in a research report—show bottom half of TRANSPARENCY 12.11 (formula and worked out example) and discuss.
   E. Show TRANSPARENCY 12.12 (effect size conventions for R^2 and relation to f and R) and discuss.

III. **Multiple Comparisons**
   A. The overall ANOVA does not test which population means are different from which.
   B. Multiple comparisons are procedures for significance testing of comparisons of specific population means.
   C. A major problem is keeping overall probability of falsely rejecting any null hypothesis at an acceptable level while testing many comparisons.
   D. Planned comparisons.
      1. A subset of all possible comparisons that the researcher specifies in advance of the study.
      2. The Bonferroni procedure: Set a more stringent significance level for each comparison, keeping overall alpha at an acceptable level.
      3. Linear contrasts test a more complex predicted relationships, such as those specifying the order of effect expected among several groups.
      4. Sometimes planned comparisons are more important than overall ANOVA's "diffuse" F test.

E. Post-hoc comparisons.
   1. All possible comparisons among groups to explore possible differences.
   2. Various procedures attempt to keep overall chance of falsely rejecting the null hypothesis low while maintaining adequate power.
   3. A common procedure with post-hoc comparisons is to report means in tables with subscripted letters such that those having the same letter are not significantly different from each other—show TRANSPARENCY 12.13 (example table with a post-hoc comparison) and discuss.

IV. **Review this Class:** Use blackboard outline.

Figure 12.1

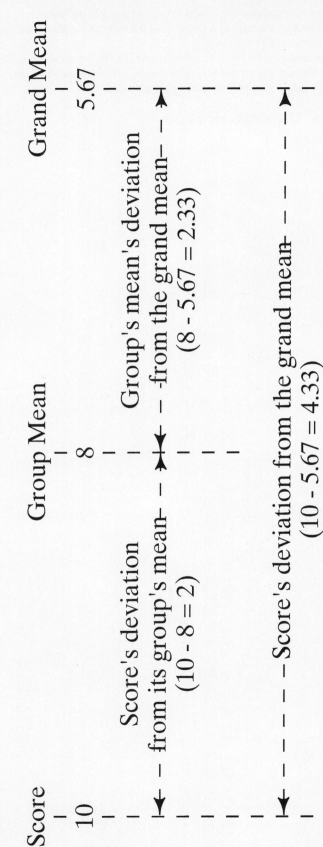

Aron/Aron
STATISTICS FOR PSYCHOLOGY

TRANSPARENCY 12.2

Formulas for structural model approach.

Addition of Sums of Squares:

$$\Sigma(\underline{X}-\underline{GM})^2 = \Sigma(\underline{X}-\underline{M})^2 + \Sigma(M-\underline{GM})^2$$

or $\underline{SS}_T = \underline{SS}_W + \underline{SS}_B$

Between-Group Population Variance Estimate:

$$\underline{S}_B{}^2 = \Sigma(\underline{M} - \underline{GM})^2/\underline{df}_B$$

or $\underline{MS}_B = \underline{SS}_B/\underline{df}_B$

With-Group Population Variance Estimate:

$$\underline{S}_W{}^2 = \Sigma(\underline{X}-\underline{M})^2/\underline{df}_W$$

or $\underline{MS}_W = \underline{SS}_W/\underline{df}_W$

Aron/Aron
STATISTICS FOR PSYCHOLOGY

TRANSPARENCY 12.3

Figure 12.2

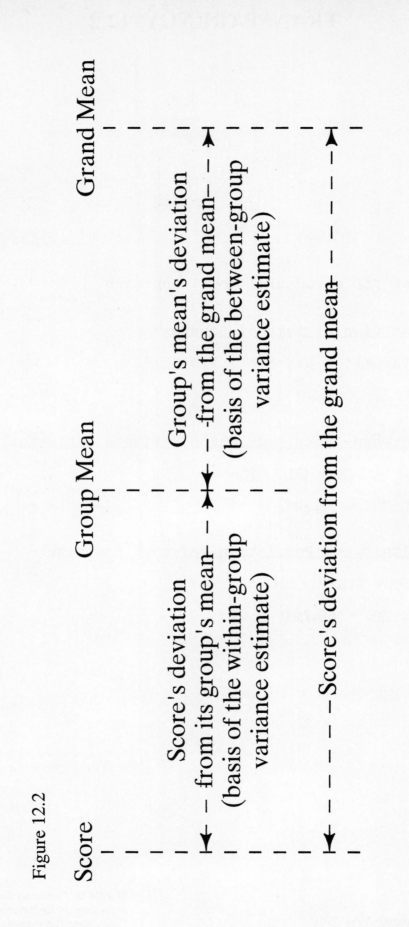

Score

Group Mean

Grand Mean

Score's deviation from its group's mean (basis of the within-group variance estimate)

Group's mean's deviation from the grand mean (basis of the between-group variance estimate)

Score's deviation from the grand mean

Aron/Aron
STATISTICS FOR PSYCHOLOGY

TABLE 12-1
Analysis of Variance for the Criminal Record Study Using the Structural Model Approach (Fictional Data)

Criminal Record Group

X	X − GM		X − M		M − GM	
	Deviation	Squared deviation	Deviation	Squared deviation	Deviation	Squared deviation
10	4.33	18.74	2	4	2.33	5.43
7	1.33	1.77	−1	1	2.33	5.43
5	− .67	.45	−3	9	2.33	5.43
10	4.33	18.74	2	4	2.33	5.43
8	2.33	5.43	0	0	2.33	5.43
$\overline{40}$		$\overline{45.13}$		$\overline{18}$		$\overline{27.14}$

$M = 40/5 = 8$

Clean Record Group

X	X − GM		X − M		M − GM	
	Deviation	Squared deviation	Deviation	Squared deviation	Deviation	Squared deviation
5	− .67	.45	1	1	−1.67	2.79
1	−4.67	21.81	−3	9	−1.67	2.79
3	−2.67	7.13	−1	1	−1.67	2.79
7	1.33	1.77	3	9	−1.67	2.79
4	−1.67	2.79	0	0	−1.67	2.79
$\overline{20}$		$\overline{33.95}$		$\overline{20}$		$\overline{13.95}$

$M = 20/5 = 4$

No Information Group

X	X − GM		X − M		M − GM	
	Deviation	Squared deviation	Deviation	Squared deviation	Deviation	Squared deviation
4	−1.67	2.79	−1	1	−.67	.45
6	.33	.11	1	1	−.67	.45
9	3.33	11.09	4	16	−.67	.45
3	−2.67	7.13	−2	4	−.67	.45
3	−2.67	7.13	−2	4	−.67	.45
$\overline{25}$		$\overline{28.25}$		$\overline{26}$		$\overline{2.25}$

$M = 25/5 = 5$

Sums of squared deviations:

$\Sigma(X - GM)^2$ or $SS_T = 45.13 + 33.95 + 28.25 = 107.33$

$\Sigma(X - M)^2$ or $SS_W = 18 + 20 + 26 = 64$

$\Sigma(M - GM)^2$ or $SS_B = 27.14 + 13.95 + 2.25 = 43.34$

Check $(SS_T = SS_W + SS_B)$: $SS_T = 107.33$; $SS_W + SS_B = 64 + 43.34 = 107.34$
   (slight difference due to rounding error)

Degrees of freedom:

$df_T = N - 1 = 15 - 1 = 14$

$df_W = df_1 + df_2 + \ldots + df_{Last} = (5 - 1) + (5 - 1) + (5 - 1) = 4 + 4 + 4 = 12$

$df_B = N_G - 1 = 3 - 1 = 2$

Check $(df_T = df_W + df_B)$: $14 = 12 + 2$

Population variance estimates:

S_W^2 or $MS_W = SS_W/df_W = 64/12 = 5.33$

S_B^2 or $MS_B = SS_B/df_B = 43.34/2 = 21.67$

F ratio: $F = S_B^2/S_W^2$ or $MS_B/MS_W = 21.67/5.33 = 4.07$

TABLE 12-5
Steps, Symbols, and Formulas for Computing an Analysis of Variance Using the Structural Model–Based Method (Equal or Unequal Sample Sizes)

Symbols Corresponding to Each Part of an Analysis of Variance Table

Source	SS	df	MS	F
Between	SS_B	df_B	MS_B (or S_B^2)	F
Within	SS_W	df_W	MS_W (or S_W^2)	
Total	SS_T	df_T		

Formulas for Each Part of an Analysis of Variance Table

Source	SS	df	MS	F
Between	$\Sigma(M - GM)^2$	$N_G - 1$	SS_B/df_B	MS_B/MS_W
Within	$\Sigma(X - M)^2$	$df_1 + df_2 + \ldots + df_{Last}$	SS_W/df_W	
Total	$\Sigma(X - GM)^2$	$N - 1$		

Definitions of Basic Symbols

Σ = sum of the appropriate numbers for all cases
M = mean of a score's group
GM = grand mean
N_G = number of groups
X = each score
N = total number of cases in the study

Aron/Aron
STATISTICS FOR PSYCHOLOGY

© 1994 by Prentice-Hall, Inc.
A Paramount Communications Company
Englewood Cliffs, New Jersey 07632

TRANSPARENCY 12.6

Responses of subjects of three attachment styles to
"Are you prone to fears?" (From class questionnaire.)

Data and computations:

$\underline{df}_B = \underline{N}_G - 1 = 3 - 1 = 2.$
$\underline{df}_W = \underline{df}_1 + \underline{df}_2 + \ldots + \underline{df}_{Last} = (55-1) + (25-1) + (17-1) = 94.$
$\underline{F}$ needed for significance at .05 level with $\underline{df} = 2,94 = 3.10.$

Secure	Avoidant	Anxious/Ambivalent
Example Score: $\underline{X}=2$	Example Score: $\underline{X}=5$	Example Score: $\underline{X} = 6$
$(\underline{X}-\underline{GM})= (2-3.67)= -1.67$	$(\underline{X}-\underline{GM})=(5-3.67)=1.33$	$(\underline{X}-\underline{GM})= (6-3.67)= 2.33$
$(\underline{X}-\underline{GM})^2 =-1.67^2 = 2.79$	$(\underline{X}-\underline{GM})^2 = 1.33^2 =1.77$	$(\underline{X}-\underline{GM})^2 = 2.33^2 = 5.43$
$(\underline{X}-\underline{M}) =(2-3.29)= -1.29$	$(\underline{X}-\underline{M}) =(5-4) =1.00$	$(\underline{X}-\underline{M}) =(6-4.41)= 1.59$
$(\underline{X}-\underline{M})^2 =-1.29^2 = 1.66$	$(\underline{X}-\underline{M})^2 = 1 =1.00$	$(\underline{X}-\underline{M})^2 = 1.59^2 = 2.53$
$(\underline{M}-\underline{GM})=(3.29-3.67)=-.38$	$(\underline{M}-\underline{GM})=(4-3.67)=-.33$	$(\underline{M}-\underline{GM})=(4.41-3.67)=.74$
$(\underline{M}-\underline{GM})^2 = -.38^2 = .14$	$(\underline{M}-\underline{GM})^2= -.33^2 = .11$	$(\underline{M}-\underline{GM})^2 = .74^2 = .55$
$\underline{M} = 181/55 = 3.29$	$\underline{M} = 100/25 = 4$	$\underline{M} = 75/17 = 4.41$

$\underline{GM} = (181+100+75)/97 = 356/97 = 3.67$

Analysis of Variance Table

Source	SS	df	MS	F
Between	19.98	2	9.99	3.92
Within	239.46	94	2.55	
Total	259.44			

Conclusion: Reject null hypothesis.

Aron/Aron
STATISTICS FOR PSYCHOLOGY

© 1994 by Prentice-Hall, Inc.
A Paramount Communications Company
Englewood Cliffs, New Jersey 07632

TRANSPARENCY 12.7

Responses of subjects of three attachment styles to
"Are you prone to fears?" (From class questionnaire.)

Steps of hypothesis testing:

1. Reframe into a research hypothesis and a null hypothesis about
   populations.

   Population 1: Individuals with a secure attachment style.
   Population 2: Individuals with an avoidant attachment style.
   Population 3: Individuals with an anxious/ambivalent attachment
                 style.

   Research hypothesis: The three populations have different means.
   Null hypothesis: The three populations have the same mean.

2. Determine the characteristics of the comparison distribution.

   F distribution with 2 and 94 degrees of freedom.

3. Determine the cutoff sample score on the comparison
   distribution at which the null hypothesis should be rejected.

   5% level, $F(2,94)$ needed = 3.10.

4. Determine score of your sample on comparison distribution.

   Within-group population variance estimate =   2.55.
   Between-group population variance estimate =   9.99.
   F ratio = 9.99/2.55 = 3.92.

5. Compare scores obtained in Steps 3 and 4 to decide whether to
   reject the null hypothesis.

   F on 4 (3.92) is more extreme than cutoff F on 3 (3.10).

   Therefore, reject the null hypothesis; the research hypothesis is
   supported (attachment style seems to be related to being prone to
   fears).

Aron/Aron
STATISTICS FOR PSYCHOLOGY

© 1994 by Prentice-Hall, Inc.
A Paramount Communications Company
Englewood Cliffs, New Jersey 07632

TRANSPARENCY 12.8

Accuracy of identifying emotion based on observing a speaker only visually (video without sound), verbally (text of what the person said without sound or video), and paraverbally (voice, distorted so that meaning cannot be detected but intonation can be). (Fictional data.)

Data and calculations:

$\underline{df}_B = \underline{N}_G - 1 = 3 - 1 = 2.$
$\underline{df}_W = \underline{df}_1 + \underline{df}_2 + . . . + \underline{df}_{Last} = 3 + 3 + 3 = 9.$
$\underline{F}$ needed for significance at .05 level with $\underline{df} = 2,9 = 4.74.$

Verbal Only							Visual Only							Paraverbal Only					
X	X$-$GM		X $-$ M		M $-$GM		X	X $-$GM		X $-$ M		M $-$GM		X	X $-$GM		X $-$ M		M $-$GM
	Dv	Dv²	Dv	Dv²	Dv	Dv²		Dv	Dv²	Dv	Dv²	Dv	Dv²		Dv	Dv²	Dv	Dv²	Dv	Dv²
1	-3	9	0	0	-3	9	7	3	9	1	1	2	4	4	0	0	-1	1	1	1
0	-4	16	-1	1	-3	9	4	0	0	-2	4	2	4	5	1	1	0	0	1	1
2	-2	4	1	1	-3	9	7	3	9	1	1	2	4	6	2	4	1	1	1	1
1	-3	9	0	0	-3	9	6	2	4	0	0	2	4	5	1	1	0	0	1	1
4		38		2		36	24		22		6		16	20		6		2		4

M = 4/4 = 1   M = 24/4 = 6   M = 20/4 = 5

GM = (4+24+20) / 12 = 48/12 = 4

$\Sigma(\underline{X}-\underline{GM})^2 = \Sigma(\underline{X}-\underline{M})^2 + \Sigma(M-\underline{GM})^2$

$\underline{SS}_T = \Sigma(\underline{X}-\underline{GM})^2 = 38 + 22 + 6 = 66$

$\underline{SS}_B = \Sigma(M-\underline{GM})^2 = 36 + 16 + 4 = 56$

$\underline{SS}_W = \Sigma(\underline{X}-\underline{M})^2 = 2 + 6 + 2 = 10$

Check: $\underline{SS}_T = \underline{SS}_W + \underline{SS}_B$; 66 = 56 + 10

Analysis of Variance Table

Source	SS	df	MS	F
Between	56	2	28.00	25.23
Within	10	9	1.11	
Total	66			

Conclusion: Reject null hypothesis.

Aron/Aron
STATISTICS FOR PSYCHOLOGY

© 1994 by Prentice-Hall, Inc.
A Paramount Communications Company
Englewood Cliffs, New Jersey 07632

TRANSPARENCY 12.9

Accuracy of identifying emotion based on observing a speaker only visually (video without sound), verbally (text of what the person said without sound or video), and paraverbally (voice, distorted so that meaning cannot be detected but intonation can be). (Fictional data.)

Steps of hypothesis testing:

1. Reframe into a research hypothesis and a null hypothesis about populations.

   Population 1: People exposed only to verbal content.
   Population 2: People exposed only to visual content.
   Population 3: People exposed only to paraverbal content.

   Research hypothesis: The three populations have different means.
   Null hypothesis: The three populations have the same mean.

2. Determine the characteristics of the comparison distribution.

   F distribution with 2 and 9 degrees of freedom.

3. Determine the cutoff sample score on the comparison distribution at which the null hypothesis should be rejected.

   5% level, $F(2,9)$ needed = 4.74.

4. Determine score of your sample on comparison distribution.

   Within-group population variance estimate =   1.11.
   Between-group population variance estimate = 28.00.
   F ratio = 28/1.11 = 25.22

5. Compare scores obtained in Steps 3 and 4 to decide whether to reject the null hypothesis.

   F on 4 (25.22) is more extreme than cutoff F on 3 (4.74).

   Therefore, reject the null hypothesis; the research hypothesis is supported (mode of information available—verbal, visual, or paraverbal—seems to affect accuracy of identifying emotion expressed).

Aron/Aron
STATISTICS FOR PSYCHOLOGY

© 1994 by Prentice-Hall, Inc.
A Paramount Communications Company
Englewood Cliffs, New Jersey 07632

TABLE 12-5
Steps, Symbols, and Formulas for Computing an Analysis of Variance Using the Structural Model–Based Method (Equal or Unequal Sample Sizes)

Hypothesis-Testing Steps

1. Reframe the question into a research hypothesis and a null hypothesis about the populations.
2. Determine the characteristics of the comparison distribution.
   a. The comparison distribution will be an F distribution.
   b. The numerator degrees of freedom is the number of groups minus 1: $df_B = N_G - 1$.
   c. The denominator degrees of freedom is the sum of the degrees of freedom in each group (the number of cases in the group minus 1): $df_W = df_1 + df_2 + \ldots + df_{Last}$.
   d. Check the accuracy of your computations by making sure that df_W and df_B sum to df_T (which is the total number of cases minus 1).
3. Determine the cutoff sample score on the comparison distribution at which the null hypothesis should be rejected.
   a. Determine the desired significance level.
   b. Look up the appropriate cutoff in an F table.
4. Determine the score of the sample on the comparison distribution. (This will be an F ratio.)
   a. Compute the mean of each group and the grand mean of all scores.
   b. Compute the following deviations for each score:
      i. Its deviation from the grand mean ($X - GM$).
      ii. Its deviation from its group's mean ($X - M$).
      iii. Its group's mean's deviation from the grand mean ($M - GM$).
   c. Square each of these deviation scores.
   d. Compute the sums of each of these three types of deviation scores (SS_T, SS_W, and SS_B).
   e. Check the accuracy of your computations by making sure that $SS_W + SS_B = SS_T$.
   f. Compute the between-group variance estimate: SS_B/df_B.
   g. Compute the within-group variance estimate: SS_W/df_W.
   h. Compute the F ratio: $F = S_B^2/S_W^2$.
5. Compare the scores in Steps 3 and 4 to decide whether to reject the null hypothesis.

Aron/Aron
STATISTICS FOR PSYCHOLOGY

TRANSPARENCY 12.11

Formulas for proportion of variance accounted for as effect size indicator in analysis of variance.

Standard Formula

$$R^2 = SS_B/SS_T$$

Criminal-Record Example:

$$R^2 = SS_B/SS_T = 43.44/107.33 = .40.$$

Formula for Computing R^2 Directly from F and df given in a research report:

$$R^2 = (F)(df_B) / ([F][df_B]+df_W)$$

Criminal Record Study Example:

$$R^2 = (F)(df_B) / ([F][df_B]+df_W)$$

$$= (4.07)(2) / ([4.07][2]+12)$$

$$= 8.14 / (8.14 + 12) = 8.14 / 20.14 = .40$$

Aron/Aron
STATISTICS FOR PSYCHOLOGY

TABLE 12-6
Cohen's Conventions for Effect Sizes in a One-Way Analysis of Variance

	Effect Size		
	Small	*Medium*	*Large*
f	.10	.25	.40
R	.10	.24	.37
R^2	.01	.06	.14

TABLE 12-7
Love Subscale Means for the Three Attachment Types (Newspaper Sample)

Scale Name	Avoidant	Anxious/ Ambivalent	Secure	$F(2, 571)$
Happiness	3.19_a	3.31_a	3.51_b	14.21***
Friendship	3.18_a	3.19_a	3.50_b	22.96***
Trust	3.11_a	3.13_a	3.43_b	16.21***
Fear of closeness	2.30_a	2.15_a	1.88_b	22.65***
Acceptance	2.86_a	3.03_b	3.01_b	4.66**
Emotional extremes	2.75_a	3.05_b	2.36_c	27.54***
Jealousy	2.57_a	2.88_b	2.17_c	43.91***
Obsessive preoccupation	3.01_a	3.29_b	3.01_a	9.47***
Sexual attraction	3.27_a	3.43_b	3.27_a	4.08*
Desire for union	2.81_a	3.25_b	2.69_a	22.67***
Desire for reciprocation	3.24_a	3.55_b	3.22_a	14.90***
Love at first sight	2.91_a	3.17_b	2.97_a	6.00**

Note. Within each row, means with different subscripts differ at the .05 level of significance according to a Scheffé test.

*$p < .05$.
**$p < .01$.
***$p < .001$.

From Hazan, C., & Shaver, P. (1987), tab. 3. Romantic love conceptualized as an attachment process. *Journal of Personality and Social Psychology, 52,* 511–524. Copyright, 1987, by the American Psychological Association. Reprinted by permission of the author.

Chapter 13
Factorial Analysis Of Variance

Instructor's Summary of Chapter

Logic of factorial designs and interaction effects. A factorial research design distributes subjects into every combination of levels of two or more independent variables permitting the simultaneous study of the influence of these variables. Such designs are efficient and provide the opportunity to examine interactions among the independent variables. Factorial designs are described according to the number of independent variables or ways ("two-way," "three-way") and by numbers of levels ("2 X 2," "3 X 4 X 3").

Interaction effects occur when the influence of one independent variable on the dependent variable differs according to the level of another independent variable.

Identifying main and interaction effects: (a) by inspection of means (main effects are indicated by differing marginal means, interactions by greater difference between cell means in one column or row than in another) or (b) graphically (main effects have different heights of lines or an average nonhorizontal slope of all lines and interaction effects are indicated when the lines are not parallel.

Computations for a two-way ANOVA follow the structural model approach. Deviations are computed for each score from its cell mean (for MS_W), between row means and the grand mean (for row main effect), between column means and the grand mean (for the column main effect), and the remainder of the deviation of the score from the grand mean (for the interaction effect). These deviations—squared, summed, and divided by their degrees of freedom—provide the respective population variance estimates for computing F ratios for testing the respective hypotheses. Each effect has its corresponding numerator population variance estimate, but all effects use the same MS_W as the denominator.

Effect size and power are computed separately for each main and interaction effect. The most useful indicator of effect size is R^2.

Extensions. Factorial ANOVA can be extended beyond two-way designs, and also to handle repeated-measures studies and unequal sample sizes in the cells.

Box 13.1. Personality and Situational Influences on Behavior: An Interaction Effect. Factorial ANOVA provides a model for thinking about relationships among variables and has influenced the structure of psychological theories. One example is that the controversy over the relative influence of personality traits and situations is sometimes resolved by seeing the relationship of the two as an interaction effect.

Lecture 13.1: The Logic and Language of Factorial Designs

Materials

Lecture outline

Transparencies 13.1 through 13.7

    (If using transparencies based on your class's questionnaires, replace 13.4 and 13.5 with 13.4R and 13.5R.)

Outline for Blackboard

I.   Review
II. Factorial Designs
III. Identifying Main and Interaction Effects from Tables of Cell Means
IV. Graphing Main and Interaction Effects
V. Additional Examples
VI. Review this Class

Instructor's Lecture Outline

I.   Review

    A. Idea of descriptive and inferential statistics.

    B. Review of types of situations addressed so far in course.

        1. One-group t test (including special case of t test for dependent means).

        2. Two-group t test—t test for independent means.

        3. More than two groups—analysis of variance.

        4. All of the above involved comparisons on one independent variable.

II. Factorial Designs: Two or more groups and two or more variables.

    A. Example: Burnkraut & Unnava (1989) examined the influence of various factors in changing attitudes, specifically:

        1. Strength of argument made.

        2. Whether subject was encouraged to think about the arguments in terms of self ("self-referencing").

    B. These researchers could have studied each separately, but instead did both together, randomly assigning subjects to one of four conditions.

    C. Show TRANSPARENCY 13.1 and discuss.

        1. Greater efficiency (two hypotheses tested with a single set of subjects).

        2. Permits testing of the interaction effect—joint effect over and above sum of two effects.

        3. Called a *factorial experiment*, in this case a *2 X 2 factorial experiment*.

        4. Indicate main and interaction effects.

III. Identifying Main and Interaction Effects from Tables of Cell Means: Show TRANSPARENCY 13.2 (tables of means from text) and discuss various examples to illustrate.

    A. Structure of table and computation of marginal means.

    B. Main effects recognized from differences in marginal means.

    C. Interaction effects indicated by variations in the pattern of differences across different rows (or columns).

IV. Graphing Main and Interaction Effects: Show TRANSPARENCY 13.3 (graphs of previous tables of means from text) and discuss various examples to illustrate.
   A. Construction of such graphs.
   B. Identifying interaction effects by nonparallel lines.
   C. Identifying one main effect by different average heights of lines.
   D. Identifying other main effect by average slope of lines being nonhorizontal.

V. Additional Examples
   A. Show TRANSPARENCY 13.4 or 13.4R (birth order X gender on crying easily and on falling in love when young, from class questionnaire) and discuss.
   B. Show TRANSPARENCY 13.5 OR 13.5R (birth order X attachment style on closeness to father and on closeness to mother, from class questionnaire) and discuss.
   C. Show TRANSPARENCIES 13.6 and 13.7 (various outcomes on mental health by nationality—French vs. U.S.—and by profession—comedians vs. actors) and discuss.

VI. Review this Class: Use blackboard outline and TRANSPARENCIES 13.2 and 13.3.

Lecture 13.2: The Two-Way Analysis of Variance

Materials

Lecture outline
Transparencies 12.3, 12.4, 12.10, 13.2, 13.3, and 13.8 through 13.17

Outline for Blackboard

I.   **Review**
II.   **Basic Logic**
III. **Structural Model**
IV.   **Effect Size (R^2)**
V.   **Review this Class**

Instructor's Lecture Outline

I.   **Review**
A. Idea of descriptive and inferential statistics.
B. Review of types of situations addressed so far in course.
   1. One-group t test (including special case of t test for dependent means).
   2. Two-group t test—t test for independent means.
   3. More than two groups—analysis of variance.
   4. More than two groups arranged in a factorial design.
C. Review of factorial designs and interaction effects.
   1. Show TRANSPARENCY 13.2 (tables of means for fictional factorial studies, from text) and discuss.
   2. Show TRANSPARENCY 13.3 (graphs of means for studies in Transparency 13.2, from text) and discuss.
D. Review of structural model of analysis of variance.
   1. Show TRANSPARENCY 12.3 (illustration of division of overall deviation from mean into two parts, from text) and discuss.
   2. Show TRANSPARENCY 12.4 (from text table of computations for criminal-record study) and discuss each step.
   3. Show TRANSPARENCY 12.10 (summary of structural model approach to ANOVA) and discuss each step.

II.   **Basic Logic**
A. In a two-way analysis of variance there are three F ratios (one for each main effect and one for the interaction effect).
B. Each F ratio represents a between-group variance estimate for its corresponding main or interaction effect divided in each case by the same within-group variance estimate, which is based on the variation within each cell.
C. Between-group variance estimates for main effects.
   1. For row independent variable, based on the variation among the row means.
   2. For the column independent variable, based on the variation among the column means.
D. Between-group variance estimate for interaction effect.
   1. Based on the variation among combinations of cells other than into columns or rows.
   2. For a 2 X 2 study, it is based on opposite diagonals. Show TRANSPARENCY 13.8 (figure showing opposite diagonals shaded, from text) and discuss.

III. Structural Model

A. Show TRANSPARENCY 13.9 (breakdown of overall deviation into four parts, from text) and discuss.

B. Show TRANSPARENCY 13.10 (formulas) and discuss each section.

C. Show TRANSPARENCY 13.11 (ANOVA table) and discuss each section.

D. Example: Show TRANSPARENCIES 13.12 through 13.16 (example based on Wong & Csikszentmihalyi study, from text) and discuss step by step.

IV. Effect Size (R^2): Show TRANSPARENCY 13.17 (formulas for R^2) and discuss logic.

V. Review this Class: Use blackboard outline.

TRANSPARENCY 13.1

Design and results of study by Burnkraut and Unnava (1989).

Design:

Dependent measure: Attitude towards product

Self-Referencing

		High	Low
Argument Strength	Strong		
	Weak		

4 Groups: A. Strong argument, high self-referencing
          B. Strong argument, low self-referencing
          C. Weak argument, high self-referencing
          D. Weak argument, low self-referencing

Results:

Self-Referencing

		High	Low	Average
Argument Strength	Strong	4.94	5.58	5.26
	Weak	4.46	3.74	4.10
Average		4.70	4.66	

Main Effects:

   Argument Strength: Means: 5.26 vs. 4.10,
                      $F(1,44) = 21.78$, $p < .001$.

   Self-Referencing: Means: 4.70 vs. 4.66,
                      $F(1,44) < 1$, ns.

Interaction Effect (Argument Strength X Self-Referencing):

                 $F(1,44) = 7.76$, $p < .01$.

Reference: Burnkraut, R. E., & Unnava, H. R. (1989). Self-
            referencing: A strategy for increasing
            processing of message content. Personality
            and Social Psychology Bulletin, 4, 628-338.

TRANPARENCY 13.2

TABLE 13-6
Possible Means for Results of a Study of the Relation of Age and Education to Income (Fictional Data, Thousands of Dollars)

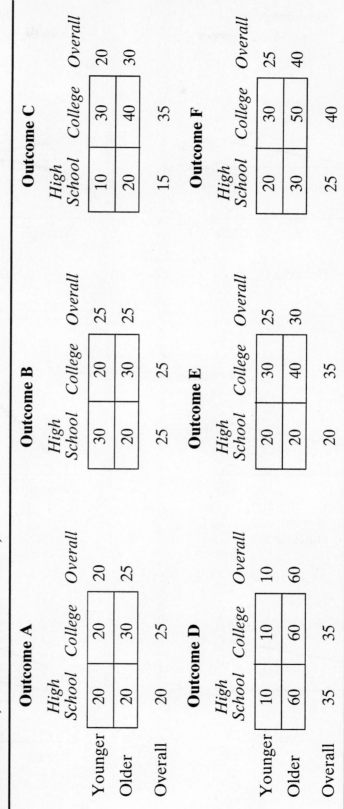

Outcome A

	High School	College	Overall
Younger	20	20	20
Older	20	30	25
Overall	20	25	

Outcome B

	High School	College	Overall
Younger	30	20	25
Older	20	30	25
Overall	25	25	

Outcome C

	High School	College	Overall
Younger	10	30	20
Older	20	40	30
Overall	15	35	

Outcome D

	High School	College	Overall
Younger	10	10	10
Older	60	60	60
Overall	35	35	

Outcome E

	High School	College	Overall
Younger	20	30	25
Older	20	40	30
Overall	20	35	

Outcome F

	High School	College	Overall
Younger	20	30	25
Older	30	50	40
Overall	25	40	

Aron/Aron
STATISTICS FOR PSYCHOLOGY

© 1994 by Prentice-Hall, Inc.
A Paramount Communications Company
Englewood Cliffs, New Jersey 07632

TRANSPARENCY 13.3

Figure 13.4

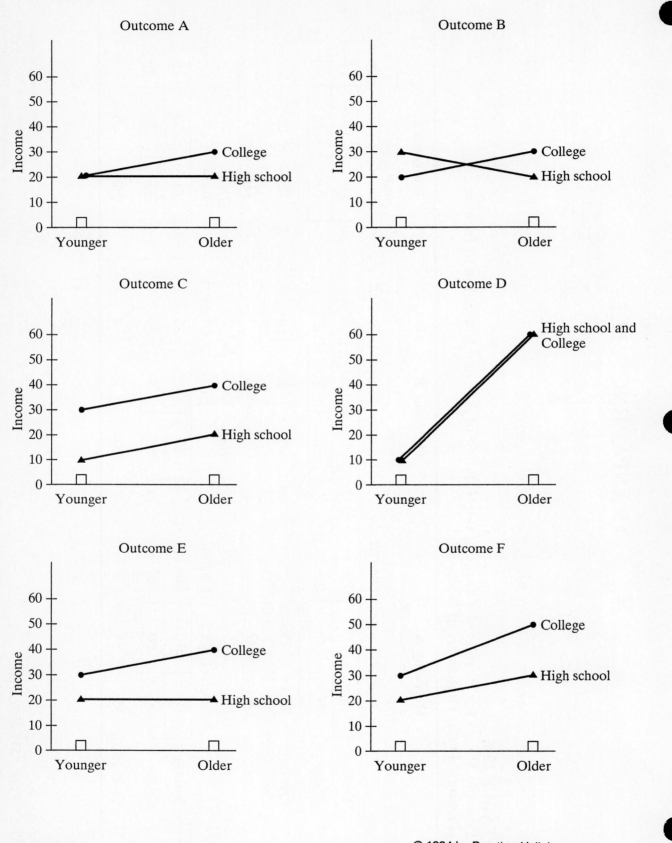

TRANSPARENCY 13.4

Examples of display of results of factorial studies.
(Data from class questionnaire.)

DV: "Do you cry easily?"

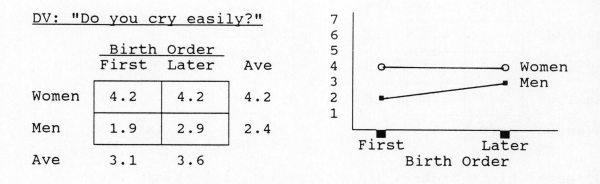

| | Birth Order | | |
	First	Later	Ave
Women	4.2	4.2	4.2
Men	1.9	2.9	2.4
Ave	3.1	3.6	

Effects: Birth Order: $F(1,93)$ =   .23, ns.
        Gender: $F(1,93)$ = 23.07, p < .0001.
        Interaction: $F(1,93)$ =   2.09, ns.

DV: "Did you tend to fall in love in your early school years?"

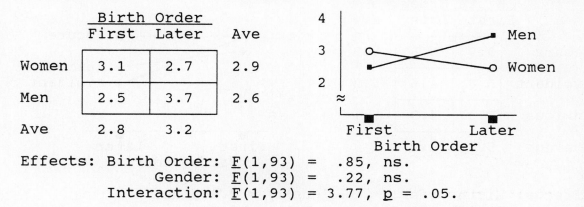

| | Birth Order | | |
	First	Later	Ave
Women	3.1	2.7	2.9
Men	2.5	3.7	2.6
Ave	2.8	3.2	

Effects: Birth Order: $F(1,93)$ = .85, ns.
        Gender: $F(1,93)$ = .22, ns.
        Interaction: $F(1,93)$ = 3.77, p = .05.

TRANSPARENCY 13.5

Examples of display of results of factorial studies.
(Data from class questionnaire.)

<u>DV</u>: "Were you close to your father?

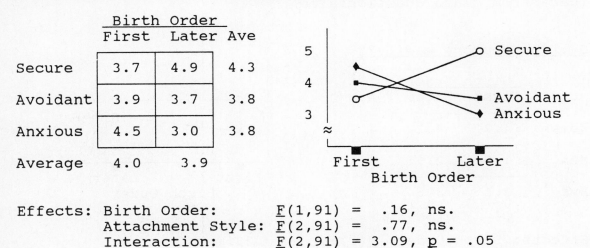

	Birth Order		
	First	Later	Ave
Secure	3.7	4.9	4.3
Avoidant	3.9	3.7	3.8
Anxious	4.5	3.0	3.8
Average	4.0	3.9	

Effects: Birth Order: $F(1,91)$ =   .16, ns.
         Attachment Style: $F(2,91)$ =   .77, ns.
         Interaction: $F(2,91)$ = 3.09, p = .05

<u>DV</u>: "Were you close to your mother?

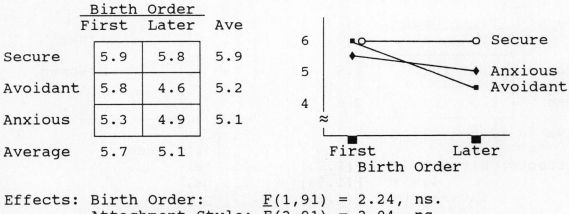

	Birth Order		
	First	Later	Ave
Secure	5.9	5.8	5.9
Avoidant	5.8	4.6	5.2
Anxious	5.3	4.9	5.1
Average	5.7	5.1	

Effects: Birth Order: $F(1,91)$ = 2.24, ns.
         Attachment Style: $F(2,91)$ = 2.04, ns.
         Interaction: $F(2,91)$ = 1.11, ns.

Aron/Aron
STATISTICS FOR PSYCHOLOGY

© 1994 by Prentice-Hall, Inc.
A Paramount Communications Company
Englewood Cliffs, New Jersey 07632

TRANSPARENCY 13.6

Possible means for mental health of French and U.S. comedians and actors. (Fictional data.)

DV: <u>Mental Health</u>

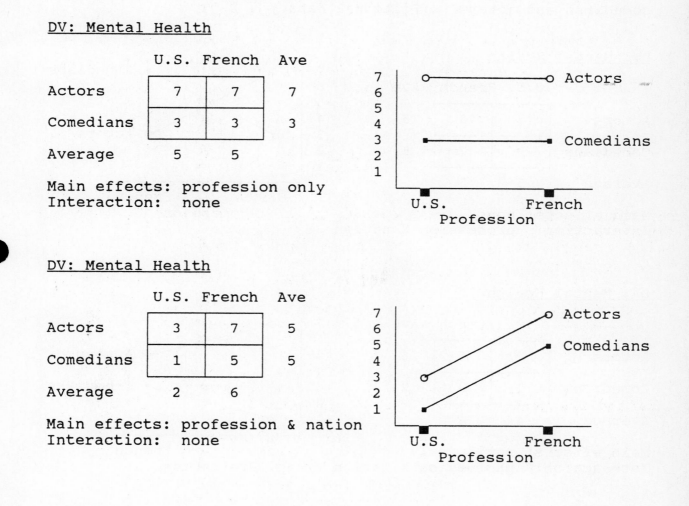

	U.S.	French	Ave
Actors	7	7	7
Comedians	3	3	3
Average	5	5	

Main effects: profession only
Interaction: none

DV: <u>Mental Health</u>

	U.S.	French	Ave
Actors	3	7	5
Comedians	1	5	5
Average	2	6	

Main effects: profession & nation
Interaction: none

Aron/Aron
STATISTICS FOR PSYCHOLOGY

TRANSPARENCY 13.7

Additional possible means for mental health of French and U.S. comedians and actors. (Fictional data.)

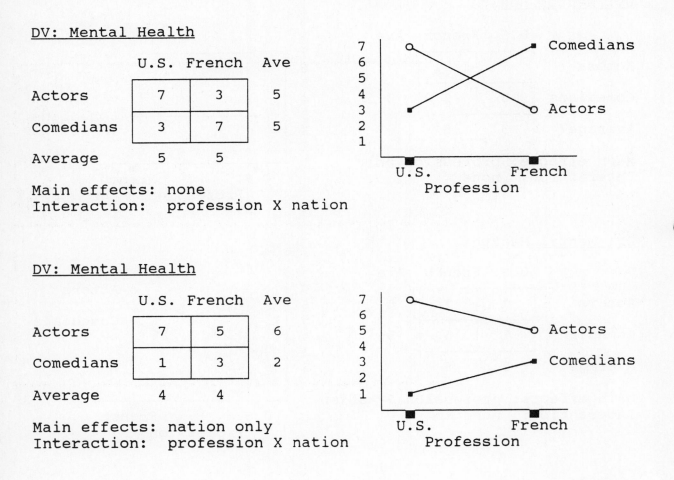

DV: Mental Health

	U.S.	French	Ave
Actors	7	3	5
Comedians	3	7	5
Average	5	5	

Main effects: none
Interaction: profession X nation

DV: Mental Health

	U.S.	French	Ave
Actors	7	5	6
Comedians	1	3	2
Average	4	4	

Main effects: nation only
Interaction: profession X nation

Aron/Aron
STATISTICS FOR PSYCHOLOGY

Figure 13.8

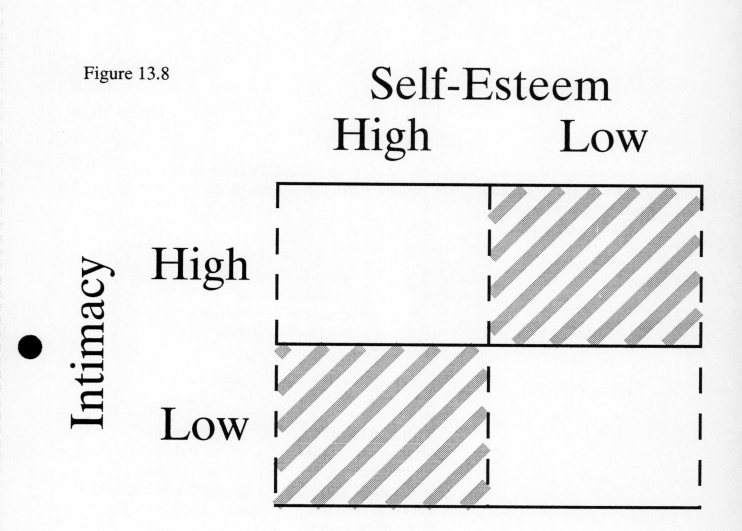

TRANSPARENCY 13.9

Figure 13.9

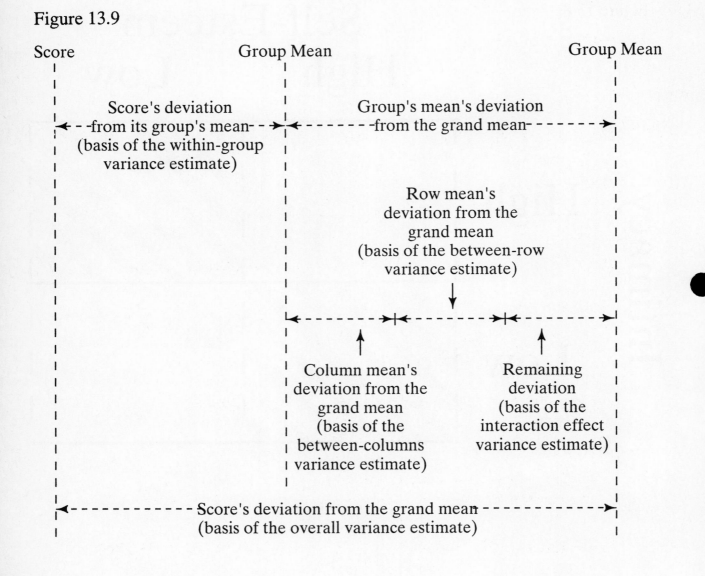

Score

Group Mean

Group Mean

Score's deviation
-from its group's mean-
(basis of the within-group
variance estimate)

Group's mean's deviation
-from the grand mean-

Row mean's
deviation from the
grand mean
(basis of the between-row
variance estimate)

Column mean's
deviation from the
grand mean
(basis of the
between-columns
variance estimate)

Remaining
deviation
(basis of the
interaction effect
variance estimate)

Score's deviation from the grand mean
(basis of the overall variance estimate)

Aron/Aron
STATISTICS FOR PSYCHOLOGY

TRANSPARENCY 13.10

Formulas for computing a two-way analysis of variance.

Within-Groups

$$S_W^2 = \frac{\Sigma(\underline{X} - \underline{M})^2}{\underline{df}_W} \quad \text{or} \quad \underline{MS}_W = \frac{\underline{SS}_W}{\underline{df}_W}$$

$$\underline{df}_W = \underline{df}_1 + \underline{df}_2 + \ldots + \underline{df}_{Last}$$

Rows:

$$S_R^2 = \frac{\Sigma(\underline{M}_R - \underline{GM})^2}{\underline{df}_R} \quad \text{or} \quad \underline{MS}_R = \frac{\underline{SS}_R}{\underline{df}_R}$$

$$\underline{df}_R = \underline{N}_R - 1 \qquad \underline{F}_R = \underline{S}_R^2/\underline{S}_W^2 \text{ or } \underline{F}_R = \underline{MS}_R/\underline{MS}_W$$

Columns:

$$S_C^2 = \frac{\Sigma(\underline{M}_C - \underline{GM})^2}{\underline{df}_C} \quad \text{or} \quad \underline{MS}_C = \frac{\underline{SS}_C}{\underline{df}_C}$$

$$\underline{df}_C = \underline{N}_C - 1 \qquad \underline{F}_C = \underline{S}_C^2/\underline{S}_W^2 \text{ or } \underline{F}_C = \underline{MS}_C/\underline{MS}_W$$

Interaction:

$$S_I^2 = \frac{\Sigma[(\underline{X}-\underline{GM})-(\underline{X}-\underline{M})-(\underline{M}_R-\underline{GM})-(\underline{M}_C-\underline{GM})]^2}{\underline{df}_C} \quad \text{or} \quad \underline{MS}_I = \frac{\underline{SS}_I}{\underline{df}_I}$$

$$\underline{df}_I = \underline{N}_{Cells} - \underline{df}_R - \underline{df}_C - 1 \qquad \underline{F}_I = \underline{S}_I^2/\underline{S}_W^2 \text{ or } \underline{F}_I = \underline{MS}_I/\underline{MS}_W$$

Aron/Aron
STATISTICS FOR PSYCHOLOGY

© 1994 by Prentice-Hall, Inc.
A Paramount Communications Company
Englewood Cliffs, New Jersey 07632

TABLE 13-8
Layout of an Analysis of Variance Table for a Two-Way Analysis

Source	SS	df	MS	F
Between:				
Columns	SS_C	df_C	MS_C	F_C
Rows	SS_R	df_R	MS_R	F_R
Interaction	SS_I	df_I	MS_I	F_I
Within	SS_W	df_W	MS_W	
Total	SS_T	df_T		

© 1994 by Prentice-Hall, Inc.
A Paramount Communications Company
Englewood Cliffs, New Jersey 07632

Aron/Aron
STATISTICS FOR PSYCHOLOGY

TABLE 13-9
Cell and Marginal Means for Number of Times Engaged in Social Activities (Data from Wong & Csikszentmihalyi, 1991)

	Affiliation		
	Low	*High*	
Boys	10.30	9.22	9.76
Girls	15.75	18.51	17.13
	13.03	13.87	13.45

Aron/Aron
STATISTICS FOR PSYCHOLOGY

TABLE 13-10
Scores, Squared Deviations, and Sums of Squared Deviations for Fictional Data Based on the Wong and Csikszentmihalyi (1991) Study

	Low Affiliation						High Affiliation				
X	$(X-GM)^2$	$(X-M)^2$	$(M_R-GM)^2$	$(M_C-GM)^2$	INT^2	X	$(X-GM)^2$	$(X-M)^2$	$(M_R-GM)^2$	$(M_C-GM)^2$	INT^2
Boys											
12.1	1.82	3.24	13.62	.18	.92	11.1	5.52	3.53	13.62	.18	.92
11.4	4.20	1.21	13.62	.18	.92	10.4	9.30	1.39	13.62	.18	.92
11.2	5.06	.81	13.62	.18	.92	10.2	10.56	.96	13.62	.18	.92
10.9	6.50	.36	13.62	.18	.92	9.8	13.32	.34	13.62	.18	.92
10.3	9.92	0.00	13.62	.18	.92	9.2	18.06	0.00	13.62	.18	.92
9.8	13.32	.25	13.62	.18	.92	9.1	18.92	.01	13.62	.18	.92
9.7	14.06	.36	13.62	.18	.92	8.9	20.70	.10	13.62	.18	.92
9.5	15.60	.64	13.62	.18	.92	8.7	22.56	.27	13.62	.18	.92
9.3	17.22	1.00	13.62	.18	.92	8.2	27.56	1.04	13.62	.18	.92
8.8	21.62	2.25	13.62	.18	.92	6.6	46.92	6.86	13.62	.18	.92
103.0	109.32	10.12	136.20	1.80	9.20	92.2	193.42	14.50	136.20	1.80	9.20
Girls											
17.4	15.60	2.74	13.54	.18	.92	22.0	73.10	2.72	13.54	.18	.92
17.1	13.32	1.82	13.54	.18	.92	20.5	49.70	3.96	13.54	.18	.92
16.8	11.22	1.10	13.54	.18	.92	19.9	41.60	1.93	13.54	.18	.92
16.7	10.56	.90	13.54	.18	.92	19.1	31.92	.35	13.54	.18	.92
15.5	4.20	.06	13.54	.18	.92	18.5	25.50	0.00	13.54	.18	.92
15.3	3.42	.20	13.54	.18	.92	17.4	15.60	1.23	13.54	.18	.92
15.0	2.40	.56	13.54	.18	.92	17.0	12.60	2.28	13.54	.18	.92
15.4	3.80	.12	13.54	.18	.92	17.1	13.32	1.99	13.54	.18	.92
14.3	.72	2.10	13.54	.18	.92	17.1	13.32	1.99	13.54	.18	.92
14.0	.30	3.06	13.54	.18	.92	16.5	9.30	4.04	13.54	.18	.92
157.5	65.54	12.64	135.40	1.80	9.20	185.1	285.96	29.95	135.40	1.80	9.20

M = mean of the score's cell
M_R = mean of the score's row
M_C = mean of the score's column
INT = score's remaining deviation for the interaction

Example of computations of deviations, using the first score in the Low Boys cell:

$$(X-GM)^2 = (12.1 - 13.45)^2 = -1.35^2 = 1.82$$
$$(X-M)^2 = (12.1 - 10.30)^2 = 1.80^2 = 3.24$$
$$(M_R-GM)^2 = (9.76 - 13.45)^2 = -3.69^2 = 13.62$$
$$(M_C-GM)^2 = (13.03 - 13.45)^2 = -0.42^2 = .18$$
$$INT^2 = [(X-GM)-(X-M)-(M_R-GM)-(M_C-GM)]^2 = [(-1.35)-(1.80)-(-3.69)-(-.42)]^2$$
$$= (-1.35 - 1.80 + 3.69 + .42)^2 = .96^2 = .92$$

$$SS_T = 109.32 + 193.42 + 65.54 + 285.96 = 654.24$$
$$SS_W = 10.12 + 14.50 + 12.64 + 29.95 = 67.21$$
$$SS_R = 136.20 + 136.20 + 135.40 + 135.40 = 543.20$$
$$SS_C = 1.80 + 1.80 + 1.80 + 1.80 = 7.20$$
$$SS_I = 9.20 + 9.20 + 9.20 + 9.20 = 36.80$$

Accuracy check: $SS_T = 654.24$; $SS_W + SS_R + SS_C + SS_I = 67.21 + 543.20 + 7.20 + 36.80 = 654.41$
(results are within rounding error).

Aron/Aron
STATISTICS FOR PSYCHOLOGY

© 1994 by Prentice-Hall, Inc.
A Paramount Communications Company
Englewood Cliffs, New Jersey 07632

TABLE 13-11
Computation of an Analysis of Variance Using Sums of Squares Based on the Wong and Csikszentmihalyi (1991) Study (Fictional Data)

F needed for Gender main effect ($df = 1, 36$; $p < .05$) = 4.12 ($df = 1, 35$ from table)
F needed for Affiliation main effect for ($df = 1, 36$; $p < .05$) = 4.12
F needed for interaction effect ($df = 1, 36$; $p < .05$) = 4.12

Source	SS	df	MS	F	
Gender	543.20	1	543.20	290.48	Reject the null hypothesis.
Affiliation	7.20	1	7.20	3.85	Do not reject the null hypothesis.
Gender × affiliation	36.80	1	36.80	19.68	Reject the null hypothesis.
Within cells	67.21	36	1.87		

Aron/Aron
STATISTICS FOR PSYCHOLOGY

Figure 13.10

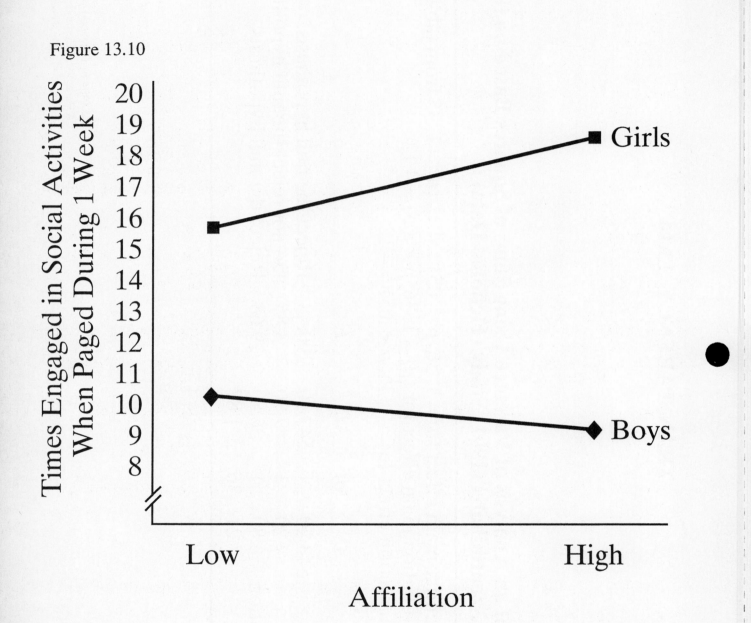

TRANSPARENCY 13.16

Wong and Csikszentmihalyi study of times engaged in social activities. (Example from text.)

Steps of hypothesis testing:

1. Reframe into a research hypothesis and a null hypothesis about populations.

   Population 1,1: Girls who are Low on desire for affiliation.
   Population 1,2: Girls who are High on desire for affiliation.
   Population 2,1: Boys who are Low on desire for affiliation.
   Population 2,2: Boys who are High on desire for affiliation.

   Rows:
   Research: Girls (1,1 & 1,2) & Boys (2,1 & 2,2) have different means.
   Null: Girls & Boys have the same mean.

   Columns:
   Research: Low (1,1 & 2,1) desire to affiliate and high desire to
             affiliate (1,2 & 2,2) have different means.
   Null: Low and high desire to affiliate have the same mean.

   Interaction:
   Research: Pattern of Girl-Boy difference varies according to whether
             there is a high or low desire to affiliate.
   Null: Difference between Girls and Boys is the same for those with
         low and high desire to affiliate.

2. Determine the characteristics of the comparison distributions.

   In all three cases, F distribution with df = 1 and 36.

3. Determine the cutoff sample score on the comparison
   distribution at which the null hypothesis should be rejected.

   In all three cases, 5% level, $F(1, 36)$ needed = 4.12.

      [actually $F(1, 35)$ used since 1,36 not in table]

4. Determine score of your sample on comparison distribution.

   F for rows (Girl-Boy) = 290.48.
   F for columns (Low-High need to affiliate) = 3.85.
   F for interaction = 19.68.

5. Compare scores obtained in Steps 3 and 4 to decide whether to
   reject the null hypothesis.

   Rows (Girl-Boy) F exceeds cutoff; reject null hypothesis.
   Columns (Low-High) F does not exceed cutoff; do not reject null.
   Interaction F exceeds cutoff; reject null hypothesis.

Aron/Aron
STATISTICS FOR PSYCHOLOGY

© 1994 by Prentice-Hall, Inc.
A Paramount Communications Company
Englewood Cliffs, New Jersey 07632

TRANSPARENCY 13.17

Formulas for proportion of variance accounted
for in a two-way analysis of variance.

$$R_R{}^2 = \frac{SS_R}{SS_T - SS_C - SS_I}$$

$$R_C{}^2 = \frac{SS_C}{SS_T - SS_R - SS_I}$$

$$R_I{}^2 = \frac{SS_I}{SS_T - SS_R - SS_C}$$

Example:

$$SS_R = 543.20$$
$$SS_C = 7.20$$
$$SS_I = 36.80$$
$$SS_T = 654.24$$

$$R_R{}^2 = \frac{SS_R}{SS_T - SS_C - SS_I} = \frac{543.20}{654.24-7.20-36.80}$$

$$= \frac{543.20}{610.24} = .89$$

Aron/Aron
STATISTICS FOR PSYCHOLOGY

© 1994 by Prentice-Hall, Inc.
A Paramount Communications Company
Englewood Cliffs, New Jersey 07632

Chapter 14
Chi-Square Tests

Instructor's Summary of Chapter

The chi-square statistic reflects the amount of discrepancy between expected and observed frequencies over several categories or combinations of categories. It is computed by finding for each category or combination the difference between observed and expected frequencies, squaring this difference (to eliminate the sign), and dividing by the expected frequency (to help make the squared differences more proportionate to the number of cases involved). The results are then added over all categories or combinations.

The distribution of the chi-square statistic is known and significance-level cutoffs for various degrees of freedom (based on the number of categories or combinations free to vary) are listed in a chi-square table.

The chi-square test for goodness of fit tests hypotheses about whether a distribution of frequencies over the levels of a nominal variable matches an expected distribution. (These expected frequencies are usually based on theory or on a known distribution in another study or circumstance). Hence, in this test, the expected frequencies are given in advance. Df = number of categories - 1.

The chi-square test of independence tests hypotheses about the relationship between two nominal variables—whether the distribution of cases over the levels of one variable has the same proportional pattern within each of the levels of the other variable. The data are set up in a contingency table, in which the two variables are crossed. The frequency expected for a given cell if the two variables are independent is the percentage of all the scores in that cell's row times the total number of cases in that cell's column. Df = number of columns minus 1 times number of rows minus 1.

Assumptions. Chi-square tests make no assumptions about normal distributions of their variables, but do require the independence of each case.

Indexes of association and effect size. The estimated effect size for a chi-square test of independence (i.e., the degree of association) for a 2 X 2 contingency table is the phi (ϕ) coefficient (the square root of the result of dividing the computed chi-square by the number of cases), and for larger tables, Cramer's phi (the square root of the result of dividing the computed chi-square by the product of the number of cases times the degrees of freedom in the smaller dimension of the contingency table). These coefficients range from 0 to 1 and can be interpreted in approximately the same way as a correlation coefficient. A phi of .10 is considered a small effect, .30 a medium effect, and .50 a large effect.

Minimum expected frequencies. The minimum acceptable frequency for a category or cell has been a subject of controversy. Currently the best advice is that even very small expected frequencies do not seriously increase the chance of a Type I error provided there are at least five times as many subjects as categories (or cells). However, low expected frequencies seriously reduce power and should be avoided if possible.

How the procedures of this chapter are reported in research articles. Chi-square tests reported in research articles often include all information about numbers in each category or cell. The computed chi-square and its significance are also usually reported, following a standard format—for example, "χ^2 (3, $N = 217$) = 8.81, $p < .05$."

Box 14.1. Karl Pearson, Inventor of Chi-Square and Center of Controversy. Summarizes the life and work of Pearson, inventor of the chi-square test.

Lecture 14.1: The Chi-Square Test of Goodness of Fit

Materials

Lecture outline

Transparencies 14.1 through 14.10

(If using transparencies based on your class's questionnaires, replace 14.1, 14.3, 14.8, and 14.9 with 14.1R, 14.3R, 14.8R, and 14.9R.)

Outline for Blackboard

I.   Review
II. Nominal Variables
III. Expected and Observed Frequencies
IV. Chi-Square Statistic
V.   Chi-Square Distribution
VI. Steps of Hypothesis Testing
VII. Review this Class

Instructor's Lecture Outline

I. Review

A. Idea of descriptive and inferential statistics.

B. Analysis of variance as looking at means over different categories.

C. Assumptions in analysis of variance.

II. Nominal Variables

A. Principle: Variable measured in levels or categories with no systematic numerical or rank order relationship among levels.

B. Examples: Hair color, religion, which school you attend, major, profession, ethnic group, attachment style.

III. Expected and Observed Frequencies: Show TRANSPARENCY 14.1 or 14.1R (distribution of attachment styles in psychology statistics students, based on class questionnaire, compared to that found in the general public, based on Hazan & Shaver, 1977) and discuss.

A. Key idea is the comparison of observed and expected frequency in each category.

B. *Observed frequency* is the number of cases in the category in the sample data.

C. *Expected frequency* is the number of cases in the category expected based on the population to which we are comparing our sample.

1. Computed by multiplying the proportion in the population to which we are comparing times the number in the sample.

2. Note: Sometimes a population distribution is not given and the comparison is to equal percentages in the categories.

IV. Chi-Square Statistic (continue showing TRANSPARENCY 14.1 or 14.1R)
    A. A measure of degree of discrepancy over all categories.
    B. The discrepancy in each category is squared, in part to eliminate the problem of signs.
    C. The squared discrepancy in each category is divided by the expected frequency to keep these discrepancies in proportion to the number of cases that would have been expected.
    D. The chi-square statistic is the sum, over the categories, of the each squared discrepancy divided by its expected frequency.

V. Chi-Square Distribution: Show TRANSPARENCY 14.2 (chi-square distribution) and discuss.
    A. Principle: If samples are randomly taken from a population and a chi-square statistic computed on each, these chi-squares follow a mathematically defined distribution (the chi-square distribution).
    B. The distribution is skewed with the tail to the right.
    C. The distribution's exact shape depends on the degrees of freedom.
    D. For a chi-square test of goodness of fit, df = number of categories - 1.
        1. This is the number of categories with frequencies free to vary, given that the frequency over all categories is fixed.
        2. In the current example, $df = 3 - 1 = 2$.
    E. The cutoff chi-square values for various significance levels are given in standard tables.
        1. A chi-square table is included in Appendix B of the text.
        2. In the example with $df = 2$, at the .05 level, cutoff chi-square is 5.992.

VI. Steps of Hypothesis Testing
    A. Show TRANSPARENCY 14.3 or 14.3R (hypothesis testing steps for attachment-style example) and discuss each step.
    B. Example 2: Show TRANSPARENCIES 14.4 and 14.5 (Mineral Q example from text) and discuss.
    C. Example 3: Show TRANSPARENCIES 14.6 and 14.7 (Fictional example of whether distribution over year in school of students who attend political demonstrations is different from the distribution of students over years in general at the same university) and discuss.
    D. Example 4: Show TRANSPARENCIES 14.8 and 14.9 or 14.8R and 14.9R (are the number of morning and night people equal, based on class questionnaire) and discuss.

VII. Review this Class: Use blackboard outline and TRANSPARENCY 14.10 (steps of conducting a chi-square test for goodness of fit).

Lecture 14.2: The Chi-Square Test of Independence

Materials

Lecture outline

Transparencies 14.10 through 14.21

(If using transparencies based on your class's questionnaires, replace 14.15, 14.16, and 14.20 with 14.15R, 14.16R, and 14.20R.)

Outline for Blackboard

I.   Review
II. Contingency Tables
III. Independence and Expected Frequencies
IV. Computing Chi-Square
V.   Degrees of Freedom
VI. Effect Size (Degree of Association)
VII. Hypothesis Testing
VIII. Review this Class

Instructor's Lecture Outline

I.   Review
   A. Idea of descriptive and inferential statistics.
   B. Idea of nominal variables.
   C. Show TRANSPARENCY 14.10 (steps of conducting a chi-square test for goodness of fit) and discuss each step.

II. Contingency Tables
   A. Breakdown of frequencies across levels of two nominal variables in which each individual is categorized on both variables.
   B. Examples.
      1. Hair color and eye color.
      2. Religious affiliation and ethnic group.
      3. Religious affiliation and eye color.
      4. Personality type of self and partner.
   C. Show top of TRANSPARENCY 14.11 (fictional example of married women with and without children who rent vs. own their home) and discuss.
      1. As example of contingency table. (Note that it does not matter which variable goes on which dimension of the table.)
      2 Cell and marginal frequencies.

III. Independence and Expected Frequencies
   A. Distribution of frequencies over categories on one variable is independent of the distribution of frequencies over categories on a second variable.
   B. In example, independence would mean that whether young married women do versus do not have children is unrelated to whether they rent or own their home. Show TRANSPARENCY 14.11 and discuss.

C. Expected frequencies are what would be expected if two variables were independent.

D. If the two variables are independent, then the expected frequencies for a given cell is the proportion of its row's observed frequency of the total observed frequency, times the observed frequency for its column.

E. Show TRANSPARENCY 14.11 and discuss.

IV. Computing Chi-Square

A. Same principle as with chi-square test for goodness of fit except that you add up over cells instead of categories.

B. Show next section of TRANSPARENCY 14.11 and discuss.

V. Degrees of Freedom

A. What is free to vary?

B. If marginal totals are known, only a small number of cells need be known to determine the rest.

C. Formula is $df = (N_R - 1)(N_C - 1)$.

B. Show next section of TRANSPARENCY 14.11 and discuss.

VI. Effect Size (Degree of Association)

A. For a 2 X 2 contingency table.

    1. $\phi = \sqrt{(\chi^2/N)}$.

    2. ϕ is like r.

    3. Show next section of TRANSPARENCY 14.11 and discuss.

B. For a larger than 2 X 2 contingency table.

    1. Cramer's $\phi = \sqrt{[\chi^2/(N)(df_L)]}$

    2. df_L is the degrees of freedom corresponding to the smaller dimension of the contingency table.

    3. Also interpreted like r.

VII. Hypothesis Testing

A. Show TRANSPARENCY 14.12 (hypothesis testing for rent-own X children-or-not example) and discuss.

B. Show TRANSPARENCY 14.13 and 14.14 (Russell, 1991, definition-of-pride example from text) and discuss.

C. Show TRANSPARENCIES 14.15 and 14.16 or 14.15R and 14.16R (abused or not as a child by gender, from class questionnaire) and discuss.

D. Show TRANSPARENCIES 14.17, 14.18, and 14.19 (fictional study of whether distribution of where a married couple originally met is related to ethnicity) and discuss.

E. Show TRANSPARENCY 14.20 or 14.20R (additional examples from class questionnaire) and discuss.

VIII. Review this Class: Use blackboard outline and TRANSPARENCY 14.21 (steps of conducting a chi-square test for independence).

TRANSPARENCY 14.1

<u>Expected Distribution</u> (from studies by Hazan & Shaver, 1987):

Secure: 56%   Avoidant: 24%   Anxious/Ambivalent: 20%

<u>Observed, Actual Frequencies</u> (from class questionnaire)

Secure: 51   Avoidant: 23   Anxious/Ambivalent: 16

<u>Expected frequencies</u> (expected percent times number in sample)

Secure: Avoidant: Anxious/Ambivalent:
(56%)(90) = 50.4   (24%)(90) = 21.6   (20%)(90) = 18

<u>Chi-Square Statistic</u> (indicates overall degree of discrepancy between observed and expected)

$$X^2 = \Sigma \frac{(O - E)^2}{E} = \frac{(51-50.4)^2}{50.4} + \frac{(23-21.6)^2}{21.6} + \frac{(16-18)^2}{18}$$

$$= \frac{.6^2}{50.4} + \frac{1.4^2}{21.6} + \frac{-2^2}{18}$$

$$= \frac{.36}{50.4} + \frac{1.96}{21.6} + \frac{4}{18}$$

$$= .007 + .091 + .222 = .32$$

Aron/Aron
STATISTICS FOR PSYCHOLOGY

TRANSPARENCY 14.2

Figure 14.1

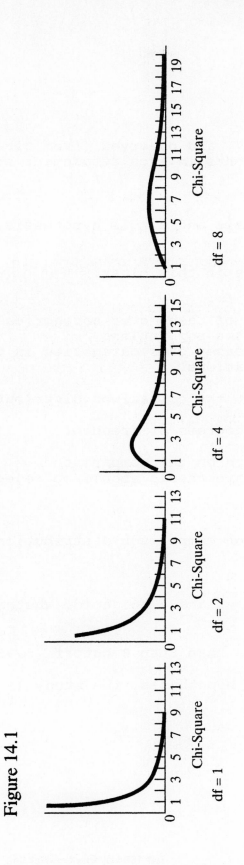

df = 1   Chi-Square

df = 2   Chi-Square

df = 4   Chi-Square

df = 8   Chi-Square

TABLE 14-2
Portion of a Chi-Square Table

	Significance Level		
df	*.10*	*.05*	*.01*
1	2.706	3.841	6.635
2	4.605	5.992	9.211
3	6.252	7.815	11.345
4	7.780	9.488	13.277
5	9.237	11.071	15.087

© 1994 by Prentice-Hall, Inc.
A Paramount Communications Company
Englewood Cliffs, New Jersey 07632

Aron/Aron
STATISTICS FOR PSYCHOLOGY

TRANSPARENCY 14.3

Expected (from Hazan & Shaver, 1987) and observed (from class questionnaire) frequencies of students in each attachment style.

Steps of hypothesis testing:

1. Reframe into a research hypothesis and a null hypothesis about populations.

   Population 1: People like students in this class.
   Population 2: General public.

   Research Hypothesis: Distribution of cases over categories differs between the two populations
   Null Hypothesis: Distribution of cases over categories is the same in the two populations.

2. Determine the characteristics of the comparison distribution.

   Chi-square distribution with two degrees of freedom.

3. Determine the cutoff sample score on the comparison distribution at which the null hypothesis should be rejected.

   .05 level, $\underline{df}$=2: $X^2 = 5.992$.

4. Determine score of your sample on comparison distribution.

   $X^2 = .32$.

5. Compare scores obtained in Steps 3 and 4 to decide whether to reject the null hypothesis.

   X^2 in Step 4 (.32) is less extreme than Step 3 cutoff (5.992).

   Therefore, do not reject the null hypothesis; the study is inconclusive.

Aron/Aron
STATISTICS FOR PSYCHOLOGY

© 1994 by Prentice-Hall, Inc.
A Paramount Communications Company
Englewood Cliffs, New Jersey 07632

TRANSPARENCY 14.4

Figure 14.2

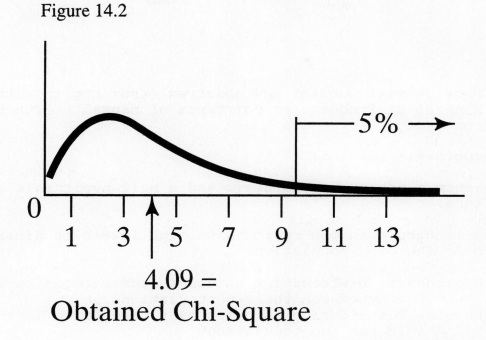

4.09 =
Obtained Chi-Square

TABLE 14-3
Observed and Expected Frequencies for Types of Mental Health Disorders in a Region High in Mineral Q Compared to the General U.S. Population and the Chi-Square Goodness of Fit Test (Fictional Data)

Condition	Observed	Expected
Anxiety disorder	134	146 (14.6% × 1,000)
Alcohol and drug abuse	160	164 (16.4% × 1,000)
Mood disorders	97	83 (8.3% × 1,000)
Schizophrenia	12	15 (1.5% × 1,000)
None of these conditions	597	592 (59.2% × 1,000)

Degrees of freedom = 5 categories − 1 = 4

Chi-square needed, df = 4, .05 level: 9.488

$$\chi^2 = \Sigma \frac{(O-E)^2}{E} = \frac{(134-146)^2}{146} + \frac{(160-164)^2}{164} + \frac{(97-83)^2}{83} + \frac{(12-15)^2}{15} + \frac{(597-592)^2}{592}$$

$$= \frac{-12^2}{146} + \frac{-4^2}{164} + \frac{14^2}{83} + \frac{-3^2}{15} + \frac{5^2}{592} = \frac{144}{146} + \frac{16}{164} + \frac{196}{83} + \frac{9}{15} + \frac{25}{592}$$

$$= .99 + .10 + 2.36 + .60 + .04 = 4.09$$

Conclusion: Do not reject the null hypothesis.

Aron/Aron
STATISTICS FOR PSYCHOLOGY

© 1994 by Prentice-Hall, Inc.
A Paramount Communications Company
Englewood Cliffs, New Jersey 07632

TRANSPARENCY 14.5

Expected (from general public) and observed (from region with high levels of Mineral Q) frequencies for types of mental health disorders. (Example from text.)

Steps of hypothesis testing:

1. Reframe into a research hypothesis and a null hypothesis about populations.

 Population 1: People in the region with high levels of Mineral Q.
 Population 2: The U.S. population.

 Research Hypothesis: Distribution of cases over categories differs between the two populations
 Null Hypothesis: Distribution of cases over categories is the same in the two populations.

2. Determine the characteristics of the comparison distribution.

 Chi-square distribution with four degrees of freedom.

3. Determine the cutoff sample score on the comparison

 .05 level, $\underline{df}$=4: X^2 = 9.488.

4. Determine score of your sample on comparison distribution.

 X^2 = 4.09.

5. Compare scores obtained in Steps 3 and 4 to decide whether to reject the null hypothesis.

 X^2 in Step 4 (4.09) is less extreme than Step 3 cutoff (9.488).

 Therefore, do not reject the null hypothesis; the study is inconclusive.

Aron/Aron
STATISTICS FOR PSYCHOLOGY

© 1994 by Prentice-Hall, Inc.
A Paramount Communications Company
Englewood Cliffs, New Jersey 07632

TRANSPARENCY 14.6

Expected (all students attending a particular university) and observed (students attending political demonstrations at that university) frequencies of those in different years. (Fictional example.)

Year	Observed Frequency	Expected Percent	Frequency	
First	30	15%	45	[(15%)(300)]
Second	20	15%	45	[(15%)(300)]
Third	90	30%	90	[(30%)(300)]
Fourth	120	30%	90	[(30%)(300)]
Graduate	40	10%	30	[(10%)(300)]
TOTAL	300	100%	300	

$\underline{df}$ = 5 categories - 1 = 4 $X^2 = \Sigma \dfrac{(O-E)^2}{E}$

$$= \frac{(30-45)^2}{45} + \frac{(20-45)^2}{45} + \frac{(90-90)^2}{90} + \frac{(120-90)^2}{90} + \frac{(40-30)^2}{30}$$

$$= \frac{225}{45} + \frac{625}{45} + \frac{0}{90} + \frac{900}{90} + \frac{100}{30}$$

$$= \quad 5 \quad + \quad 13.9 \quad + \quad 0 \quad + \quad 10 \quad + \quad 3.3 \quad = \quad 32.2$$

Aron/Aron
STATISTICS FOR PSYCHOLOGY

© 1994 by Prentice-Hall, Inc.
A Paramount Communications Company
Englewood Cliffs, New Jersey 07632

TRANSPARENCY 14.7

Expected (all students attending a particular university) and observed (students attending political demonstrations at that university) frequencies of those in different years. (Fictional example.)

Steps of hypothesis testing:

1. Reframe into a research hypothesis and a null hypothesis about populations.

   Population 1: Students at this university who attend political demonstrations.
   Population 2: Students at this university in general.

   Research Hypothesis: Distribution of cases over categories differs between the two populations
   Null Hypothesis: Distribution of cases over categories is the same in the two populations.

2. Determine the characteristics of the comparison distribution.

   Chi-square distribution with four degrees of freedom.

3. Determine the cutoff sample score on the comparison distribution at which the null hypothesis should be rejected.

   .01 level, $\underline{df}$=2: X^2 = 13.277.

4. Determine score of your sample on comparison distribution.

   X^2 = 32.2.

5. Compare scores obtained in Steps 3 and 4 to decide whether to reject the null hypothesis.

   X^2 in Step 4 (32.2) is more extreme than Step 3 cutoff (13.277).

   Therefore, reject the null hypothesis; the research hypothesis is supported (those attending political demonstrations appear to have a different distribution of year in school than do students at this university in general).

Aron/Aron
STATISTICS FOR PSYCHOLOGY

© 1994 by Prentice-Hall, Inc.
A Paramount Communications Company
Englewood Cliffs, New Jersey 07632

TRANSPARENCY 14.8

Expected (even distribution) and observed (from class questionnaire) frequencies of students who are morning and night people.

Question: "To what extent are you a "morning person?"
Those in class answering 5-7 (high) vs 1-3 (low)

Answer	Observed Frequency	Expected Percent	Frequency	
Morning (5-7)	27	50%	42	[(50%)(84)]
Night (1-3)	57	50%	42	[(50%)(84)]
TOTAL	84	100%		

$\underline{df}$ = 2 categories - 1 = 1

$$X^2 = \Sigma \frac{(O-E)^2}{E} = \frac{(27-42)^2}{42} + \frac{(57-42)^2}{42} = \frac{225}{42} + \frac{225}{42} = 10.72$$

Aron/Aron
STATISTICS FOR PSYCHOLOGY

© 1994 by Prentice-Hall, Inc.
A Paramount Communications Company
Englewood Cliffs, New Jersey 07632

TRANSPARENCY 14.9

Expected (even distribution) and observed (from class questionnaire) frequencies of students who are morning and night people.

Steps of hypothesis testing:

1. Reframe into a research hypothesis and a null hypothesis about populations.

   Population 1: Students like those in class.
   Population 2: Students who are equally divided between morning and night people.

   Research Hypothesis: Distribution of cases over categories differs between the two populations
   Null Hypothesis: Distribution of cases over categories is the same in the two populations.

2. Determine the characteristics of the comparison distribution.

   Chi-square distribution with one degree of freedom.

3. Determine the cutoff sample score on the comparison distribution at which the null hypothesis should be rejected.

   .05 level, $\underline{df}$=1: X^2 = 3.841.

4. Determine score of your sample on comparison distribution.

   X^2 = 10.72.

5. Compare scores obtained in Steps 3 and 4 to decide whether to reject the null hypothesis.

   X^2 in Step 4 (10.72) is less extreme than Step 3 cutoff (3.841).

   Therefore, reject the null hypothesis; the research hypothesis is supported (there appear to be more night than morning people among students like those in this class).

Aron/Aron
STATISTICS FOR PSYCHOLOGY

© 1994 by Prentice-Hall, Inc.
A Paramount Communications Company
Englewood Cliffs, New Jersey 07632

TRANSPARENCY 14.10

How to conduct a chi-square test for goodness of fit.

1. Reframe the question into a research hypothesis and a null hypothesis about populations.

   A. Populations:

      1. Population 1 are people like those in the study.
      2. Population 2 are people who have the hypothesized distribution over categories.

   B. Hypotheses:

      1. Research hypothesis: The two populations have different distributions of cases over categories.
      2. Null Hypothesis: The two populations have the same distribution of cases over categories.

2. Determine the characteristics of the comparison distribution:

   A. Chi-square distribution.
   B. <u>df</u> = Number of categories minus 1.

3. Determine the cutoff sample score on the comparison distribution at which the null hypothesis should be rejected.

   A. Determine the desired significance level.
   B. Look up the appropriate cutoff on a chi-square table, using the degrees of freedom calculated above.

4. Determine the score of your sample on the comparison distribution.

$$X^2 = \Sigma \frac{(O-E)^2}{E}$$

5. Compare the scores obtained in Steps 3 and 4 to decide whether to reject the null hypothesis.

Aron/Aron
STATISTICS FOR PSYCHOLOGY

© 1994 by Prentice-Hall, Inc.
A Paramount Communications Company
Englewood Cliffs, New Jersey 07632

TRANSPARENCY 14.11

Young married women who have versus do not have children (first nominal variable) and rent or own their home (second nominal variable). (Fictional data.)

Contingency Table:

	Rent	Own	Total	
Have Children	5	11	16	40%
No Children	20	4	24	60%
Total	25	15	40	

Expected Frequencies:

$$E = (R/\underline{N})(C) \text{ or } E = (\text{Row's Percent})(\text{Column Total})$$

Have Children-Rent = (40%)(25) = 10
Have Children-Own = (40%)(15) = 6
No Children-Rent   = (60%)(25) = 15
No Children-Own   = (60%)(15) = 9

Table with Observed (and Expected) Frequencies:

	Rent	Own	Total
Have Children	5(10)	11(6)	16(16)
No Children	20(15)	4(9)	24(24)
Total	25(25)	15(15)	40(40)

$$X^2 = \Sigma \frac{(O-E)^2}{E} = \frac{(5-10)^2}{10} + \frac{(11-6)^2}{6} + \frac{(20-15)^2}{15} + \frac{(4-9)^2}{9}$$

$$= \frac{25}{10} + \frac{25}{6} + \frac{25}{15} + \frac{25}{9}$$

$$= 2.5 + 4.17 + 1.67 + 2.78 = 11.12$$

Degrees of Freedom:

	Rent	Own	Total
Have Children	5		16
No Children			24
Total	25	15	40

NOTE: Given $\underline{N}$ of $\underline{one}$ cell and marginal totals, all other cells' $\underline{N}$s are determined.

$\underline{df}$=(Number of Rows-1)(Number of Columns-1)=(2-1)(2-1)=1

$\phi = \sqrt{(X^2/\underline{N})} = \sqrt{(11.12/40)} = \sqrt{.28} = .53$, large effect size.

Aron/Aron
STATISTICS FOR PSYCHOLOGY

© 1994 by Prentice-Hall, Inc.
A Paramount Communications Company
Englewood Cliffs, New Jersey 07632

TRANSPARENCY 14.12

Young married women who have versus do not have children (first nominal variable) and either rent or own their home (second nominal variable). (Fictional data.)

Steps of hypothesis testing:

1. Reframe into a research hypothesis and a null hypothesis about populations.

   Population 1: Young married women like those surveyed.
   Population 2: Young married women for whom renting or owning their home is independent of whether they have children.

   Research Hypothesis: Two populations are different (renting or owning is not independent of having children).
   Null Hypothesis: Two populations are the same (renting or owning is independent of having children).

2. Determine the characteristics of the comparison distribution.

   Chi-square distribution with one degree of freedom.

3. Determine the cutoff sample score on the comparison

   .05 level, <u>df</u>=1: X^2 = 3.841.

4. Determine score of your sample on comparison distribution.

   X^2 = 11.12.

5. Compare scores obtained in Steps 3 and 4 to decide whether to reject the null hypothesis.

   X^2 in Step 4 (11.12) is more extreme than Step 3 cutoff (3.841).

   Therefore, reject the null hypothesis; the research hypothesis is supported (among young married women like those surveyed, renting or owning appears not to be independent of having children).

6. Examine effect size and interpret.

   ϕ = .53, a large effect size.

Aron/Aron
STATISTICS FOR PSYCHOLOGY

© 1994 by Prentice-Hall, Inc.
A Paramount Communications Company
Englewood Cliffs, New Jersey 07632

TRANSPARENCY 14.13

TABLE 14-9
Results and Computation of the Chi-Square Test for Independence Comparing Whether Subjects Rate Pride as an Emotion Differently According to the Definition of Pride Given

		Response		Total
		Yes	*No*	
Definition	*Satisfaction*	47 (43.3)	14 (17.7)	61 (52.14%)
	Sense of dignity	36 (39.7)	20 (16.3)	56 (47.86%)
		83	34	117

Degrees of freedom $= (N_C - 1)(N_R - 1) = (2 - 1)(2 - 1) = (1)(1) = 1$.

Chi-square needed, $df = 1$, .05 level: 3.841.

$$x^2 = S\frac{(O - E)^2}{E} = \frac{(47 - 43.3)^2}{43.3} + \frac{(14 - 17.7)^2}{17.7} + \frac{(36 - 39.7)^2}{39.7} + \frac{(20 - 16.3)^2}{16.3}$$

$$= .32 + .77 + .34 + .84 = 2.27.$$

Conclusion: Do not reject the null hypothesis.

Note. Data from Russell (1991).

© 1994 by Prentice-Hall, Inc.
A Paramount Communications Company
Englewood Cliffs, New Jersey 07632

Aron/Aron
STATISTICS FOR PSYCHOLOGY

TRANSPARENCY 14.14

Rating of whether or not pride is an emotion (first nominal variable) and definition of pride given (second nominal variable). (From Russell, 1991)

Steps of hypothesis testing:

1. Reframe into a research hypothesis and a null hypothesis about populations.

   Population 2: People like those surveyed.
   Population 1: People for whom whether or not pride is an emotion is independent of how it is defined.

   Research Hypothesis: Two populations are different.
   Null Hypothesis: Two populations are the same.

2. Determine the characteristics of the comparison distribution.

   Chi-square distribution with one degree of freedom.

3. Determine the cutoff sample score on the comparison.

   .05 level, $\underline{df}$=1: X^2 = 3.841.

4. Determine score of your sample on comparison distribution.

   X^2 = 2.27.

5. Compare scores obtained in Steps 3 and 4 to decide whether to reject the null hypothesis.

   X^2 in Step 4 (2.27) is less extreme than Step 3 cutoff (3.841).

   Therefore, do not reject the null hypothesis; the study is inconclusive.

6. Examine effect size and interpret.

   $\phi = \sqrt{(X^2/\underline{N})} = \sqrt{(2.27/117)} = \sqrt{(.0194)} = .14$, small effect size.

   From tables, power to achieve a large effect size is near 100%. However, power for a medium effect size is only 56%.

Aron/Aron
STATISTICS FOR PSYCHOLOGY

© 1994 by Prentice-Hall, Inc.
A Paramount Communications Company
Englewood Cliffs, New Jersey 07632

TRANSPARENCY 14.15

Students of both genders (first nominal variable) who report either having been abused somewhat or not having been abused at all as a child (second nominal variable). (From class questionnaire.)

	Women	Men	Total	
Not Abused	50	33	83	80.6%
Abused	18	2	20	19.4%
Total	68	35	103	100.0%

Expected frequencies: Expected Freq = (Row %) X (Column Total)

Not Abused-Women = (80.6%)(68) = 54.8
Not Abused-Men   = (80.6%)(35) = 28.2
Abused-Women = (19.4%)(68) = 13.2
Abused-Men   = (19.4%)(35) = 6.8

Table showing Observed (and Expected) Frequencies

	Women	Men	Total
Not Abused	50(54.8)	33(28.2)	83(83)
Abused	18(13.2)	2(6.8)	20(20)
Total	68(68.0)	35(35.0)	103(103)

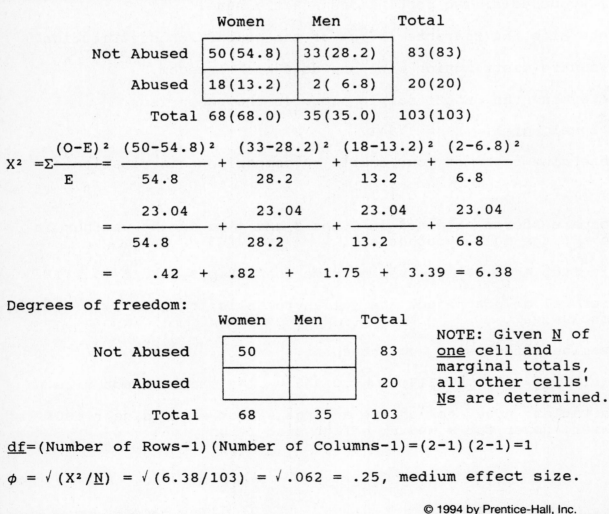

$$X^2 = \Sigma \frac{(O-E)^2}{E} = \frac{(50-54.8)^2}{54.8} + \frac{(33-28.2)^2}{28.2} + \frac{(18-13.2)^2}{13.2} + \frac{(2-6.8)^2}{6.8}$$

$$= \frac{23.04}{54.8} + \frac{23.04}{28.2} + \frac{23.04}{13.2} + \frac{23.04}{6.8}$$

$$= .42 + .82 + 1.75 + 3.39 = 6.38$$

Degrees of freedom:

	Women	Men	Total
Not Abused	50		83
Abused			20
Total	68	35	103

NOTE: Given <u>N</u> of <u>one</u> cell and marginal totals, all other cells' <u>N</u>s are determined.

<u>df</u>=(Number of Rows-1)(Number of Columns-1)=(2-1)(2-1)=1

$\phi = \sqrt{(X^2/\underline{N})} = \sqrt{(6.38/103)} = \sqrt{.062} = .25$, medium effect size.

TRANSPARENCY 14.16

Students of both genders (first nominal variable) who report either having been abused somewhat or not having been abused at all as a child (second nominal variable). (From class questionnaire.)

Steps of hypothesis testing:

1. Reframe into a research hypothesis and a null hypothesis about populations.

   Population 1: Psychology students in general.
   Population 2: Psychology students for whom gender is independent of whether they were abused.

   Research Hypothesis: Two populations are different (gender is not independent of having been abused).
   Null Hypothesis: Two populations are the same (gender is independent of having been abused).

2. Determine the characteristics of the comparison distribution.

   Chi-square distribution with one degree of freedom.

3. Determine the cutoff sample score on the comparison

   .05 level, $\underline{df}$=1: X^2 = 3.841.

4. Determine score of your sample on comparison distribution.

   X^2 = 6.38.

5. Compare scores obtained in Steps 3 and 4 to decide whether to reject the null hypothesis.

   X^2 in Step 4 (6.38) is less extreme than Step 3 cutoff (3.841).

   Therefore, reject the null hypothesis; the research hypothesis is supported (among psychology students gender does seem to be related to having been abused as a child).

6. Examine effect size and interpret.

   ϕ = .25, medium effect size.

Aron/Aron
STATISTICS FOR PSYCHOLOGY

© 1994 by Prentice-Hall, Inc.
A Paramount Communications Company
Englewood Cliffs, New Jersey 07632

TRANSPARENCY 14.17

Where a married couple met (first nominal variable) and
ethnicity (second nominal variable). (Fictional data.)

Data and computations:

Where Met	Ethnic Group A	B	C	Total	
Parent's home	6	24	20	50	28%
School/college	8	12	30	50	28%
Friend's home	20	10	0	30	17%
At work	16	4	30	50	28%
Total	50	50	80	180	100%

Expected Frequency: (Row %)(Column Total)

Parent's home A =(28%)(50) = 14   Friend's home A =(17%)(50)= 8.5
Parent's home B =(28%)(50) = 14   Friend's home B =(17%)(50)= 8.5
Parent's home C =(28%)(80) = 22.4   Friend's home C =(17%)(80)= 13.6
School/college A =(28%)(50)= 14   At work   A =(28%)(50)= 14
School/college B =(28%)(50)= 14   At work   B =(28%)(50)= 14
School/college C =(28%)(80)= 22.4   At work   C =(28%)(80)= 22.4

Where Met	Ethnic Group A	B	C	Total	
Parent's home	6(14)	24(14)	20(22.4)	50	28%
School/college	8(14)	12(14)	30(22.4)	50	28%
Friend's home	20(8.5)	10(8.5)	0(13.6)	30	17%
At work	16(14)	4(14)	30(22.4)	50	28%
Total	50	50	80	180	100%

$$X^2 = (6-14)^2/14 + (24-14)^2/14 + (20-22.4)^2/22.4 + (8-14)^2/14$$

$$+ (12-14)^2/14 + (30-22.4)^2/22.4 + (20-8.5)^2/8.5 +$$

$$+ (10-8.5)^2/8.5 + (0-13.6)^2/13.6 + (16-14)^2/14$$

$$+ (4-14)^2/14 + (30-22.44)^2/22.4 = 69.1$$

Aron/Aron
STATISTICS FOR PSYCHOLOGY

TRANSPARENCY 14.18

Where a married couple met (first nominal variable) and ethnicity (second nominal variable). (Fictional data.)

Data and computations (continued):

Degrees of Freedom:

Where Met	Ethnic Group			Total
	A	B	C	
Parent's home	6	24		50
School/college	8	12		50
Friend's home	20	10		30
At work				50
Total	50	50	80	180

   NOTE: Given the $\underline{N}$s in $\underline{six}$ cells and the marginal totals, all other cells' $\underline{N}$s are determined.

$\underline{df}$ = ($\underline{N}_R$-1)($\underline{N}_C$-1) = (4-1)(3-1) = (3)(2) = 6

Cramer's ϕ = $\sqrt{(X^2/[\underline{N}][\underline{df}_L])}$ = $\sqrt{(69.1/[180][2])}$ = $\sqrt{(69.1/360)}$

           = $\sqrt{1.92}$ = .44, large effect size.

Aron/Aron
STATISTICS FOR PSYCHOLOGY

TRANSPARENCY 14.19

Where a married couple met (first nominal variable) and
ethnicity (second nominal variable). (Fictional data.)

Steps of hypothesis testing:

1. Reframe into a research hypothesis and a null hypothesis about
   populations.

   Population 1: Married couples in general.
   Population 2: Married couples for whom ethnicity is independent
                 of where they initially met.

   Research Hypothesis: Two populations are different (ethnic group is
                        not independent of where initially met).
   Null Hypothesis: Two populations are the same (ethnic group is
                    independent of where initially met).

2. Determine the characteristics of the comparison distribution.

   Chi-square distribution with six degrees of freedom.

3. Determine the cutoff sample score on the comparison.

   .01 level, $\underline{df}$=6: X^2 = 16.812.

4. Determine score of your sample on comparison distribution.

   X^2 = 69.1.

5. Compare scores obtained in Steps 3 and 4 to decide whether to
   reject the null hypothesis.

   X^2 in Step 4 (69.1) is more extreme than Step 3 cutoff (16.812).

   Therefore, reject the null hypothesis; the research hypothesis is
   supported (among married couples like those studied, where they met
   seems to be related to their ethnicity).

6. Examine effect size and interpret.

   Cramer's ϕ = .44, large effect size.

Aron/Aron
STATISTICS FOR PSYCHOLOGY

More examples of the chi-square test for independence. (From class questionnaire.)

Gender X whether answered "Not at All" to item "Was alcoholism a problem in your immediate family when you were growing up?"

	Women	Men	Total	
Alcoholism not a problem	43	25	68	66%
Alcoholism was a problem	25	10	35	34%
Total	68	35	103	100.0%

X^2(1, $\underline{N}$ = 103) =.691, $\underline{ns}$; ϕ = .08, small effect size.

Attachment style X whether abused or not as a child:

	Secure	Avoidant	Anxious	Total	
Not Abused	50	18	11	79	80.6%
Abused	5	7	7	19	19.4%
Total	55	25	18	98	100.0%

X^2(2, $\underline{N}$ = 98) = 9.30, $\underline{p}$ < .01; Cramer's ϕ = .31, medium effect size.

Aron/Aron
STATISTICS FOR PSYCHOLOGY

© 1994 by Prentice-Hall, Inc.
A Paramount Communications Company
Englewood Cliffs, New Jersey 07632

TRANSPARENCY 14.21

How to Conduct a Chi-Square Test for Independence

1. Reframe the question into a research hypothesis and a null hypothesis about populations.

   A. Populations:

      1. Population 1 are people like those in the study.
      2. Population 2 are people whose distribution of cases over categories on the first variable is independent of the distribution of cases over categories for the second variable.

   B. Hypotheses.

      1. Research hypothesis: Two populations have different distributions of cases over categories.
      2. Null Hypothesis: Two populations have the same distributions of cases over categories.

2. Determine the characteristics of the comparison distribution:

   A. Chi-square distribution.
   B. $\underline{df}$ = $(\underline{N}_C-1)(\underline{N}_R-1)$.

3. Determine the cutoff sample score on the comparison distribution at which the null hypothesis should be rejected.

   A. Determine the desired significance level.
   B. Look up the appropriate cutoff on a chi-square table, using the degrees of freedom calculated above.

4. Determine the score of your sample on the comparison distribution.

$$X^2 = \Sigma \frac{(O-E)^2}{E}$$

5. Compare the scores obtained in Steps 3 and 4 to decide whether to reject the null hypothesis.

6. Examine effect size and interpret.

   A. If a 2 X 2 table:

      1. $\phi = \sqrt{(X^2/\underline{N})}$
      2. Rules of thumb: small=.10, medium=.30, large=.50.

   B. If greater than a 2 X 2 table

      1. Cramer's $\phi = \sqrt{(X^2/[\underline{N}][\underline{df}_L])}$
      2. Rules of thumb depend on $\underline{df}_L$.

Aron/Aron
STATISTICS FOR PSYCHOLOGY

Chapter 15
Strategies When Population Distributions Are Not Normal:
Data Transformations, Rank-Order Tests, and Computer-Intensive Methods

Instructor's Summary of Chapter

Normal distribution assumption in standard parametric tests. The *t* test, analysis of variance, and the significance tests for correlation and regression all assume that populations are normally distributed. While these parametric statistical tests seem to be robust over many types of moderate violations of this assumption, when the population is severely nonnormal they can permit too many Type I or Type II errors. It is difficult to assess from the sample whether a population is normal, but outliers are clearly a problem, as is extreme skewness or kurtosis.

Data transformations. One approach when the populations appear nonnormal is to transform the scores mathematically, such as taking the square root, log, or inverse of each score so that the distribution of transformed scores appears to represent a normally distributed population. The ordinary hypothesis-testing procedures can then be applied.

Rank-order methods. Another approach is to rank all of the subjects based on their scores on a variable. Special rank-order statistical tests ("nonparametric" or "distribution-free" tests) exist which use simple principles of probability to determine the chance of the ranks being unevenly distributed across experimental groups. However, in many cases, using the rank-transformed data in an ordinary parametric test produces an acceptable approximation.

Computer-intensive methods. An example of a computer-intensive method is a randomization test, which considers every possible rearrangement of the scores obtained in a study to determine the probability of the obtained arrangement (in terms, for example, of the difference in means between groups) arising by chance.

Comparison of methods. Data transformations permit the use of familiar parametric techniques but can not always be applied and may distort the meaning of the data. Rank-order methods can be applied to any data set, are especially appropriate with rank or rank-like data, and have a straightforward conceptual foundation. But rank-order techniques have not been developed for many complex data-analysis situations and like other kinds of data transformations, information may be lost or meaning distorted. Computer-intensive methods are widely applicable—sometimes to situations in which no other method exists—and have an appealing direct logic. But they are unfamiliar to researchers; being new, their possible limitations are not well worked out, and they can be difficult to set up as they are not provided on standard computer programs. There is little consensus about how the various techniques compare in terms of the relative risks of Type I and Type II errors when populations are not normal.

The procedures of this chapter as described in research articles. Data transformations are usually described and justified at the beginning of the Results sections, and rank-order methods are described much like any other kind of hypothesis test. Computer-intensive methods, being less familiar, are typically described in more than the usual detail.

Box 15.1. Where do Random Numbers Come From? Describes the history of the production of lists of random numbers, including how pseudorandom numbers are generated today by computer.

Lecture 15.1: Nonnormal Distributions and Data Transformations

Materials

Lecture outline

Transparencies 15.1 through 15.11

    (If using transparencies based on your class's questionnaires, replace 15.4, 15.5, 15.6, and 15.10 with 15.4R, 15.5R, 15.6R, and 15.10R.)

Outline for Blackboard

  I.   Review
  II. Assumptions for Parametric Tests
  III. Solutions to Suspected Nonnormal Populations
  IV. Data Transformations
  V.   Rank-Order Tests
  VI. Review this Class

Instructor's Lecture Outline

I.   Review

  A. Idea of descriptive and inferential statistics.

  B. Major inferential statistical techniques (parametric tests).

     1. t test.

     2. Analysis of variance.

     3. Significance test for correlation and regression.

II. Assumptions for Parametric Tests

  A. Assumptions are requirements for the statistical procedure to work properly. They are built into the logic or mathematics.

  B. Two main assumptions.

     1. Normal populations—if not met, then comparison distribution (t or F) is not the shape it is supposed to be (and thus tabled values are wrong).

     2. Equal population variances for each group—if not met, then the idea of pooling does not work and we can not compute a single estimate.

  C. Almost never met exactly in practice, but close is good enough—in most cases, these techniques are robust.

  D. Main emphasis has been on normal distribution assumption and the problems of skew, kurtosis, and outliers.

  E. How to know if a population is normal based on sample data.

     1. Show TRANSPARENCY 15.1 (sample histograms from normally distributed populations from text) and discuss.

     2. Principle: If sample not highly nonnormal, we assume population is normal. (Innocent until proven guilty.)

III. Solutions to Suspected Nonnormal Populations

  A. Data transformations (to be discussed today).

  B. Rank-order methods (also called "nonparametric tests," to be discussed today).

  C. Computer-intensive methods (to be discussed next class).

IV. Data Transformations

A. Principle: If the distribution is not normal, change the scale of the scores so that the distribution is closer to normal.

B. Example: Show TRANSPARENCY 15.2 (History vs. English majors on love for classical music) and discuss.
1. Need for transformation.
2. Method of carrying out transformation.
4. Result of transformation.
5. Next step would be to carry out the t test for independent means in the usual way.

C. Legitimacy.
1. Underlying numerical meaning of scale is arbitrary.
2. Done in advance of data analysis and to all scores in all groups, so there is no systematic bias in favor of the researcher's hypotheses.
3. Does not alter the order of the scores.

D. Methods of transformation.
1. Each of the major methods we consider is used when there is positive skew. (When there is negative skew, other transformations can be used, or the scores can be first reflected.)
2. Show TRANSPARENCY 15.3 (effects of transformations on distribution shape, from text) and discuss three kinds of transformations covered.
   a. Square root, to correct a moderate positive skew.
   b. Log, to correct a strong positive skew.
   c. Inverse, to correct a very strong positive skew.

E. Additional examples.
1. Show TRANSPARENCIES 15.4 and 15.5 or 15.4R and 15.5R (ages of women who do or do not make a point of avoiding watching violent TV, from class questionnaire) and discuss.
2. Show TRANSPARENCY 15.6 or 15.6R (ratings of tending to fall in love as a child for those who do or do not make a point of avoiding watching violent TV, from class questionnaire) and discuss.
3. Show TRANSPARENCY 15.7 (number of books read by highly sensitive and not highly sensitive children, from text) and discuss.

V. Rank-Order Tests

A. Basic approach.
1. Transform scores to ranks.
   a. Ignoring which group a subject is in if it is a comparison of means of groups (as in a t test or ANOVA situation).
   b. Within each variable in a correlation situation.
2. This creates a distribution with known characteristics (it is rectangular, with mean and variance depending only on how many cases are ranked).
3. Thus, the possibility of any particular distribution of ranks (such as a particular mean or ranks in one group, or a particular high correlation of ranks) can be computed exactly by principles of probability.
4. Specific tests exist for the computations involved in computing the ranks and the cutoff means or sums or correlations of ranks for various hypothesis testing situations. Show TRANSPARENCY 15.8 (table of rank-order tests from text) and discuss.

B. Simply applying standard parametric statistics to ranked data provides acceptable approximations.

C. Examples.
1. Show TRANSPARENCY 15.9 (number of books read by highly sensitive and not highly sensitive children, from text) and discuss.
2. Show TRANSPARENCY 15.10 or 15.10R (ages of women who do or do not make a point of avoiding watching violent TV, from class questionnaire) and discuss.
3. Show TRANSPARENCY 15.11 (History vs. English majors on number of records listened to) and discuss.

VI. Review this Class: Use blackboard outline.

Lecture 15.2: Computer-Intensive Methods

Materials

Lecture outline

Transparencies 15.12 through 15.16

    (If using transparencies based on your class's questionnaires, replace 15.15 with 15.15R.)

Outline for Blackboard

I.   Review
II. Randomization Tests
III. Approximate Randomization Tests
IV. Comparison of Methods with Nonnormal Population Distributions
V.   Review this Class

Instructor's Lecture Outline

I.   Review

    A. Idea of descriptive and inferential statistics.

    B. Assumptions of standard parametric tests.

        1. Normal population distribution and equal population variances.

        2. Robustness fails when there is extreme skewness or kurtosis or outliers.

        3. Difficulty of determining whether population meets assumptions based on samples.

    C. Data transformation methods.

    D. Rank-order tests.

II. Randomization Tests

    A. These are an example of computer-intensive methods—other examples are bootstrap tests, the jackknife, and Monte Carlo procedures.

    B. Principle: Take every possible reorganization of scores in your sample and find what percent of results are higher than the one obtained with the actual organization.

    C. Examples.

        1. Show TRANSPARENCIES 15.12 and 15.13 (sensitive children study from text) and discuss.

        2. Show TRANSPARENCY 15.14 (History vs. English majors on number of records listened to) and discuss.

    D. Procedure.

        1. Determine number of organizations and how many more extreme combinations would be possible for your sample to be in the top 5% (or 1%, etc.).

        2. Lay out all the possible organizations.

           a. For comparisons of group means, reorder all scores, ignoring which group they were originally in, keeping number in each group fixed.

           b. For repeated-measures, reorder scores within each testing, but do not shuffle scores between testings.

           c. For correlation, reorder scores within each variable, but do not shuffle between variables.

        3. Compute the difference between means or other statistic (such as the correlation) for each organization.

        4. Order these from lowest to highest.

        5. Locate where your sample's actual difference (or correlation, etc.) falls in this order.

        6. Compare Step 5 to Step 1 cutoff to determine statistical significance.

III. Approximate Randomization Tests

    A. Used because, with reasonable numbers of subjects, regular randomization tests become unwieldy (too many different ways to reorganize the data).

    B. Principle: Randomly select a large number (such as 1,000) of the possible reorganizations of the data and see if your actual sample falls in the top 5% (or 1%, etc.) of these.

    C. Example: Show TRANSPARENCY 15.15 or 15.15R (ages of women who do or do not make a point of avoiding watching violent TV, from class questionnaire) and discuss.

IV. Comparison of Methods with Nonnormal Population Distributions: Show TRANSPARENCY 15.16 (table of comparisons) and discuss.

VI. Review this Class: Use blackboard outline.

TRANSPARENCY 15.1

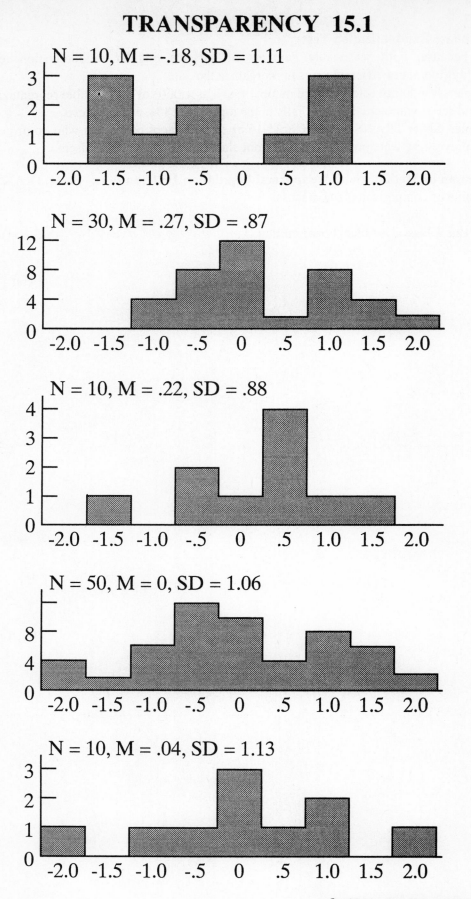

N = 10, M = -.18, SD = 1.11

N = 30, M = .27, SD = .87

N = 10, M = .22, SD = .88

N = 50, M = 0, SD = 1.06

N = 10, M = .04, SD = 1.13

Aron/Aron
STATISTICS FOR PSYCHOLOGY

© 1994 by Prentice-Hall, Inc.
A Paramount Communications Company
Englewood Cliffs, New Jersey 07632

TRANSPARENCY 15.2

Data transformation example.   (Fictional data.)

IV: Major
DV: Number of classical music records
      listened to in last week

	History	English
	4	36
	16	9
	9	0
	9	9
Σ	38	54
M	9.5	13.5

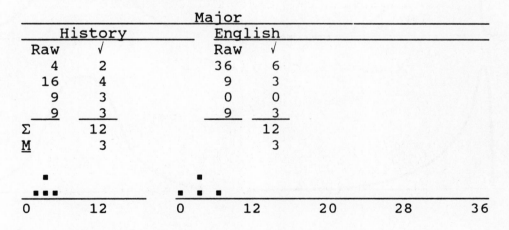

Square-root transformation:

	History		English	
	Raw	√	Raw	√
	4	2	36	6
	16	4	9	3
	9	3	0	0
	9	3	9	3
Σ		12		12
M		3		3

TRANSPARENCY 15.3

Before   After

(a)

(b)

(c)

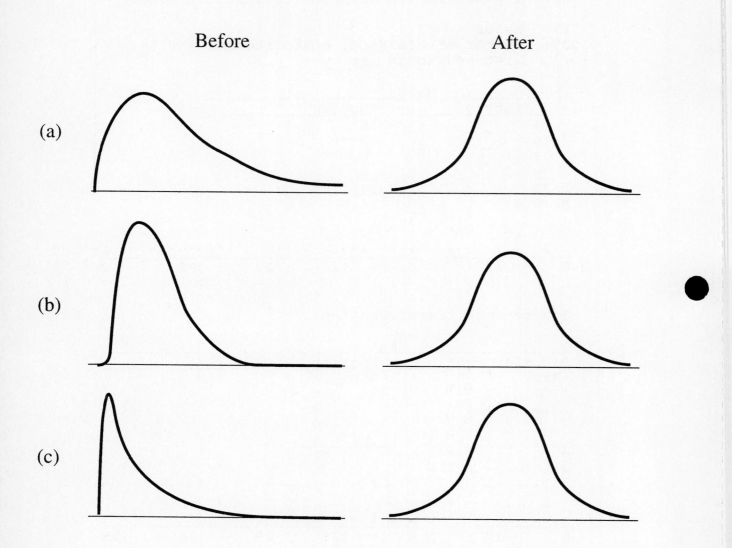

Aron/Aron
STATISTICS FOR PSYCHOLOGY

© 1994 by Prentice-Hall, Inc.
A Paramount Communications Company
Englewood Cliffs, New Jersey 07632

TRANSPARENCY 15.4

Example of data suggesting nonnormal distributions.
(From class questionnaire.)

Sample: Students in class
IV: Do or do not make a point of avoiding watching violent TV
DV: Ages of women in class

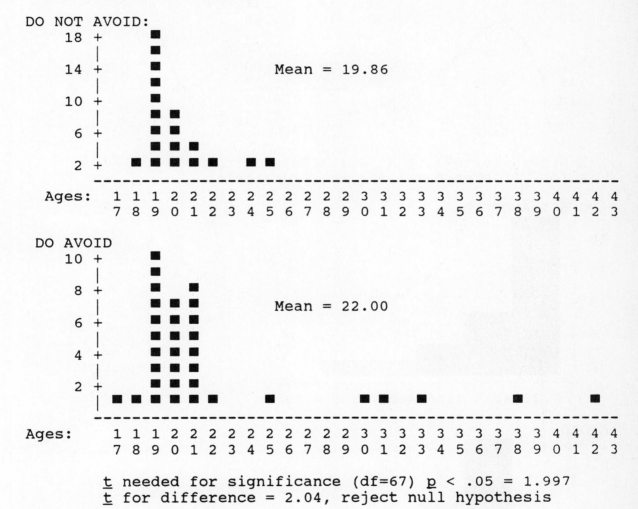

DO NOT AVOID:

Mean = 19.86

DO AVOID

Mean = 22.00

t needed for significance (df=67) p < .05 = 1.997
t for difference = 2.04, reject null hypothesis

TRANSPARENCY 15.5

Example of applying square root transformation to data appearing to have nonnormal distributions. (From class questionnaire.)

Sample: Students in class
DV: Ages of women
IV: Do or do not make a point of avoiding watching violent TV

Square root transformation:

Do not avoid				Avoid		
S#	Age	√Age		S#	Age	√Age
1	21	4.58		36	21	4.58
2	25	5.00		37	38	6.16
3	20	4.47		38	30	5.48
.	.	.		.	.	.
.	.	.		.	.	.
.	.	.		68	25	5.00
34	19	4.36		69	31	5.57
35	19	4.36				158.10
Σ =		155.75				4.65
M =		4.45				

$\underline{t}$ needed (df=67), $\underline{p}$ < .05: 1.997
$\underline{t}$ for difference = 1.93; not significant.

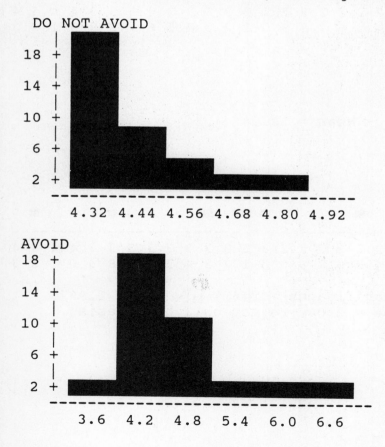

DO NOT AVOID

(x-axis: 4.32 4.44 4.56 4.68 4.80 4.92)

AVOID

(x-axis: 3.6 4.2 4.8 5.4 6.0 6.6)

Aron/Aron
STATISTICS FOR PSYCHOLOGY

TRANSPARENCY 15.6

Example of applying data transformation to data appearing to have nonnormal distributions. (From class questionnaire.)
DV: "Did you tend to fall in love in your early school years?"
IV: Do or do not make a point of avoiding watching violent TV

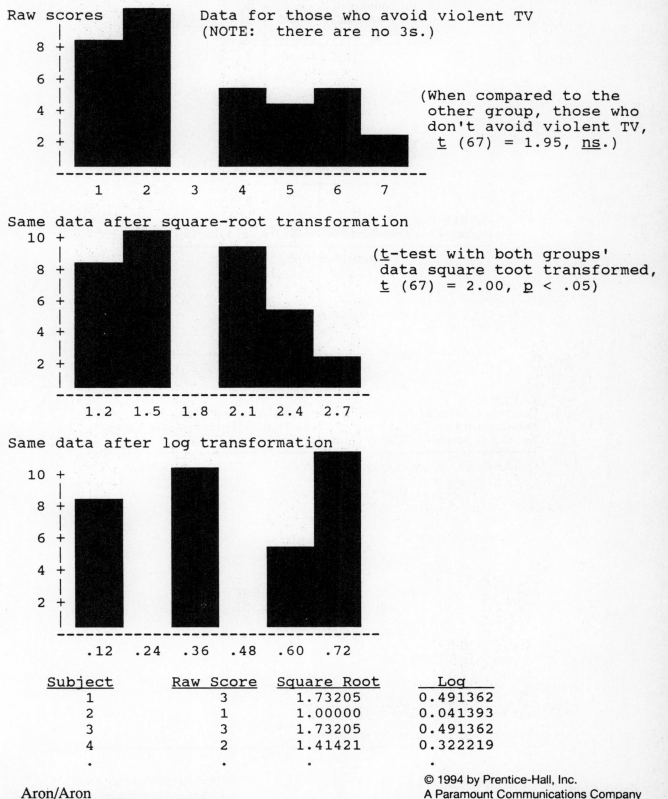

Raw scores   Data for those who avoid violent TV
                    (NOTE: there are no 3s.)

(When compared to the other group, those who don't avoid violent TV, t (67) = 1.95, ns.)

Same data after square-root transformation

(t-test with both groups' data square toot transformed, t (67) = 2.00, p < .05)

Same data after log transformation

Subject	Raw Score	Square Root	Log
1	3	1.73205	0.491362
2	1	1.00000	0.041393
3	3	1.73205	0.491362
4	2	1.41421	0.322219
.	.	.	.

Aron/Aron
STATISTICS FOR PSYCHOLOGY

TABLE 15-1
Results of a Study Comparing Highly Sensitive and Not Highly Sensitive Children on the Number of Books Read in the Past Year (Fictional Data)

	Highly Sensitive	
	No	*Yes*
	0	17
	3	36
	10	45
	22	75
Σ:	35	173
$M =$	8.75	43.25
$S^2 =$	95.58	584.00

TABLE 15-2
Square-Root Transformation of the Data in Table 15-1

Highly Sensitive			
No		*Yes*	
X	$\div X$	X	$\div X$
0	0.00	17	4.12
3	1.73	36	6.00
10	3.16	45	6.71
22	4.69	75	8.66

TABLE 15-3
Computations for a *t* Test for Independent Means Using Square-Root-Transformed Data for the Study of Books Read by Highly Sensitive Versus Not Highly Sensitive Children (Fictional Data)

t needed for .05 significance level, $df = (4-1) + (4-1) = 6$, one tailed = 1.943.

	Highly Sensitive	
	No	*Yes*
	0.00	4.12
	1.73	6.00
	3.16	6.71
	4.69	8.66
Σ:	9.58	25.49
$M =$	9.58/4 = 2.40	25.49/4 = 6.37
$S^2 =$	12.03/3 = 4.01	10.56/3 = 3.52

$$S_P^2 = 3.77$$

$S_M^2 =$	3.77/4 = .94	3.77/4 = .94

$S_{DIF}^2 = .94 + .94 = 1.88$
$S_{DIF} = \div 1.88 = 1.37$
$t = (6.37 - 2.40)/1.37 = 2.90$

Conclusion: Reject the null hypothesis.

Aron/Aron
STATISTICS FOR PSYCHOLOGY

© 1994 by Prentice-Hall, Inc.
A Paramount Communications Company
Englewood Cliffs, New Jersey 07632

TABLE 15-7
Computations for a Wilcoxin Rank-Sum Test for the Study of Books Read by Highly Sensitive Versus Not Highly Sensitive Children (Fictional Data)

Cutoff for significance: Maximum sum of ranks in the not highly sensitive group for significance at the .05 level, one-tailed (from a standard table) = 11.

Highly Sensitive

No			Yes	
X	Rank		X	Rank
0	1		17	4
3	2		36	6
10	3		45	7
22	5		75	8
Σ:	11			

Comparison to cutoff: Sum of ranks of group predicted to have lower scores, 11, equals but does not exceed cutoff for significance.

Conclusion: Reject the null hypothesis.

© 1994 by Prentice-Hall, Inc.
A Paramount Communications Company
Englewood Cliffs, New Jersey 07632

TABLE 15-8
Computations for a *t* test for Independent Means Using Ranks Instead of Raw Scores for the Study of Books Read by Highly Sensitive Versus Not Highly Sensitive Children (Fictional Data)

t needed for .05 significance level, $df = (4-1) + (4-1) = 6$, one-tailed $= -1.943$

Highly Sensitive

	No	Yes
	1	4
	2	6
	3	7
	5	8
Σ	11	25
$M =$	$11/4 = 2.75$	$25/4 = 6.25$
$S^2 =$	$8.75/3 = 2.92$	$8.75/3 = 2.92$

$$S_P^2 = 2.92$$

$S_M^2 =$	$2.92/4 = .73$	$2.92/4 = .73$

$S_{DIF}^2 = .73 + .73 = 1.46$
$S_{DIF} = \div 1.46 = 1.21$
$t = (2.75 - 6.25)/1.21 = -2.89$

Conclusion: Reject the null hypothesis.

TABLE 15-9
Randomization Test Computations for the Study Comparing Highly Sensitive and Not Highly Sensitive Children on the Number of Books Read in the Past Year (Fictional Data)

Actual Results:

Highly Sensitive

	No	Yes
	0	17
	3	36
	10	45
	22	75
Σ	35	173
$M =$	8.75	43.25

Actual difference $= M_{Yes} - M_{No} = 34.5$
Needed to reject the null hypothesis: This mean difference must be in top 5% of mean differences. With 70 mean differences, it must be among the three highest differences.

All Possible Divisions (70) of the Eight Scores into Two Groups of Four Each:

	Actual		No	Yes		No	Yes		No	Yes		No	Yes		No	Yes		No	Yes
	No	Yes		0	22		0	22		0	22		0	22		0	10		0	10
	0	17		3	36		3	17		3	17		3	17		3	36		3	17
	3	36		10	45		10	45		10	36		10	36		22	45		22	45
	10	45		17	75		36	75		45	75		75	45		17	75		36	75
	22	75																		
$M_{Yes} - M_{No}$	34.5			37			27.5			23			8			31			21.5	

Seventy Differences Ordered from Lowest (Most Negative) to Highest:

−37, −37, −34.5, −32, −27.5, −27.5, −26, −21.5, −24, −23, −20.5, −16, −16.5, −17, −18, −19, −14.5, −13.5, −13, −12, −10, −10, −10, −9.5, −8.5, −8, −6.5, −5, −5, −5, −4.5, −1.5, −3, −2, −1, 1, 1.5, 2, 3, 4.5, 5, 5, 5, 6.5, 8, 8.5, 9.5, 10, 10, 10, 12, 13, 13.5, 14.5, 16, 16.5, 17, 18, 19, 20.5, 21.5, 23, 24, 26, 27.5, 27.5, 31, 34.5, 37, 37

Conclusion: Actual mean difference is among the three highest. Reject the null hypothesis.

Example of applying rank transformation to data suggesting
nonnormal distributions. (From class questionnaire.)

DV: Ages of women in class
IV: Do or do not make a point of avoiding watching violent TV

Rank transformation:

Do not avoid				Avoid		
S#	Age	Rank		S#	Age	Rank
1	21	52.5		36	21	52.5
2	25	63.5		37	38	68.0
3	20	39.5		38	30	65.0
·	·	·		·	·	·
·	·	·		·	·	·
·	·	·		68	25	63.5
34	19	18.5		69	31	66.0
35	19	18.5				1344.0
Σ =		1071.0				39.5
M =		30.6				

t needed (df=67), p < .05: 1.997.
t for difference = 1.97. Not Significant.

Aron/Aron
STATISTICS FOR PSYCHOLOGY

Example of study employing rank transformation.
(Fictional data.)

IV: Major
DV: Number of classical music records
     listened to in last week

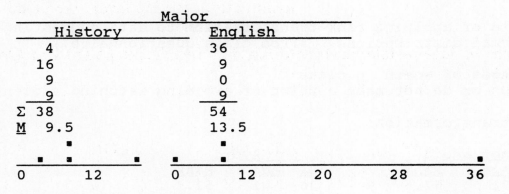

Major	
History	English
4	36
16	9
9	0
9	9
Σ 38	54
M 9.5	13.5

Rank transformation:

(NOTE: ranks from lowest score to highest, ignoring
which group a subject is in; tied ranks--3, 4, 5, 6--
get average rank, 4.5.)

Major			
History		English	
Raw	Rank	Raw	Rank
4	2	36	8
16	7	9	4.5
9	4.5	0	1
9	4.5	9	4.5
Σ	18.0		18.0
M	4.5		4.5

TABLE 15-10
Randomization Test Computation for the Study Correlating Grade Level and Algebra Test Scores (Fictional Data)

Needed to reject the null hypothesis: The actual correlation must be the highest of the 24 possible correlations to reject null hypothesis at the 5% level, one-tailed.

Correlations of All Possible Pairings of Algebra Test Scores (ATS) with Grade Level (GL)

Actual

ATS	GL
1	4
4	6
10	9
95	7
$r = .25$	

ATS	GL
1	6
4	9
10	7
95	4
$r = -.79$	

ATS	GL
1	9
4	7
10	4
95	6
$r = -.24$	

ATS	GL
1	7
4	4
10	6
95	9
$r = .79$	

ATS	GL
1	9
4	6
10	7
95	4
$r = -.82$	

ATS	GL
1	7
4	6
10	9
95	4
$r = -.76$	

ATS	GL
1	9
4	4
10	6
95	7
$r = .12$	

ATS	GL
1	7
4	6
10	4
95	9
$r = .75$	

ATS	GL
1	4
4	7
10	6
95	9
$r = .82$	

ATS	GL
1	4
4	9
10	7
95	6
$r = .52$	

Aron/Aron
STATISTICS FOR PSYCHOLOGY

© 1994 by Prentice-Hall, Inc.
A Paramount Communications Company
Englewood Cliffs, New Jersey 07632

Table 15.10 (Continued)

Randomization Test Computation for the Study Correlating Grade Level and Algebra Test Scores (Fictional Data)

ATS	GL		ATS	GL		ATS	GL		ATS	GL		ATS	GL
1	7		1	4		1	4		1	6		1	6
4	4		4	9		4	7		4	4		4	4
10	9		10	6		10	9		10	9		10	7
95	6		95	7		95	6		95	7		95	9
$r = -.11$			$r = .18$			$r = -.08$			$r = .22$			$r = .82$	

ATS	GL		ATS	GL		ATS	GL		ATS	GL		ATS	GL
1	6		1	6		1	9		1	9		1	7
4	9		4	7		4	4		4	7		4	9
10	4		10	4		10	7		10	6		10	6
95	7		95	9		95	6		95	4		95	4
$r = .11$			$r = .76$			$r = .18$			$r = -.84$			$r = .82$	

ATS	GL		ATS	GL		ATS	GL		ATS	GL
1	4		1	6		1	9		1	7
4	6		4	9		4	7		4	9
10	7		10	4		10	6		10	4
95	9		95	7		95	4		95	6
$r = .84$			$r = .11$			$r = -.84$			$r = -.22$	

Correlations from Lowest to Highest:

−.84, −.84, −.82, −.79, −.76, −.24, −.22, −.18, −.11, −.08, .11, .11, .12, .18, .22, .25, .52, .75, .76, .79, .82, .82, .82, .84

Conclusion: Do not reject the null hypothesis.

Aron/Aron
STATISTICS FOR PSYCHOLOGY

© 1994 by Prentice-Hall, Inc.
A Paramount Communications Company
Englewood Cliffs, New Jersey 07632

Example of randomization test. (Fictional data.)

IV: Major
DV: Number of classical music records

	Major	
History	English	
4	36	
16	9	
9	0	
9	9	
Σ 38	54	
M 9.5	13.5	

Reorganization of data: 70 possibilities. Examples:

Hist	Eng		Hist	Eng		Hist	Eng		Hist	Eng
4	36		36	4		4	36		4	36
16	9		16	9		9	16		16	9
9	0		9	0		9	0		0	9
9	9		9	9		9	9		9	9
38	54		70	22		31	61		29	63
9.5	13.5		7.5	5.5		2.75	15.25		7.25	15.75
DIF=-4			DIF=12			DIF=-12.5			DIF=-8.5	

. . . .

© 1994 by Prentice-Hall, Inc.
A Paramount Communications Company
Englewood Cliffs, New Jersey 07632

TRANSPARENCY 15.15

Example of randomization test. (From class questionnaire.)

DV: Ages of women
IV: Do or do not make a point of avoiding watching violent TV

Actual Sample Data			
Avoid			
Do Not		Do	
S#	Age	S#	Age
1	21	36	21
2	25	37	38
3	20	38	30
.	.	.	.
.	.	68	25
34	19	69	31
35	19		
M	19.8		20.4

Random Reorganization 1			
Avoid			
Do Not		Do	
S#	Age	S#	Age
32	18	6	22
12	25	28	17
43	19	59	21
.	.	.	.
.	.	37	38
3	20	1	21
14	19		
	20.1		20.0

Random Reorganization 2			
Avoid			
Do Not		Do	
S#	Age	S#	Age
69	31	15	33
42	23	7	24
38	30	44	23
.	.	.	.
.	.	21	19
58	29	39	22
5	18		
	19.3		21.7

...

Aron/Aron
STATISTICS FOR PSYCHOLOGY

© 1994 by Prentice-Hall, Inc.
A Paramount Communications Company
Englewood Cliffs, New Jersey 07632

TRANSPARENCY 15.16

Comparison of methods.

Method	Advantages	Disadvantages
Parametric tests	♦ Most power when assumptions are met ♦ Most familiar	♦ Give wrong results when assumptions not met
Transformations	♦ Permit use of familiar parametric tests	♦ May distort meaning of data ♦ Can not always be applied
Rank-Order Methods	♦ Do not distort data (instead just ignore aspects of data) ♦ Easy hand computation ♦ Population distribution known ♦ Can use ordinal data	♦ Loses information ♦ May be less powerful ♦ Less familiar than parametric
Computer-Intensive Methods	♦ Simple basic logic ♦ Very flexible, can be applied to unusual data analysis situations	♦ Unfamiliar ♦ Computer packages for using them not widely available ♦ New; thus impact on power and other issues not all worked out yet

Aron/Aron
STATISTICS FOR PSYCHOLOGY

© 1994 by Prentice-Hall, Inc.
A Paramount Communications Company
Englewood Cliffs, New Jersey 07632

Chapter 16
Integrating What You Have Learned: The General Linear Model

Instructor's Summary of Chapter

The general linear model equates the value of a variable for any individual case with the sum of (a) a constant, (b) the partial, weighted influence of each of several other variables, and (c) error. Correlation and multiple regression/correlation (and associated significance tests), the *t* test, and the analysis of variance are all special cases of the general linear model.

Multiple regression and the general linear model are nearly identical.

Bivariate regression/correlation is the special case of multiple regression/correlation in which there is only one predictor variable.

The t test for independent means as a special case of the analysis of variance. The former can be mathematically derived from the latter. There are many similarities in computation: the *t* score is the square root of the *F* ratio; the numerators of both are built on the differences between group means; the denominators of both are built on the variance within the groups; the denominator of *t* involves dividing by the number of subjects, and the numerator of *F* involves multiplying by the number of subjects; the *t* and *F* denominator degrees of freedom are the same.

The t test for independent means as a special case of the significance test for the correlation coefficient. A correlation measures the degree of association of a predictor or independent variable with a dependent variable. In the same way, by showing a difference between group means, the *t* test identifies an association of the variable on which the groups are divided, the independent or predictor variable, with the dependent variable. If you assign a score of 1 to each subject in one of the two groups and a 2 to each subject in the other group (or any two different numbers) and then compute a correlation of these scores with the dependent variable, the significance of that correlation will be the same as the *t* test would produce. Drawing a scatter diagram of these data would make a column of scores for each group, with the regression line passing through the means of each group. The more the means are different, the greater the proportionate reduction in error from using the grand mean and the greater the *t* score based on a comparison of the two groups' means.

Analysis of variance as a special case of correlation and regression. In both regression and analysis of variance, SS_T is about the deviation of each score from the mean of all the dependent-variable scores. Since a score's group's mean in an analysis of variance is the predicted score for each case in regression, SS_E and SS_W are the same. The reduction in squared error ($SS_T - SS_E$) is the same as the sum of squared deviations of scores' group means from the grand mean (SS_B) in analysis of variance. Finally, regression's proportionate reduction in error (r^2) is the same as the proportion of variance accounted for (R^2), an indicator of effect size in the analysis of variance.

Nominal coding. Any analysis of variance can be set up as a multiple regression by making the categories that represent the different groups into one or more dichotomous numerical variables.

Assumptions. All of these methods share the same assumptions that population distributions are normal and have constant variance over levels of the predictor variable.

Choice of method. Conventional practice (and sometimes confusion) leads to mathematically identical procedures being used in different research contexts as if they were actually different.

Controversy: Views of causality. The regularity view identifies *X* as a cause of *Y* if *X* and *Y* are associated, *X* precedes *Y*, and no other third factors precede *X* that could cause them both. The generative view argues that in addition there must be a clear understanding of the mechanism by which *X* affects *Y*.

Box 16.1. The Golden Age of Statistics: Four Guys around London. Discusses the relations among Galton, Gossett, Pearson, and Fisher, and the circumstances that may have contributed to the rapid development of statistics.

Box 16.2. Two Women Make Their Point about Gender and Statistics. An interview with Linda Fidell, coauthor of an influential multivariate statistics text. The focus is on the role of women in statistics.

Lecture 16.1: Introduction to the General Linear Model and the *t* Test as a Special Case of Analysis of Variance and of Correlation

Materials

Lecture outline

Transparencies 16.1 through 16.13

(If using transparencies based on your class's questionnaires, replace 16.6, 16.10 and 16.11 with 16.6R, 16.10R, and 16.11R.)

Outline for Blackboard

I. **Review of Multiple Regression/Correlation (MRC)**

II. **The General Linear Model**

III. *r* **as a Special Case of MRC**

IV. *t* **as a Special Case of ANOVA**

V. *t* **as a Special Case of** *r*

VI. **Review this Class**

Instructor's Lecture Outline

I. **Review of Multiple Regression/Correlation (MRC)**

   A. Show TRANSPARENCY 16.1 (multiple regression formula and worked out example from text).

   B. The correlation between the set of independent variables and the dependent variable is called a multiple correlation (R).

   C. R^2 is the proportionate reduction in squared error gained by using the multiple-regression prediction rule compared to simply predicting the dependent variable from its mean.

   D. R and R^2 can be tested against the null hypothesis that the population value is 0.

II. **The General Linear Model**

   A. Show TRANSPARENCY 16.2 (general linear model formula) and discuss.

   B. This formula is similar to a raw-score multiple regression formula.

       1. The as, bs and Xs are the same.

       2. However:

           a. The formula is for the actual, not the predicted, score on the dependent variable.

           b. The formula includes error, all influences on the dependent variable not included in the prediction rule.

   C. It is called a linear model because the equation does not include any squared or higher power terms.

   D. Show TRANSPARENCY 16.3 (diagram of relations among procedures) and discuss.

III. *r* **as a Special Case of MRC:** It is the special case in which there is only one predictor variable.

IV. *t* **as a Special Case of ANOVA**

    A. The *t* test for independent means is equivalent to ANOVA for two groups.

    B. Both *t* and *F* can be understood as ratios of signal (numerators of both are based on difference or variation among means of groups) to noise (denominators of both are based on variation within groups).

    C. Show TRANSPARENCY 16.4 (table showing *t* and ANOVA similarities, from text) and discuss.

    D. Examples.

       1. Show TRANSPARENCY 16.5 (table showing *t* and ANOVA parallel computations using job-skills-program example, from text) and discuss.

       2. Show TRANSPARENCY 16.6 or 16.6R (table showing parallel computations for first vs. later borns on avoiding crowds, stat class data) and discuss.

       3. Show TRANSPARENCY 16.7 (table showing parallel computations for example of those reading neutral vs. depressing stories on helping behavior) and discuss.

V. *t* **as a Special Case of** *r*

    A. The significance test for *r*.

       1. Tests the null hypothesis that in the population $r = 0$.

       2. The comparison distribution is a *t* distribution with degrees of freedom equal to number of subjects minus 2.

       3. The score on the comparison distribution is a *t* score, with $t = (r)(\sqrt{[N-2]}) / \sqrt{(1-r^2)}$. (Write formula on board.)

    B. A comparison of means of groups (the focus of a *t* test for independent means) can be thought of as a nominal variable with two levels.

    C. Representing a nominal variable with two levels as a numerical variable, using any two arbitrary numbers to stand for the two levels, permits one to conduct a correlation between that variable and the dependent variable.

       1. Which two numbers you use is arbitrary, since when computing a correlation each variable is converted to Z score, and with just two numbers, whatever the two numbers are, the Z scores always come out the same.

       2. Show TRANSPARENCY 16.8 (nominal coding and worked out correlation for job-skills-program example, from text), discuss and compare to TRANSPARENCY 16.4 *t* test result.

    D. For analyses conducted on the same data.

       1. Both methods give the same *t*.

       2. Both methods give the same degrees of freedom.

       3. Thus, both methods yield the same cutoff *t* score and the same conclusions regarding significance.

    E. Show TRANSPARENCY 16.9 (scatter diagram for the job-skills-program example, from text) and discuss.

       1. The mean of each group is the same as the predicted score for each of the levels of the two-level numerical variable based on the regression equation for it.

       2. The variation between the two means is equivalent to the slope of the regression line—the greater this is, the more likely the *t* is significant.

       3. The variation within each of the groups is equivalent to the spread around each predicted value in the scatter diagram—the smaller this is, the more likely the *t* is significant.

    F. Additional examples.

       1. Show TRANSPARENCIES 16.10 or 16.10R (nominal coding and worked out correlation for first vs. later borns on avoiding crowds, from class questionnaire) and 16.11 or 16.11R (scatter diagram for same example) and discuss.

       2. Show TRANSPARENCIES 16.12 (nominal coding and worked out correlation for example of those reading neutral vs. depressing stories on helping behavior) and 16.13 (scatter diagram for same example) and discuss.

VII. Review this Class: Use blackboard outline and TRANSPARENCIES 16.2, 16.4, 16.5. 16.8, and 16.9.

Lecture 16.2: The Analysis of Variance as a Special Case of Multiple Regression

Materials

Lecture outline

Transparencies 16.1 through 16.4, 16.8, 16.9, and 16.14 through 16.20

(If using transparencies based on your class's questionnaires, replace 16.16 with 16.16R.)

Outline for Blackboard

I. Review
II. ANOVA for Two Groups as a Special Case of *r*
III. ANOVA for More than Two Groups as a Special Case of MRC
IV. Choice of Test when Results Would Be Equivalent
V. Causality
VI. Review this Class

Instructor's Lecture Outline

I. Review

A. Review of MRC. Show TRANSPARENCY 16.1 (multiple regression formula and worked out example from text) and discuss.

B. The general linear model. Show TRANSPARENCY 16.2 (general linear model formula) and discuss, including its relation to the raw-score multiple regression formula.

C. Show TRANSPARENCY 16.3 (diagram of relations among procedures) and discuss.

D. *t* as a special case of ANOVA. Show TRANSPARENCY 16.4 (table showing similarities from text) and discuss.

E. *t* as a special case of *r*.

    1. A comparison of means of groups (the focus of a *t* test for independent means) can be thought of as a nominal variable with two levels, which can be represented as a numerical variable with any two arbitrary values.

    2. Show TRANSPARENCY 16.8 (nominal coding and worked out correlation for job-skills-program example, from text), discuss—note that the *t* is the same as with a *t* test for independent means for the same data.

    3. Show TRANSPARENCY 16.9 (scatter diagram for the job-skills-program example, from text) and discuss.

II. ANOVA for Two Groups as a Special Case of *r*

A. Logic of going from a difference between groups to an association is same as with *t* test for independent means and correlation—the required intermediary is a set of two-level numerical variables.

B. Clearest link is between computations of R^2 in ANOVA using the structural model approach and the r^2 in bivariate regression. Show TRANSPARENCY 16.14 (links between *SS*s in ANOVA and regression) and discuss each step.

C. Examples.

    1. Show TRANSPARENCY 16.15 (table showing parallel computations using job-skills program example, from text) and discuss.

    2. Show TRANSPARENCY 16.16 or 16.16R (table showing parallel computations for first vs. later borns on avoiding crowds, from class questionnaire) and discuss.

    3. Show TRANSPARENCY 16.17 (table showing parallel computations for example of those reading neutral vs. depressing stories on helping behavior) and discuss.

III. ANOVA for More than Two Groups as a Special Case of MRC

A. A comparison of means of several groups = nominal independent variable with as many levels as there are groups.

B. However, such a many-leveled nominal variable can not be made into a single numerical variable because the numerical levels assigned would imply specific ordered and quantitative relations among the levels.

C. Nominal coding.

1. Such a many-leveled nominal variable can be made into a set of two-level numerical variables, such that each represents being in or not being in one of the groups.

2. It takes one less two-leveled variable than there are groups because the last group is represented by a case not being in any of the preceding groups.

3. Show TRANSPARENCY 16.18 (nominal coding of nationality information, from text) and discuss.

4. Show TRANSPARENCY 16.19 (nominal coding of five-different-memory-cues example) and discuss.

D. ANOVA = multiple correlation of dependent variable with set of two-level nominally coded numerical variables.

1. R^2 comes out the same.

2. F comes out the same.

IV. Choice of Test when Results Would Be Equivalent

A. Based in part on tradition and what people are used to.

B. Based in part on confusing a correlational research design with a correlational statistic.

1. Correlational research design involves no random assignment to groups.

2. Correlational research designs usually involve predictor variables of more than two levels (e.g., income level, marital satisfaction), so the correlation statistics are usually used.

3. However, correlational research designs can also be used when the predictor variable is a nominal variable and a t test or ANOVA is an appropriate statistical approach. Examples:

a. t test of first born vs. later born.

b. ANOVA comparing people whose native language is French, Spanish, German, or Italian.

4. Experimental research designs involve true random assignment to groups.

5. Experimental research designs usually involve random assignment to two or a small number of discrete groups, so t tests and analysis of variance are usually used.

6. However, experimental research designs can also randomly assign subjects to different levels of a numerical independent variable, in which case correlation and regression statistics are appropriate.

V. Causality

A. The regularity theory of causality.

1. Associated with philosophers Hume and Mill.

2. Show TRANSPARENCY 16.20 (diagram of requirements for causality) and discuss regularity theory part.

B. The generative theory of causality.

1. Associated with philosophers Aristotle, Aquinas, and Kant.

2. Show TRANSPARENCY 16.20 (diagram of requirements for causality) and discuss generative theory part.

VII. Review this Class: Use blackboard outline and TRANSPARENCIES 16.2, 16.3 16.14, 16.15, 16.18, and 16.20.

TRANSPARENCY 16.1

Multiple Regression Formula:

$$\hat{Y} = a + (b_1)(X_1) + (b_2)(X_2) + (b_3)(X_3) + \cdots$$

Example Prediction Rule:

Predicted Stress = -4.70 + (.56)(Number Supervised)
                        + (.06)(Noise in Decibels)
                        + (.86)(Number of Deadlines per Month)

Example Prediction of a Single Case:

Manger: 4 people supervised
        50 decibel work area
         1 deadline per month

Predicted Stress = -4.70 + (.56)(4) + (.06)(50) + (.86)(1)

                 = -4.70 + 2.24   +   3   +   .86

                 = 1.40

Aron/Aron
STATISTICS FOR PSYCHOLOGY

© 1994 by Prentice-Hall, Inc.
A Paramount Communications Company
Englewood Cliffs, New Jersey 07632

TRANSPARENCY 16.2

General Linear Model

$$\underline{Y} = \underline{a} + (\underline{b}_1)(\underline{X}_1) + (\underline{b}_2)(\underline{X}_2) + (\underline{b}_3)(\underline{X}_3) + \cdot \ \cdot \ \cdot \ \cdot + \underline{e}$$

$\underline{Y}$ is a person's actual score on some dependent variable.

$\underline{a}$ is the fixed influence that applies to all individuals.

$\underline{b}_1$ is the degree of influence of the first predictor variable.

$\underline{e}$ is the error, the sum of all other influences on the person's
score on $\underline{Y}$

Aron/Aron
STATISTICS FOR PSYCHOLOGY

© 1994 by Prentice-Hall, Inc.
A Paramount Communications Company
Englewood Cliffs, New Jersey 07632

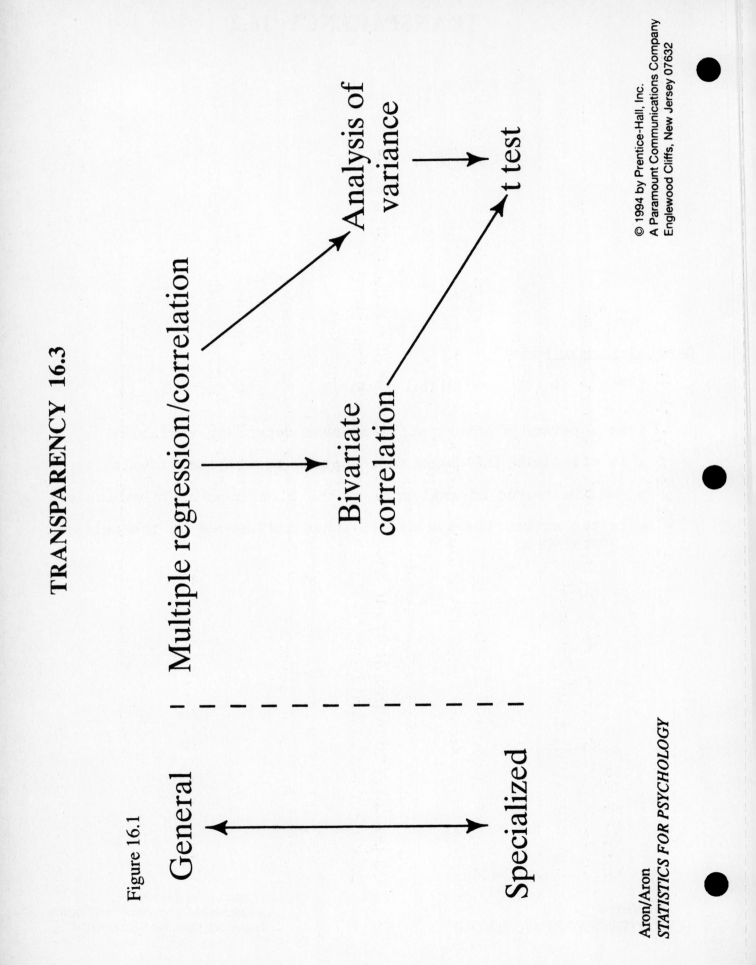

TRANSPARENCY 16.3

Figure 16.1

Multiple regression/correlation

Analysis of variance

Bivariate correlation

t test

General ←――――――――→ Specialized

© 1994 by Prentice-Hall, Inc.
A Paramount Communications Company
Englewood Cliffs, New Jersey 07632

Aron/Aron
STATISTICS FOR PSYCHOLOGY

TABLE 16-1
Some Links Between the *t* Test for Independent Means and the Analysis of Variance

t Test	Analysis of Variance
Numerator of *t* is the difference between the means of the two groups.	Numerator of *F* is partly based on variation between the means of the two or more groups.
Denominator of *t* is partly based on pooling the population variance estimates computed from each group.	Denominator of *F* is computed by pooling the population variance estimates computed from each group.
Denominator of *t* involves dividing by total number of subjects.	Numerator of *F* involves multiplying by total number of subjects. (Multiplying a numerator by a number has the same effect as dividing the denominator by that number.)
When using two groups, $t = \div F$	When using two groups, $F = t^2$

$$df = (N_1 - 1) + (N_2 - 1)$$

Aron/Aron
STATISTICS FOR PSYCHOLOGY

© 1994 by Prentice-Hall, Inc.
A Paramount Communications Company
Englewood Cliffs, New Jersey 07632

TRANSPARENCY 16.5

TABLE 16-2
t **Test and Analysis of Variance Computations for an Experiment Examining the Effectiveness of a New Job Skills Program for People Who Have Previously Not Been Able to Hold Jobs (Fictional Data)**

	Experimental Group (Special Program)			Control Group (Standard Program)		
	X_1	$X_1 - M_1$	$(X_1 - M_1)^2$	X_2	$X_2 - M_2$	$(X_2 - M_2)^2$
	6	0	0	6	3	9
	4	−2	4	1	−2	4
	9	3	9	5	2	4
	7	1	1	3	0	0
	7	1	1	1	−2	4
	3	−3	9	1	−2	4
	6	0	0	4	1	1
Σ	42	0	24	21	0	26

$M_1 = 6$ $S_1^2 = 24/6 = 4$ $\qquad$ $M_2 = 3$ $S_2^2 = 26/6 = 4.33$
$N_1 = 7$ $df_1 = N_1 - 1 = 6$ $\qquad$ $N_2 = 7$ $df_2 = N_2 - 1 = 6$

t test Computations	ANOVA Computations

Numerator

Mean difference = 6.00 − 3.00 = 3.00

$$df_B = N_G - 1 = 2 - 1 = 1$$

$$GM = (6 + 3)/2 = 9/2 = 4.5$$

$$\Sigma(M - GM)^2 = (6 - 4.5)^2 + (3 - 4.5)^2$$
$$= 1.5^2 + -1.5^2$$
$$= 2.25 + 2.25 = 4.5$$

$$S_B^2 \text{ or } MS_B = \left(\frac{\Sigma(M - GM^2)}{df_B}\right)(n) = \left(\frac{4.5}{1}\right)(7) = 31.5$$

Denominator

$$S_P^2 = \left(\frac{df_1}{df_T}\right)(S_1^2) + \left(\frac{df_2}{df_T}\right)(S_2^2) = \left(\frac{6}{12}\right)(4) + \left(\frac{6}{12}\right)(4.33)$$
$$= (.5)(4) + (.5)(4.33) = 2.00 + 2.17 = 4.17$$

$$S_{DIF}^2 = S_{M1}^2 + S_{M2}^2 = (S_P^2/N_1) + (S_P^2/N_2)$$
$$= (4.17/7) + (4.17/7)$$
$$= .60 + .60 = 1.20$$

$$S_{DIF} = \div S_{DIF}^2 = \div 1.20 = 1.10$$

Degrees of Freedom

$$df_T = df_1 + df_2 = 6 + 6 = 12$$
$$df_W = df_1 + df_2 \ldots df_{Last} = 6 + 6 = 12$$

Cutoff

Needed *t* with *df* = 12 at 5% level, two-tailed = ±2.179

Needed *F* with *df* = 1, 12 at 5% level = 4.75

Score on Comparison Distribution

$$t = (M_1 - M_2)/S_{DIF} = (6.00 - 3.00)/1.10 = 3.00/1.10 = 2.73$$
$$F = S_B^2/S_W^2 \text{ or } MS_B/MS_W = 31.5/4.17 = 7.55$$

Conclusions

Reject the null hypothesis; the research hypothesis is supported.

Reject the null hypothesis; the research hypothesis is supported.

Aron/Aron
STATISTICS FOR PSYCHOLOGY

© 1994 by Prentice-Hall, Inc.
A Paramount Communications Company
Englewood Cliffs, New Jersey 07632

TRANSPARENCY 16.6

First versus later born's responses to "Do you avoid crowds?"
(From class questionnaire.)

	First Borns				Later Borns	
$\underline{N}$	Mean	S^2		$\underline{N}$	Mean	S^2
17	2.53	2.64		17	3.82	3.03

t-Test Computations	ANOVA Computations

Numerator

$\underline{df}_B = \underline{N}_G - 1 = 2 - 1 = 1$

$\underline{GM} = (2.53+3.82)/2 = 6.35/2 = 3.18$

$\Sigma(\underline{M}-\underline{GM})^2 = (2.53-3.18)^2 + (3.82-3.18)^2$

$\qquad\qquad = \quad -.65^2 \quad + \quad .64^2$

$\qquad\qquad = \quad .42 \quad + \quad .41 = .83$

$\underline{S}_B^2$ or $\underline{MS}_B = [\Sigma(\underline{M}-\underline{GM})/\underline{df}_B][\underline{n}]$

Mean Difference $= 11.90-10.74 = 1.16$

$\qquad\qquad\qquad\quad = [.83 \quad 1][17] = 14.11$

Denominator

$$S_P^2 = \frac{df_1}{df_T} S^2_1 + \frac{df_2}{df_T} S^2_2$$

$$\underline{S}_W^2 \text{ or } \underline{MS}_W = \frac{\underline{S}_1^2 + \underline{S}_1^2 + ... + \underline{S}_{Last}^2}{\underline{N}_G}$$

$$= \frac{16}{32} 2.64 + \frac{16}{32} 3.03 = 2.84$$

$$= \frac{2.64 + 3.03}{2} = 2.84$$

$\underline{S}_{DIF}^2 = \underline{S}_M^2{}_1 + \underline{S}_M^2{}_2 = (\underline{S}_P^2/\underline{N}_1) + (\underline{S}_P^2/\underline{N}_2)$

$\quad = (2.84/17) + (2.84/17) = .17 + .17 = .34$

$\underline{S}_{DIF} = \sqrt{\underline{S}_{DIF}^2} = \sqrt{.34} = .58$

Degrees of Freedom

$\underline{df}_T = \underline{df}_1 + \underline{df}_2 = 16 + 16 = 32$

$\underline{df}_W = \underline{df}_1 + \underline{df}_2 + ... + \underline{df}_{Last} = 16 + 16 = 32$

Cutoff

Needed $\underline{t}(32), \underline{p}<.05, 2\text{-tailed} = \pm 2.043$
($\sqrt{4.17} = 2.042$)

Needed $\underline{F}(1,32), \underline{p}<.05 = 4.17$

Score on Comparison Distribution

$\underline{t} = (\underline{M}_1 - \underline{M}_2)/\underline{S}_{DIF} = (2.53-3.82)/.58$

$\quad = -1.29/.58 = 2.22$

$\quad = -1.29/.58 = -2.22$

$\underline{F} = \underline{S}_B^2/\underline{S}_W^2$ or $\underline{MS}_B/\underline{MS}_W = 14.11/2.84$

$\quad = 4.97$ (NOTE: $\sqrt{\underline{F}} = \sqrt{4.97} = 2.23$)

Conclusion

Reject null hypothesis;
research hypothesis is supported

Reject null hypothesis;
research hypothesis is supported.

Aron/Aron
STATISTICS FOR PSYCHOLOGY

TRANSPARENCY 16.7

Helping behavior following induction of depressed vs. neutral mood.
(Fictional data.)

Scores	Neutral Mood			Depressed Mood		
Neutral: 31,36,34,31	N	Mean	S^2	N	Mean	S^2
Depressed: 38,41,39,42	4	33.00	6.00	4	40.00	3.33

t-Test Computations	ANOVA Computations

Numerator

$$\underline{df}_B = \underline{N}_G-1 = 2-1 = 1$$
$$\underline{GM} = (33+40)/2 = 73/2 = 36.5$$
$$\Sigma(\underline{M}-\underline{GM})^2=(33-36.5)^2+(40-36.5)^2$$
$$= -3.5^2 \quad + \quad 3.5^2 = 24.5$$
$$\underline{S}_B^2 \text{ or } \underline{MS}_B =[\Sigma(\underline{M}-\underline{GM})/\underline{df}_B][\underline{n}]$$
$$=[\ 24.5\ /\ 1\ ][4] = 98$$

Mean Difference=33-40=-7

Denominator

$$S_P^2 = \frac{df_1}{\underline{df}_T} S^2_1 + \frac{df_2}{\underline{df}_T} S^2_2$$

$$\underline{S}_W^2 \text{ or } \underline{MS}_W = \frac{\underline{S}_1^2+\underline{S}_1^2+\ldots+\underline{S}_{Last}^2}{\underline{N}_G}$$

$$= \frac{3}{6}\ 6.00 + \frac{3}{6}\ 3.33 = 4.67$$

$$= \frac{6.00 + 3.33}{2} = 4.67$$

$$\underline{S}_{DIF}^2=\underline{S}_M^2{}_1+\underline{S}_M^2{}_2=(S_P^2/\underline{N}_1)+(S_P^2/\underline{N}_2)$$
$$=(4.67/4)+(4.67/4)=1.17+1.17=2.34$$
$$\underline{S}_{DIF} = \sqrt{\underline{S}_{DIF}^2} = \sqrt{2.34} = 1.53$$

Degrees of Freedom

$$\underline{df}_T = \underline{df}_1 + \underline{df}_2 = 3 + 3 = 6$$

$$\underline{df}_W = \underline{df}_1+\underline{df}_2+\ldots+\underline{df}_{Last} = 3+3 = 6$$

Cutoff

Needed $\underline{t}(6),\underline{p}<.05,2\text{-tailed}=\pm2.447$
$$(\sqrt{4.17}=2.447)$$

Needed $\underline{F}(1,6),\underline{p}<.05=5.99$

Score on Comparison Distribution

$\underline{t} = (\underline{M}_1-\underline{M}_2)/\underline{S}_{DIF} = 7/1.53 = 4.58$
21

$\underline{F} = \underline{S}_B^2/\underline{S}_W^2$ or $\underline{MS}_B/\underline{MS}_W =98/4.67=$

(NOTE: $\sqrt{\underline{F}} = \sqrt{21} = 4.58$)

Conclusion

Reject null hypothesis;
research hypothesis is supported

Reject null hypothesis;
research hypothesis is supported.

Aron/Aron
STATISTICS FOR PSYCHOLOGY

© 1994 by Prentice-Hall, Inc.
A Paramount Communications Company
Englewood Cliffs, New Jersey 07632

TRANSPARENCY 16.8

TABLE 16-3
Computation of the Correlation Coefficient and a Hypothesis Test of the Correlation Coefficient Using the Data From Table 10-3 and Converting the Predictor (Independent) Variable Into a Numerical Variable Having Values of 1 (for the Experimental Group) or 2 (for the Control Group)

Predictor Variable (Experimental Versus Control)		Dependent Variable (Employer's Rating)		Cross-Product
Raw	Z_X	Raw	Z_Y	$Z_X Z_Y$
1	−1	6	.62	− .62
1	−1	4	− .21	.21
1	−1	9	1.87	−1.87
1	−1	7	1.04	−1.04
1	−1	7	1.04	−1.04
1	−1	3	− .62	.62
1	−1	6	.62	− .62
2	1	6	.62	.62
2	1	1	−1.45	−1.45
2	1	5	.21	.21
2	1	3	− .62	− .62
2	1	1	−1.45	−1.45
2	1	1	−1.45	−1.45
2	1	4	− .21	− .21
S 21	0	63	0	−8.71
M = 1.5	0	4.5	0	r = − .62
(SD = .5)		(SD = 2.41)		

$df = N - 2 = 14 - 2 = 12.$

t needed with $df = 12$ at 5% level, two-tailed $= \pm 2.179$.

$t = r\sqrt{N-2}/\sqrt{1-r^2} = -.62\sqrt{14-2}/\sqrt{1-(-.62)^2} = -.62\sqrt{12}/\sqrt{1 - .38} = -.62(3.46)/\sqrt{.62} = -2.15/.79 = -2.72.$

Conclusion: Reject the null hypothesis; the research hypothesis is supported.

Aron/Aron
STATISTICS FOR PSYCHOLOGY

Figure 16.2

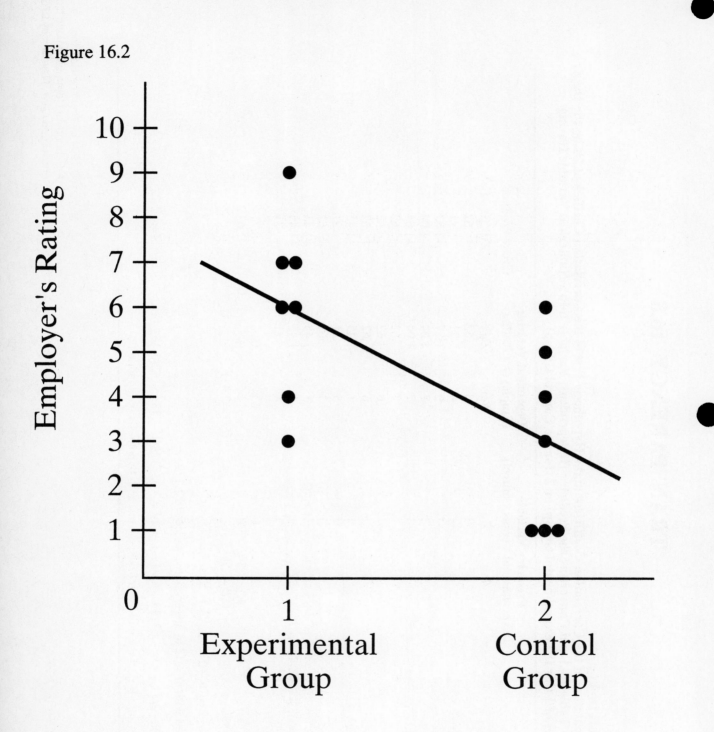

Aron/Aron
STATISTICS FOR PSYCHOLOGY

© 1994 by Prentice-Hall, Inc.
A Paramount Communications Company
Englewood Cliffs, New Jersey 07632

Responses of first and later borns to "Do you avoid crowds?"
(From class questionnaire.)

Predictor Variable (First vs. Later Born)		Dependent Variable (Avoiding Crowds)		Cross-Product
Raw	Z	Raw	Z	$Z_X Z_Y$
1	-1	1	-1.24	1.24
1	-1	5	1.04	-1.04
1	-1	4	.47	- .47
.	.	.	.	.
.	.	.	.	.
.	.	.	.	.
2	1	3	- .10	- .10
2	1	6	1.61	1.61

Σ	51	0	108	0		12.52
M	1.5	0	3.18	0	$r =$	.368
SD	.5		3.09			

$df = N - 2 = 34 - 2 = 32$

t needed with df = 32 at 5% level, two-tailed = ±2.043

$t = (r)(\sqrt{[N-2]}) / \sqrt{(1-r^2)} = (.368)(\sqrt{[34-2]}) / \sqrt{(1-.368^2)}$

  $= (.368)(\sqrt{[32]})/\sqrt{(1-.135)} = (.368)(5.66)/\sqrt{(.865)} = 2.08/.93 = 2.24$

Conclusion: Reject null hypothesis; research hypothesis supported.

Aron/Aron
STATISTICS FOR PSYCHOLOGY

TRANSPARENCY 16.11

Responses of first and later borns to "Do you avoid crowds?"
(From class questionnaire.)

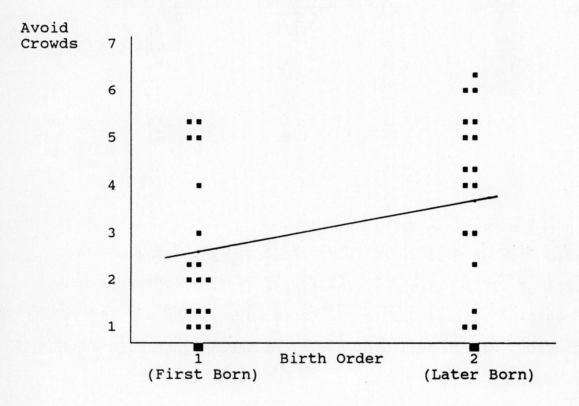

TRANSPARENCY 16.12

Helping behavior following induction of depressed vs. neutral mood:
Computation of correlation. (Fictional data.)

Predictor Variable

Mood Neutral=1 Depressed=2		Dependent Variable (Helping Behavior)		Cross-Product
Raw	Z	Raw	Z	$Z_X Z_Y$
1	-1	31	-1.38	1.38
1	-1	36	- .13	.13
1	-1	34	- .63	.63
1	-1	31	-1.39	1.39
2	1	38	.38	.38
2	1	41	1.13	1.13
2	1	39	.63	.63
2	1	42	1.39	1.39

Σ	12	0	292	0	7.06
M	1.5	0	36.5	0	r = .883
SD	.5		3.97		

$\underline{df} = \underline{N} - 2 = 8 - 2 = 6$

$\underline{t}$ needed with $\underline{df}$ = 6 at 5% level, two-tailed = ±2.447

$\underline{t} = (\underline{r})(\sqrt{[\underline{N}-2]}) / \sqrt{(1-\underline{r}^2)} = (.883)(\sqrt{[8-2]}) / \sqrt{(1-.883^2)}$

$= (.883)(\sqrt{[6]}) / \sqrt{(1-.78)} = (.883)(2.45) / \sqrt{(.22)} = 2.16/.47 = 4.60$

Conclusion: Reject null hypothesis; research hypothesis supported.

Aron/Aron
STATISTICS FOR PSYCHOLOGY

TRANSPARENCY 16.13

Helping behavior following induction of depressed vs. neutral mood:
Scatter diagram. (Fictional data)

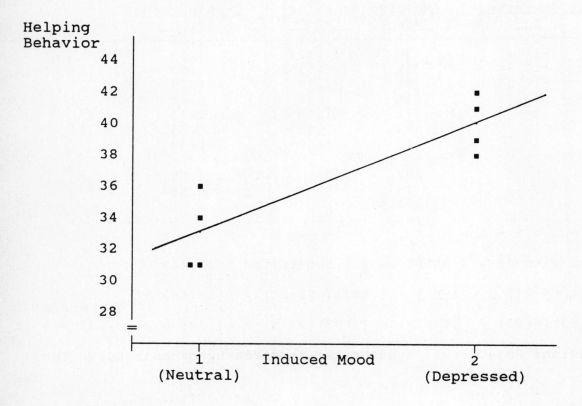

Aron/Aron
STATISTICS FOR PSYCHOLOGY

TRANSPARENCY 16.14

TRANSPARENCY 16.14

Relation of R^2 in ANOVA to r^2 in bivariate regression.

SS_W in ANOVA = SS_E in regression
1. SS_W = $\Sigma(X-M)^2$
2. SS_E = $\Sigma(X-\hat{Y})^2$
3. When X has only two levels, each $\hat{Y}$ = M

SS_T in ANOVA = SS_T in regression
1. SS_T in ANOVA   = $\Sigma(X\ GM)^2$
2. SS_T in regression = $\Sigma(X-M)^2$
3. GM in ANOVA = M (of DV) in regression

SS_B in ANOVA = reduction in squared error in regression
1. SS_B = SS_T - SS_W
2. Reduction in squared error in regression = SS_T - SS_E
3. As noted above, SS_W = SS_E

R^2 in ANOVA = r^2 in regression
1. In ANOVA, R^2 = SS_B/SS_T
2. In regression, r^2 = reduction in squared error / SS_T.
3. As noted above:
    a. SS_B in ANOVA = reduction in squared error in regression
    b. SS_T in ANOVA = SS_T in regression

F in ANOVA = F in regression
1. F in ANOVA   = SS_B/SS_W
2. F in regression = reduction in squared error / SS_E
3. As noted above:
    a. SS_B in ANOVA = reduction in squared error in regression
    b. SS_W in ANOVA = SS_E in regression

Aron/Aron
STATISTICS FOR PSYCHOLOGY

© 1994 by Prentice-Hall, Inc.
A Paramount Communications Company
Englewood Cliffs, New Jersey 07632

TRANSPARENCY 16.15

TABLE 16-4
Computation of the Proportionate Reduction in Error With Raw Scores and Analysis of Variance, Structural Model Approach, Using the Data From Table 10-3

PROPORTIONATE REDUCTION IN ERROR

Predictor Variable (Experimental Versus Control)	Dependent Variable (Employer's Rating)			
Raw	Score	Predicted	Difference	Squared Difference
1	6	6	0	0
1	4	6	−2	4
1	9	6	3	9
1	7	6	1	1
1	7	6	1	1
1	3	6	−3	9
1	6	6	0	0
2	6	3	3	9
2	1	3	−2	4
2	5	3	2	4
2	3	3	0	0
2	1	3	−2	4
2	1	3	−2	4
2	4	3	1	1

$\Sigma = SS_E = 50$

Sum of squared error using the overall mean as a prediction rule (computation not shown): $SS_T = 81.34$

$$\text{Proportionate reduction in squared error} = \frac{SS_T - SS_E}{SS_T} = \frac{81.34 - 50}{81.34} = \frac{31.34}{81.34} = .39$$

$r^2 = .39; \quad r = \div r^2 = \div .39 = \pm.62.$

ONE-WAY ANALYSIS OF VARIANCE STRUCTURAL MODEL CALCULATION

$GM = 4.5$

	Experimental Group (Special Program)							Control Group (Standard Program)						
X_1	$X - GM$		$X - M$		$M - GM$		X	$X - GM$		$X - M$		$M - GM$	
	Dev	Dev²	Dev	Dev²	Dev	Dev²		Dev	Dev²	Dev	Dev²	Dev	Dev²
6	1.5	2.25	0	0	1.5	2.25	6	1.5	2.25	3	9	−1.5	2.25
4	−.5	.25	−2	4	1.5	2.25	1	−3.5	12.25	−2	4	−1.5	2.25
9	4.5	20.25	3	9	1.5	2.25	5	.5	.25	2	4	−1.5	2.25
7	2.5	6.25	1	1	1.5	2.25	3	−1.5	2.25	0	0	−1.5	2.25
7	2.5	6.25	1	1	1.5	2.25	1	−3.5	12.25	−2	4	−1.5	2.25
3	−1.5	2.25	−3	9	1.5	2.25	1	−3.5	12.25	−2	4	−1.5	2.25
6	1.5	2.25	0	0	1.5	2.25	4	−.5	.25	1	1	−1.5	2.25
Σ:		39.75		24		15.75			41.75		26		15.75

Note: Dev = Deviation; Dev² = Squared deviation

Sums of squared deviations:
$\Sigma(X - GM)^2$ or $SS_T = 39.75 + 41.75 = 81.5$
$\Sigma(X - M)^2$ or $SS_W = 24 + 26 = 50$
$\Sigma(M - GM)^2$ or $SS_B = 15.75 + 15.75 = 31.5$
Check $(SS_T = SS_W + SS_B)$: $81.5 = 50 + 31.5$
Degrees of freedom:
$df_T = N - 1 = 14 - 1 = 13$
$df_W = df_1 + df_2 + \ldots + df_{Last} = 6 + 6 = 12$
$df_B = N_G - 1 = 2 - 1 = 1$
Check $(df_T = df_W + df_B)$: $13 = 12 + 1$

Population variance estimates:
S_T^2 or $MS_T = SS_T/df_T = 81.5/13 = 6.27$
S_W^2 or $MS_W = SS_W/df_W = 50/12 = 4.17$
S_B^2 or $MS_B = SS_B/df_B = 31.5/1 = 31.5$
F ratio: $F = S_B^2/S_W^2$ or $MS_B/MS_W = 31.5/4.17 = 7.55$
$R^2 = SS_B/SS_T = 31.5/81.5 = .39$

Aron/Aron
STATISTICS FOR PSYCHOLOGY

TRANSPARENCY 16.16

Responses of first and later borns to "Do you avoid crowds?"
(From class questionnaire.)

<u>Proportionate Reduction in Error (regression analysis)</u>:

Predictor Variable   Dependent Variable ($\underline{Y}$)
(First Born = 1;
<u>Later Born = 2)</u>   <u>(Rating for Avoiding Crowds)</u>

$\underline{X}$	$\underline{Y}$	$\hat{Y}$	$(\underline{Y}-\hat{\underline{Y}})$	$(\underline{Y}-\hat{\underline{Y}})^2$
1	1	2.53	-1.53	2.34
1	5	2.53	2.47	6.10
1	4	2.53	1.47	2.16
.	.	.	.	.
.	.	.	.	.
.	.	.	.	.
2	3	3.82	- .82	.67
2	6	3.82	2.18	4.75
			$\underline{SS}_E =$	90.71

$\underline{SS}_T = 104.94$ (computation not shown)

Reduction in squared error = $\underline{SS}_T - \underline{SS}_E$ = 104.94-90.71 = 14.23

$\underline{r}^2$ = reduction in squared error/$\underline{SS}_T$ = 14.23/104.94 = .136

      ($\underline{r} = \sqrt{\underline{r}^2} = \sqrt{.136}$ = .37)

<u>ANOVA (Structural Model Approach)</u>:   GM = 3.18; $\underline{M}_1$=2.53; $\underline{M}_2$=3.82

(First Borns)							(Later Borns)					
$\underline{X}_1$	X - GM		X - M		M - GM		$\underline{X}_2$	X - GM		X - M		M - GM	
	Dev	Dev²	Dev	Dev²	Dev	Dev²		Dev	Dev²	Dev	Dev²	Dev	Dev²
1	2.18	4.75	-1.53	2.34	-.65	.42	2	-1.18	1.39	-1.82	3.31	.64	.41
5	1.82	3.31	2.47	6.10	-.65	.42	6	2.82	7.95	2.18	4.75	.64	.41
.	.	.	.	.	.	.	.	.	.	.	.	.	.
.	.	.	.	.	.	.	.	.	.	.	.	.	.
4	.82	.67	1.47	2.16	-.65	.42	3	-.18	.03	-.82	.67	.64	.41
2	-1.18	1.39	-.53	.28	-.65	.42	6	2.82	7.95	2.18	4.75	.64	.41
Σ		49.35		42.24		7.14			55.46		48.47		6.97

$\Sigma(\underline{X}-\underline{GM})^2$ or $\underline{SS}_T$ = 49.35+55.46 = 104.81
$\Sigma(\underline{X}-\underline{M})^2$ or $\underline{SS}_W$ = 42.24+48.47 = 90.71 | $R^2 = \underline{SS}_B/\underline{SS}_T$ = 14.11/104.81
$\Sigma(\underline{M}-\underline{GM})^2$ or $\underline{SS}_B$ = 7.14+ 6.97 = 14.11 | = .135

Aron/Aron
STATISTICS FOR PSYCHOLOGY

Helping behavior following induction of depressed vs. neutral mood. (Fictional data.)

Proportionate Reduction in Error (regression analysis):

Predictor Variable (Neutral = 1; Depressed = 2)	Dependent Variable ($\underline{Y}$) (Helping Behavior)			
$\underline{X}$	$\underline{Y}$	$\hat{Y}$	$(\underline{Y}-\hat{Y})$	$(\underline{Y}-\hat{Y})^2$
1	31	33	-2	4
1	36	33	3	9
1	34	33	1	1
1	31	33	-2	4
2	38	40	-2	4
2	41	40	1	1
2	39	40	-1	1
2	42	40	2	4
				$SS_F = 28$

$\underline{SS}_T = 126$ (computation not shown)

Reduction in squared error = $\underline{SS}_T - \underline{SS}_F = 126 - 28 = 98$

$\underline{r}^2$ = reduction in squared error/$\underline{SS}_T = 98/126 = .778$

$\quad$ ($\underline{r} = \sqrt{\underline{r}^2} = \sqrt{.778} = .88$)

ANOVA (Structural Model Approach): $\underline{GM} = 36.5$; $\underline{M}_1 = 33$; $\underline{M}_2 = 40$

	(Neutral Mood Group)							(Depressed Mood Group)					
X_1	X - GM Dev	Dev²	X - M Dev	Dev²	M - GM Dev	Dev²	X_2	X - GM Dev	Dev²	X - M Dev	Dev²	M - GM Dev	Dev²
31	-5.5	30.25	-2	4	-3.5	12.25	38	1.5	2.25	-2	4	3.5	12.25
36	-.5	.25	3	9	-3.5	12.25	41	4.5	20.25	1	1	3.5	12.25
34	-2.5	6.25	1	1	-3.5	12.25	39	2.5	6.25	-1	1	3.5	12.25
31	-5.5	30.25	-2	4	-3.5	12.25	42	5.5	30.25	2	4	3.5	12.25
Σ		67		18		49			59		10		49

$\Sigma(\underline{X}-\underline{GM})^2$ or $\underline{SS}_T = 67 + 59 = 126$

$\Sigma(\underline{X}-\underline{M})^2$ or $\underline{SS}_W = 18 + 10 = 28$

$\Sigma(\underline{M}-\underline{GM})^2$ or $\underline{SS}_B = 49 + 49 = 98$

$\underline{R}^2 = \underline{SS}_B/\underline{SS}_T = 98/126 = .778$

Aron/Aron
STATISTICS FOR PSYCHOLOGY

© 1994 by Prentice-Hall, Inc.
A Paramount Communications Company
Englewood Cliffs, New Jersey 07632

TABLE 16-5
Example of Nominal Coding for Ten Subjects' Nationality in a Fictional Study of Subjects of Four European Nationalities

Subject	Nationality	Variable 1 French or Not	Variable 2 Spanish or Not	Variable 3 Italian or Not
1	Spanish	0	1	0
2	Italian	0	0	1
3	German	0	0	0
4	Italian	0	0	1
5	French	1	0	0
6	French	1	0	0
7	German	0	0	0
8	Italian	0	0	1
9	French	1	0	0
10	Spanish	0	1	0

© 1994 by Prentice-Hall, Inc.
A Paramount Communications Company
Englewood Cliffs, New Jersey 07632

Aron/Aron
STATISTICS FOR PSYCHOLOGY

TRANSPARENCY 16.19

Subjects randomly assigned to five different memory cues.
(Fictional example.)

<u>Nominal coding</u>

S	Dependent Variable (Recall)	Independent Variable (Cue)	Nominally Coded 2-Level Numerical Vars			
			A or Not	B or Not	C or Not	D or Not
1	17	A	1	0	0	0
2	14	A	1	0	0	0
3	12	A	1	0	0	0
4	8	B	0	1	0	0
5	6	B	0	1	0	0
6	8	B	0	1	0	0
7	10	C	0	0	1	0
8	14	C	0	0	1	0
9	12	C	0	0	1	0
10	8	D	0	0	0	1
11	11	D	0	0	0	1
12	9	D	0	0	0	1
13	15	E	0	0	0	0
14	14	E	0	0	0	0
15	11	E	0	0	0	0

Aron/Aron
STATISTICS FOR PSYCHOLOGY

© 1994 by Prentice-Hall, Inc.
A Paramount Communications Company
Englewood Cliffs, New Jersey 07632

TRANSPARENCY 16.20

Analysis of causality.

Regularity Theory Requirements X is a cause of Y if:	How Psychology Research Attempts to Meet these Requirements
1. X and Y regularly associated.	1. Significant correlation obtained between measures of X and Y in a representative sample.
2. X precedes Y.	2. a. Experiment which manipulates X before Y is presented. b. Longitudinal study where X is measured before Y.
3. And no third causes that precede X that might cause X and Y.	3. a. Experiment where exposure to a particular level of X is randomly determined (so no previous cause could affect it). b. Statistical control for or matching on potential third causes.

Generative Theory Requirements X is a cause of Y if:	How Psychology Research Attempts to Meet these Requirements
1. All requirement of regularity theory are met.	1. See above.
2. And a plausible explanation for mechanism through which X causes Y	2. Theory and research results on intervening processes.

Aron/Aron
STATISTICS FOR PSYCHOLOGY

Chapter 17
Making Sense of Advanced Statistical Procedures in Research Articles

Instructor's Summary of Chapter

This chapter introduces various widely-used advanced procedures in order to give the student an appreciation of the basic idea and issues sufficient to make sense of research articles employing these procedures.

Hierarchical multiple regression. Predictor variables are included in a prediction rule in a planned sequential fashion, permitting the researcher to determine the relative contribution of each successive variable over and above those already included.

Stepwise multiple regression. An exploratory procedure in which potential predictor variables are searched in order to find the best predictor, then the remaining variables are searched for the predictor which in combination with the first produces the best prediction. This process continues until adding the last remaining variable does not provide a significant improvement.

Partial correlation. The degree of correlation between two variables while holding one or more other variables constant.

Reliability coefficients. Measures of the extent to which scores on a test are internally consistent (e.g., Cronbach's alpha) or consistent over time (test-retest reliability).

Factor analysis. Identifies groupings of variables which correlate maximally with each other and minimally with other variables.

Path analysis. Examines whether the correlations among a set of variables is consistent with a systematic model of the pattern of causal relationships among them. A diagram describes these relationships with arrows pointing from cause to effect, each with a path coefficient indicating the influence of the hypothesized causal variable on the hypothesized effect variable. Path coefficients are standardized regression coefficients from a multiple-regression prediction rule in which the variable at the end of the arrow is the dependent variable and the variable at the start of the arrow is the predictor, along with all other variables to that dependent variable.

Latent variable modeling. A sophisticated version of path analysis which includes paths involving latent, unmeasured theoretical variables (each of which consists of the common elements of several measured variables). It also permits a kind of significance test and provides measures of the overall fit of the data to the hypothesized causal pattern.

Analysis of covariance. An analysis of variance which controls for one or more variables.

Multivariate analysis of variance. An analysis of variance with more than one dependent variable; multivariate analysis of covariance is an analysis of covariance with more than one dependent variable.

Controversy: Should statistics be controversial? Statistical methods are usually taught in psychology as if they were well-established truths. However, what is usually taught is a hybrid of competing approaches that some feel are incompatible.

How to read results involving unfamiliar statistical techniques. It is usually possible to extract the general meaning of an unfamiliar statistical technique by its context; in general a *p* is reported if it is a significance test, and most indicators of effect size fall on the scale of 0 to 1 of a correlation.

Box 17.1. The Forced Marriage of Fisher and Neyman-Pearson. Briefly describes the issues and personalities involved in the controversies between Fisher and the team of Neyman and Pearson about the fundamentals of statistics in the social sciences.

Lecture 17.1: Advanced Procedures I

Materials

Lecture outline
Transparencies 17.1 through 17.10
(If using transparencies based on your class's questionnaires, replace 17.3, 17.6, and 17.8 with 17.3R, 17.6R, and 17.8R.)

Outline for Blackboard

I.   Review of Multiple Regression/Correlation
II.   Types of Advanced Statistical Techniques
III. Hierarchical Multiple Regression
IV. Stepwise Multiple Regression
V.   Partial Correlation
VI. Reliability Coefficients
VII. Review this Class

Instructor's Lecture Outline

I.   Review of Multiple Regression/Correlation
   A. Show TRANSPARENCY 17.1 (multiple regression formula and worked-out example, from text).
   B. The correlation between the set of independent variables and the dependent variable is called a multiple correlation (R).
   C. R^2 is the proportionate reduction in squared error gained by using the multiple-regression prediction rule compared to simply predicting the dependent variable from its mean.
   D. R and R^2 can be tested against the null hypothesis that the population value is 0.

II.   Types of Advanced Statistical Techniques
   A. Those that focus on associations among variables (these are variations and extensions of correlation and regression).
   B. Those that focus on differences among groups (these are variations and extensions of analysis of variance).

III. Hierarchical Multiple Regression
   A. Show TRANSPARENCY 17.2 (outline of procedure, and falling-in-love fictional example) and discuss.
   B. Show TRANSPARENCY 17.3 or 17.3R (analysis of hypothesized predictors of tendency to avoid crowds, from class questionnaire) and discuss.

IV. Stepwise Multiple Regression
   A. Purpose: Exploratory procedure to determine which predictor variables of many that have been measured usefully contribute to the prediction.
   B. Show TRANSPARENCY 17.4 (outline of stepwise procedure) and discuss.

C. Show TRANSPARENCY 17.5 (predictors of job success at ABC Enterprises) and discuss.

D. Show TRANSPARENCY 17.6 or 17.6R (childhood-related variables as possible predictors of score on High Sensitivity Scale, from class questionnaire) and discuss.

E. Caution: The prediction formula that results is the optimal small set of variables for predicting the dependent variable, *as determined from the sample studied*—when tried with a new sample, somewhat different combinations often result.

V.   Partial Correlation

A. The degree of association between two variables, over and above the influence of one or more other variables.

B. Variable(s) over and above which the partial correlation is computed are said to be *held constant*, *partialed out*, or *controlled for*. (These terms are interchangeable.)

C. You can think of a partial correlation as the average of the correlations between two variables, each correlation computed among just those subjects at each level of the variable being controlled for.

D. Partial correlation is often used to help sort out alternative explanations in a correlational study.

   1. If correlation between two variables dramatically drops or is eliminated when a third variable is partialled out, it suggests that the third variable was behind the correlation.

   2. If correlation between two variables is largely unaffected when a third variable is partialled out, it suggests that the third variable is not behind the correlation.

E. Show TRANSPARENCY 17.7 (associations among marital satisfaction, passionate love, and marriage length— fictional data based on pattern of results in Tucker & Aron, 1993) and discuss.

F. Show TRANSPARENCY 17.8 or 17.8R (prone to fears, tense or worried by nature, and High Sensitivity Scale scores, from class questionnaires) and discuss.

VI.   Reliability Coefficients

A. Reliability, the accuracy and consistency of a measure, is the extent to which, if you were to give the same measure again to the same person under the same circumstances, you would obtain the same result.

B. Reports of computations of reliability of a measurement are very common in research articles.

C. Test-retest reliability.

   1. The correlation between the scores of the same people who take a measure twice.

   2. Often impractical or not appropriate, since having taken the test once would influence the second taking.

   3. Show TRANSPARENCY 17.9 (IOS Scale study example) and discuss.

E. Reliability as internal consistency.

   1. Split-half reliability is the correlation between two halves of the same test.

   2. Cronbach's alpha can be thought of as the average correlation for all possible divisions of a test into halves.

F. In general, a test should have a reliability of at least .7, and preferably closer to .9, to be considered useful.

G. If a measure has low reliability, it tends to reduce the correlation between it and any other variable. (This can be adjusted for in bivariate correlation using a correction for attenuation.)

H. Show TRANSPARENCY 17.10 (examples of effect of unreliability on corrections) and discuss.

VII. Review this Class: Use blackboard outline.

Lecture 17.2: Advanced Procedures II

Materials

Lecture outline

Transparencies 17.11 through 17.19

(If using transparencies based on your class's questionnaires, replace 17.11, 17.15, and 17.18 with 17.11R, 17.15R, and 17.18R.)

Outline for Blackboard

I. Review
II. Factor Analysis
III. Causal Modeling
IV. ANCOVA
V. MANOVA and MANCOVA
VI. Review this Class

Instructor's Lecture Outline

I. **Review**
   A. Advanced statistical techniques that focus on associations among variables (these are variations and extensions of correlation and regression).
      1. Hierarchical multiple regression.
      2. Stepwise multiple regression.
      3. Partial correlation.
      4. Reliability coefficients.
      5. Factor analysis (to be covered this class).
      6. Causal models (to be covered this class).
   B. Those that focus on differences among groups—variations and extensions of analysis of variance (to be covered this class).

II. **Factor Analysis**
   A. Widely used procedure applied when a researcher has measured people on a large number of variables.
   B. Identifies groupings of variables (called *factors*) such that those within each group correlate with each other and not with variables in other groupings.
   C. The correlation between a variable and a factor is called the variable's *factor loading* on that factor.
   D. Show TRANSPARENCY 17.11 (IOS Scale factor analysis example from Aron et al., 1992) and discuss.
   E. Show TRANSPARENCY 17.12 or 17.12R (factor analysis of items regarding sensitivity and childhood experiences, from class questionnaire) and discuss.

III. Causal Modeling

A. Path analysis.

   1. Focuses on a diagram with arrows connecting the variables, indicating the hypothesized pattern of causal relations among them.
   2. Show TRANSPARENCY 17.13 (religious well-being, social desirability, and marital satisfaction, from Leong, 1989) and discuss.

B. Latent variable models.

   1. Also widely known as *structural equation modeling* or as *LISREL*.
   2. An extension of path analysis with several advantages.
      a. Produces an overall measure of how good the model fits the data.
      b. Includes a significance test—null hypothesis is that the model fits.
      c. Permits modeling of latent variables (as assessed by a set of manifest variables).
   3. Path diagram.
      a. Manifest shown in squares.
      b. Latent variables shown in circles.
      c. The measurement model, the relation of the manifest to the latent variables they assess, usually involves arrows from the each latent to its associated manifest variables.
      d. The causal model usually involves arrows showing the relations among the latent variables.
   4. Show TRANSPARENCY 17.14 (confirmatory factor analysis for measures of closeness in cross-validation sample, from Aron et al., 1992) and discuss.
   5. Limitations.
      a. A well fitting model that is not a significantly bad fit is still only one of the possible models that could fit the data.
      b. Shares limitations of all methods ultimately based on the correlation coefficient:
         i. Association does not demonstrate direction of causality.
         ii. Only takes into account linear relationships.
         iii. Results can be distorted by restriction in range.

IV. Analysis of Covariance (ANCOVA)

A. Same as an ordinary ANOVA, except one or more variables are partialed out.

B. The variable partialed out is called a covariate.

C. The rest of the results are interpreted like any other analysis of variance.

D. The analysis of covariance is generally used in one of two cases.

   1. Analysis of a random-assignment experiment in which some nuisance variable is partialed out.
   2. Show TRANSPARENCY 17.15 or 17.15R (gender by birth order on abused as a child, partialing out alcoholism in family, from class questionnaire) and discuss.
   3. In a study in which it is not possible to employ random assignment, variables on which groups may differ are partialed out.
   4. Show TRANSPARENCY 17.16 (attractiveness of communicator on attitude change, posttest scores controlling for pretest) and discuss.

E. Assumes that correlation between the covariates and the dependent variable is the same in all the cells.

V. Multivariate Analysis of Variance (MANOVA) and Multivariate Analysis of Covariance (MANCOVA)

A. Multivariate statistical techniques involve more than one dependent variable. (Like ANOVA, etc., these can have one or more IVs.)

B. The most widely used multivariate techniques are MANOVA and MANCOVA.

C. MANOVA is simply an analysis of variance in which there is more than one dependent variable.

D. MANOVA tests each main and interaction effect of the independent variables on the combination of dependent variables.

E. Show TRANSPARENCY 17.17 (closest relationship, IV, and measures of closeness for other in that relationship, DVs; from Aron et al., 1992) and discuss.

F. Show TRANSPARENCY 17.18 or 17.18R (relation of adult attachment style, IV, to the combination of close to mother as a child and mother fond of children, DVs; from class questionnaire) and discuss.

G. MANCOVA is a MANOVA in which one or more variables are partialed out of the analysis.

VII. Review this Class: Use blackboard outline and show TRANSPARENCY 17.19 (table of methods from text).

TRANSPARENCY 17.1

Multiple regression formula.

$$\hat{Z}_y = (\beta_1)(Z_{\underline{x}1}) + (\beta_2)(Z_{\underline{x}2}) + (\beta_3)(Z_{\underline{x}3})$$

Example:

$$\hat{Z}_{Stress} = \quad (.51)(Z_{NumberSupervised})$$
$$+ \quad (.11)(Z_{Noise})$$
$$+ \quad (.33)(Z_{Decisions/Month})$$

Aron/Aron
STATISTICS FOR PSYCHOLOGY

TRANSPARENCY 17.2

Hierarchical regression.

<u>Procedure</u> (based on hypothesized order of influence):

Step 1: R^2 for Y with X_1

Step 2: R^2 for Y with X_1 and X_2

    What is the improvement in prediction over X_1 alone?

    Is the improvement significant?

Step 3: R^2 for Y with X_1, X_2, and X_3

    What is the improvement in prediction over X_1 and X_2 alone?

    Is the improvement significant?

Etc.

<u>Example</u>:

Study of influences on falling in love (fictional):

    Predicted order of influence:

      X_1 = Readiness to fall in love

      X_2 = Other having desirable characteristics

      X_3 = Perception that other likes self

Step 1: R^2 for Y with X_1 = .08 (p < .05)

Step 2: R^2 for Y with X_1 and X_2 = .17

    Increment = .09 (p < .01)

Step 3: R^2 for Y with X_1, X_2, and X_3 = .23

    Increment = .06 (p < .05)

Aron/Aron
STATISTICS FOR PSYCHOLOGY

© 1994 by Prentice-Hall, Inc.
A Paramount Communications Company
Englewood Cliffs, New Jersey 07632

Study of hypothesized predictors of tendency to avoid crowds.
(From class questionnaire.)

Dependent Variable: $\underline{Y}$ = Tendency to avoid crowds

Predicted order of influence:

X_1 = "Are you a tense or worried person by nature?"

X_2 = "Do you make it a high priority to arrange your
life to avoid upsetting or overwhelming
situations?"

X_3 = "Are you made uncomfortable by loud noise?"

Hierarchical Multiple Regression ($\underline{N}$ = 102)

Predictor	Betas			Overall Model		Increment	
Variables	β_1	β_2	β_3	R^2	F	R	F
X_1	-.01			.002	.02	.002	.02
X_1 and X_2	-.04	.32		.999	5.50*	.097	10.96**
X_1, X_2, and X_3	-.05	.19	.34	.201	8.26**	.102	3.96*

*$\underline{p}$ < .05; **$\underline{p}$ < .01

Aron/Aron
STATISTICS FOR PSYCHOLOGY

© 1994 by Prentice-Hall, Inc.
A Paramount Communications Company
Englewood Cliffs, New Jersey 07632

TABLE 17-2
The Process of a Stepwise Multiple Regression

Step 1: Search all potential predictor variables and find the best bivariate correlation with the dependent variable.

Step 2: Test significance.

    If not significant, $\longrightarrow$ STOP.

    If significant, include this variable in all further steps, and $\longrightarrow$ CONTINUE.

Step 3: Search all remaining potential predictor variables for the best single variable to combine with those already included for predicting the dependent variable.

    If no addition is significant, $\longrightarrow$ STOP.

    If an addition is significant, include this variable in all further steps, and $\longrightarrow$ REPEAT STEP 3 TO SEARCH FOR THE NEXT BEST REMAINING PREDICTOR VARIABLE.

Aron/Aron
STATISTICS FOR PSYCHOLOGY

TRANSPARENCY 17.5

Predictors of job success at ABC Enterprises.   (Fictional study.)

Dependent Variable: Y = Rating of job success after 3 months of employment for clerical employees.

Potential predictors:

1. Typing test score
2. General clerical skills test score
3. Language test score
4. Psychological stability test score
5. Previous job history (rated by personnel officer)
6. Letters of recommendation (rated by personnel officer)
7. Social skills (rated by interviewer)
8. Personal integrity (rated by interviewer)
9. Efficiency/energy (rated by interviewer)

Stepwise Regression for Predicting Job Success

Step	Variable Added	Overall Model R^2	Overall Model F	Increment R	Increment F
1	Previous job history	.09	4.20**	.09	5.20**
2	Typing test score	.16	5.98**	.07	4.11**
3	Social skills	.24	6.21**	.05	3.88*
4	Language test score	.26	6.35**	.02	1.39

*p < .05; **p < .01

Aron/Aron
STATISTICS FOR PSYCHOLOGY

TRANSPARENCY 17.6

Exploratory analysis of childhood related variables as possible predictors of score on High Sensitivity Scale. (From class questionnaire.)

Dependent Variable: $\underline{Y}$ = High Sensitivity Scale score.

Potential predictors:

1. Close to mother as a child
2. Close to father as a child
3. Father involved in your family
4. Mother fond of infants
5. Fell in love as a child
6. Abused as a child
7. Alcohol a problem in your family
8. Prone to hide as a child

Stepwise Regression for Predicting Score on High Sensitivity Scale

Step	Variable Added	Overall Model		Increment	
		R^2	F	R	F
1	Fell in love as a child	.046	5.16*	.046	5.16*
2	Close to mother as a child	.091	5.23**	.044	5.10*
3	Mother fond of infants	.105	4.06**	.014	1.67

*$\underline{p}$ < .05; **$\underline{p}$ < .01

NOTE: Only the first two variables added a significant increment to the prediction. Third variable came close. None of the other variables added any meaningful increment.

Aron/Aron
STATISTICS FOR PSYCHOLOGY

© 1994 by Prentice-Hall, Inc.
A Paramount Communications Company
Englewood Cliffs, New Jersey 07632

TRANSPARENCY 17.7

Marital satisfaction, passionate love, and length of marriage. (Fictional data based on pattern of results in Tucker & Aron, 1993.)

Correlation:

    Marital satisfaction and marriage length: $r = -.31$, $p < .01$.

Partial correlation:

    Marital satisfaction and marriage
    length controlling for passionate love: Partial $r = -.03$, ns

Correlation:

    Passionate love and marriage length: $r = -.34$, $p < .01$.

Partial correlation:

    Passionate love and marriage length
    controlling for marital satisfaction: Partial $r = -.27$, $p < .01$.

Implications:

1. The correlation between marital satisfaction and marriage length is mainly accounted for by passionate love.

2. The correlation between passionate love and marriage length is not mainly accounted for by marital satisfaction.

3. Therefore, passionate love is the primary variable associated with length of marriage.

Aron/Aron
STATISTICS FOR PSYCHOLOGY

© 1994 by Prentice-Hall, Inc.
A Paramount Communications Company
Englewood Cliffs, New Jersey 07632

TRANSPARENCY 17.8

Being prone to fears, being a tense or worried person
by nature, and score on the High Sensitivity Scale.
(From class questionnaire.)

Correlation of prone to fears and
tense or worried by nature ($\underline{N}$=104): $\underline{r}$ = .53, $\underline{p}$ < .01

Correlation of prone to fears and
tense or worried by nature,
partialing out scores on the
High Sensitivity Scale ($\underline{N}$=104): $\underline{r}$ = .49, $\underline{p}$ < .01

Implication:

The correlation between being prone to fears and being
tense or worried by nature is not primarily a result of
the correlation of each of these with scores on the High
Sensitivity Scale.

Aron/Aron
STATISTICS FOR PSYCHOLOGY

TRANSPARENCY 17.9

Inclusion of Other in Self (IOS) Scale administered to undergraduate students twice over a 2-week period, each time describing their relationship with the person with whom they have the closest, most intimate relationship. (From Aron, Smollan, & Aron, 1992.)

Please circle the picture below which best describes your relationship

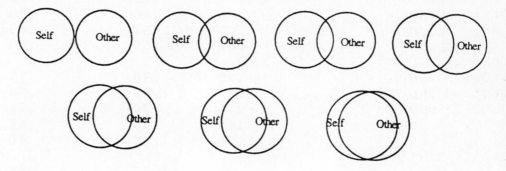

Correlation between two testings (test-retest reliability):

Those rating closeness to a family member (n = 13): r = .85.

Those rating closeness to a nonromantic friend: (n = 31): r = .86.

Those rating closeness to a romantic partner: (n = 48): r = .85.

Aron/Aron
STATISTICS FOR PSYCHOLOGY

© 1994 by Prentice-Hall, Inc.
A Paramount Communications Company
Englewood Cliffs, New Jersey 07632

TRANSPARENCY 17.10

Effect of unreliable measures on correlations:
Comparison of correlations obtained with perfectly
reliable measures versus measures with less than
perfect reliability.

Correlation of X and Y, each measured with perfect reliability	Reliability (Alpha)		Obtained Correlation
	X	Y	
1.00	1.00	1.00	1.00
1.00	.80	1.00	.89
1.00	1.00	.80	.89
1.00	.80	.80	.80
1.00	.60	.80	.69
1.00	.60	.60	.60
.80	.90	.90	.72
.80	.80	.80	.64
.80	.70	.70	.56
.80	.60	.60	.48
.80	.50	.50	.40
.50	.90	.90	.45
.50	.80	.80	.40
.50	.70	.70	.35
.50	.60	.60	.30
.50	.50	.50	.25
.20	.90	.90	.18
.20	.80	.80	.16
.20	.70	.70	.14
.20	.60	.60	.12
.20	.50	.50	.10

Aron/Aron
STATISTICS FOR PSYCHOLOGY

TRANSPARENCY 17.11

Closeness measures administered to undergraduate students, describing their relationship with the person with whom they have the closest, most intimate relationship. (From Aron, Smollan, & Aron, 1992.)

Correlations Among Closeness Measures (N = 208)

	RCI			Subjective	Sternberg
	Frequency	Diversity	Strength	Closeness	Intimacy
IOS Scale	.09	.16*	.36**	.34**	.45**
RCI-Frequency		.71**	.18**	-.01	-.04
RCI-Diversity			.27**	.08	.05
RCI-Strength				.26**	.13
Subjective Closeness					.64**

*p < .05; **p <.01

Factor Loadings for Closeness Measures
(Principal Factors Extraction, Oblique Rotation)

	Feeling Close	Behaving Close
Sternberg Intimacy Scale	.78	-.09
Subjective Closeness Index	.77	.00
IOS Scale (sense of interconnectedness)	.72	.21
RCI-Strength (other's influence on me)	.55	.43
RCI-Frequency (Time spent together)	-.00	.89
RCI-Diversity (number of shared activities)	.13	.91

Aron/Aron
STATISTICS FOR PSYCHOLOGY

TRANSPARENCY 17.12

Exploratory factor analysis of questionnaire items regarding
sensitivity and childhood experiences. (From class questionnaire.)

Factor Loadings
(Principle Components Analysis, Varimax Rotation)

Questionnaire Item	Sensitive	Relation with Father	Relation with Mother
Prone to fears	.82	-	-
Bothered by loud noises	.62	-	-
Fall in love very hard	.55	-	-
Avoid upsetting situations	.54	-	-
Tense or worried person	.53	-	-
Cry easily	.49	-	-
Father involved in family	-	.87	-
Close to father as child	-	.87	-
Close to mother as child	-	-	.83
Mother fond of infants	-	-	.68
Proportion of total variance	.22	.18	.14

Note: Loadings below .40 are indicated by a dash

TRANSPARENCY 17.13

Path analysis for predicted pattern in which religious well-being is seen as a cause of marital satisfaction, and two correlated aspects of social desirability (measured by the Crowne-Marlowe and Edmonds tests) are seen as causes of both religious well-being and marital satisfaction. (From Leong, 1989.)

Correlations Among Measures (N = 56)

	Social Desirability Crowne-Marlowe	Edmonds	Religious Well-Being	Marital Satisfaction
Crowne-Marlowe (CM)				
Edmonds (E)	.10			
Religious Well-Being (RW)	.03	.16		
Marital Satisfaction (MS)	.14	.76**	.41**	

**p < .01

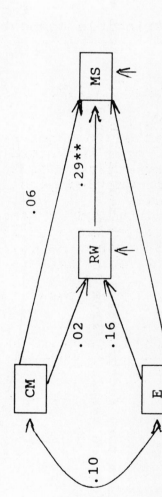

Aron/Aron
STATISTICS FOR PSYCHOLOGY

TRANSPARENCY 17.14

Confirmatory factor analysis for measures of closeness in cross-validation sample. (From Aron et al., 1992.)

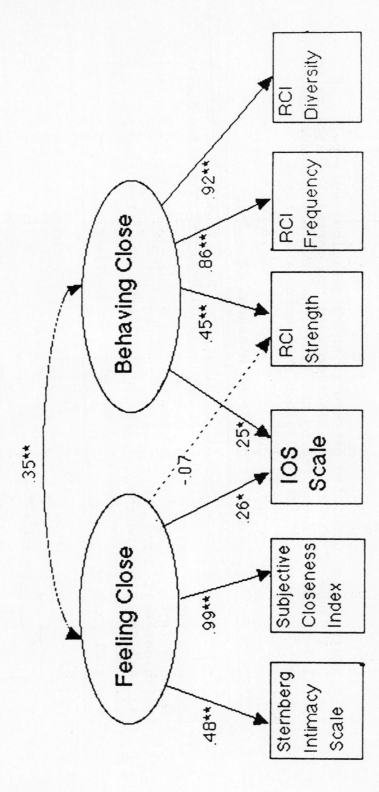

Aron/Aron
STATISTICS FOR PSYCHOLOGY

Relation of gender and birth order (first vs. later born) to reported abuse (log transformed) as a child. (From class questionnaire.)

Analysis of variance:

Effect	Significance	Effect Size
Gender	$F(1,98) = 5.85$, $p < .05$	$R^2 = .056$
Birth-Order	$F(1,98) = 2.12$, ns	$R^2 = .021$
Interaction	$F(1,98) = .14$, ns	$R^2 = .001$

		Gender		
		Women	Men	
Birth Order	First Born	1.02	.00	.63
	Later Born	1.94	.54	1.53
		1.58	.27	

Analysis of covariance, controlling for alcoholism in the family:

Effect	Significance	Effect Size
Gender	$F(1,97) = 5.04$, $p < .05$	$R^2 = .049$
Birth-Order	$F(1,97) = 1.72$, ns	$R^2 = .017$
Interaction	$F(1,97) = .23$, ns	$R^2 = .002$

ADJUSTED MEANS:

		Gender		
		Women	Men	
Birth Order	First Born	1.01	.13	.57
	Later Born	1.90	.54	1.22
		1.45	.34	

Aron/Aron
STATISTICS FOR PSYCHOLOGY

© 1994 by Prentice-Hall, Inc.
A Paramount Communications Company
Englewood Cliffs, New Jersey 07632

TRANSPARENCY 17.16

Influence of attractive versus nonattractive communicators on attitude change. (Fictional data.)

Attractive Communicator		
	Attitude	
S	Pre	Post
A	4	7
F	1	2
G	5	8
J	2	5
K	4	6
M	3.2	5.6

Unattractive Communicator		
	Attitude	
S	Pre	Post
C	3	4
D	2	1
I	5	6
M	2	2
O	5	6
M	3.4	3.8

Control (No Communicator)		
	Attitude	
S	Pre	Post
B	4	3
E	5	6
H	1	1
L	3	3
N	7	6
M	4.0	3.8

Analysis of variance for post-test attitude ratings:

$$F(2,12) = 1.07, \text{ ns}, R^2 = .15$$

Posttest Means		
Attractive	Unattractive	Control
5.60	3.80	3.80

Analysis of covariance for post-test attitude ratings, with pretest ratings as the covariate:

$$F(2,11) = 12.73, \text{ } p < .01, R^2 = .70$$

Adjusted Posttest Means		
Attractive	Unattractive	Control
5.98	3.27	3.95

Aron/Aron
STATISTICS FOR PSYCHOLOGY

TRANSPARENCY 17.17

Relation of type of closest relationship (family, friend, romantic; independent variable) to the combination of three questionnaire measures of closeness about that relationship (intimacy, subjective closeness, and interconnectedness; the dependent variables). (From data collected in Aron et al., 1992.)

Univariate analysis of variance for Sternberg Intimacy Scale:

$$\underline{F}(2,194) = 1.20, \underline{ns}, \underline{R}^2 = .012$$

Closest Relationship

Family	Friend	Romantic
6.35	6.23	6.10

Univariate analysis of variance for Subjective Closeness Scale:

$$\underline{F}(2,194) = 0.02, \underline{ns}, \underline{R}^2 = .000$$

Closest Relationship

Family	Friend	Romantic
6.03	6.00	6.02

Univariate analysis of variance for interconnectedness (IOS Scale):

$$\underline{F}(2,194) = 2.41, \underline{p} = .09, \underline{R}^2 = .024$$

Closest Relationship

Family	Friend	Romantic
4.35	4.58	4.94

Multivariate analysis of variance for all three dependent variables taken together:

Wilks' $\underline{F}(6,384) = 2.37$, $\underline{p} < .05$, effect size (1-Lambda) = .070

Aron/Aron
STATISTICS FOR PSYCHOLOGY

© 1994 by Prentice-Hall, Inc.
A Paramount Communications Company
Englewood Cliffs, New Jersey 07632

TRANSPARENCY 17.18

Relation of adult attachment style (independent variable) to the combination of close to mother as a child and mother fond of children (dependent variables). (From class questionnaire.)

Univariate analysis of variance for close to mother as a child:

$$\underline{F}(2,94) = 2.70, \underline{p} = .07, \underline{R}^2 = .054$$

Adult Attachment Style		
Secures	Avoidants	Anxious
3.77	3.04	2.86

Univariate analysis of variance for mother fond of children:

$$\underline{F}(2,94) = 2.93, \underline{p} = .06, \underline{R}^2 = .059$$

Adult Attachment Style		
Secures	Avoidants	Anxious
3.32	3.29	2.11

Multivariate analysis of variance for both dependent variables taken together.

Wilks' $\underline{F}(4,186) = 2.44$, $\underline{p} < .05$, effect size (1-Lambda) = .097

Aron/Aron
STATISTICS FOR PSYCHOLOGY

© 1994 by Prentice-Hall, Inc.
A Paramount Communications Company
Englewood Cliffs, New Jersey 07632

TRANSPARENCY 17.19

TABLE 17-4
Major Statistical Techniques

Association or Difference	Number of Independent Variables	Number of Dependent Variables	Any Variables Controlled?	Name of Technique
Association	1	1	No	Bivariate correlation/regression
Association	Any number	1	No	Multiple regression (including hierarchical and stepwise regression)
Association	1	1	Yes	Partial correlation
Association	Many, not differentiated		No	Reliability coefficients Factor analysis
Association	Many, with specified causal patterns			Path analysis Latent variable modeling
Difference	1	1	No	One-way analysis of variance; *t* test
Difference	Any number	1	No	Analysis of variance
Difference	Any number	1	Yes	Analysis of covariance
Difference	Any number	Any number	No	Multivariate analysis of variance
Difference	Any number	Any number	Yes	Multivariate analysis of covariance

Aron/Aron
STATISTICS FOR PSYCHOLOGY